THE EVOLUTION OF HUMAN SOCIETIES

Allen W. Johnson & Timothy Earle

The Evolution of Human Societies

*From Foraging Group
to Agrarian State*

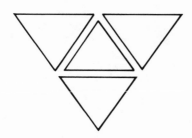

Stanford University Press · Stanford, California · 1987

Stanford University Press
Stanford, California

© 1987 by the Board of Trustees of
Leland Stanford Junior University
Printed in the United States of America

CIP data appear at the end of the book

Dedicated to
Marvin Harris and Marshall Sahlins
For their inspiration and disputation

Preface

THE PRESENT WORK, a synthesis of what we currently know about the evolution of economy in society, has undergone an evolution of its own. For several years we had both been teaching courses at UCLA entitled "Economic Anthropology" and "Cultural Ecology," when it occurred to us that we might profit by teaching them jointly for a few years. That experience resulted in two discoveries. First, the subjects of cultural ecology and economic anthropology were not adequately served by the division into two separate and unrelated courses; the theories and data of one are profoundly relevant to the other, the practitioners of both subdisciplines are often the same people, and a theory unifying both is not only desirable but necessary. And second, the existing literature is either outdated or too specialized to provide an adequate introduction and overview for scholars and students of anthropology and related disciplines. To be sure, the literature is not lacking in excellent works; but each carves out only a piece from the pie of economic and ecological theory and cross-cultural data. Readers approaching the subject by way of some number of these works are often dismayed by the differences between them in their discourse and miss what we believe to be the underlying unity of thought that they should share.

We therefore decided to write a theoretically integrated overview of the fields of economic and ecological anthropology. We had no intention at the outset of writing a book on social or cultural evolution. Theories of sociocultural evolution are not popular at the moment in anthropology. But we feel that this turning away from evolutionism, after a period of creativity a generation ago, is not warranted by the evidence. In fact, most researchers on problems of ecology and economy in other cultures work within an implicit evolutionary typology—usually that of Forager, Horticulturalist, Agriculturalist (or Peasant), and Industrialist/Proletarian, with Pastoralists and Fishers somehow appended. Furthermore, the idea of "intensification of production" that has become so basic to our field is itself an inherently evolutionary concept, as we shall argue throughout the book. Thus the point arrived in our work when we could no longer deceive ourselves: in order to present an orderly and synthetic treatment of our subject matter, we had to acknowledge that it could only be done using evolutionary theory and an evolutionary typology.

In a work of this scope we naturally have more intellectual debts than we could hope to acknowledge. Most of our anthropological colleagues, whether in their contributions to ecological and economic theory in anthropology or in their criticisms of it, have sharpened the tools that were at hand when we set out to write this book. Two people, Marvin Harris and Marshall Sahlins, who often express opposing views on key issues, have provided unusual inspiration to us. We dedicate our book to them in acknowledgment not only of their separate contributions, but also of the importance of their dialogue to us. We have had much direct help evaluating our argument. Robert McC. Netting read the first draft and gave us pages of detailed comments and criticisms, greatly contributing to what is of value in the present version. The following scholars read chapters or parts of chapters and commented on their accuracy and reasonableness: Robert Bettinger, Napoleon Chagnon, Myron Cohen, Terence D'Altroy, Walter Goldschmidt, Daniel Gross, Patrick Kirch, Richard Lee, Cherry Lowman, Mervin Meggitt, Philip Newman, Wendell Oswalt, Melanie Patton, Nazif Shahrani, David Hurst Thomas, Jan Weinpahl, Lynn White, Jr., and Johannes Wilbert. Despite such expert and generously given help, errors of fact and judgment no doubt remain, the responsibility for which must be entirely ours. The maps were drawn by Amalie Orme and reflect her creative input.

Any co-authored book must list one author first, implying some kind of seniority. In this case the first-named author took the early initiative, but thereafter the book was a collaboration in the fullest sense. Hardly a paragraph remains that has not been revised by both of us. Certainly, none of the major ideas and arguments survived our discussions in the form we had separately envisioned. What we each learned from the other, whether facts or ways of thinking, has reinforced our appreciation of the benefits of collaboration across specialties and subdisciplines in anthropology.

A.W.J.

T.E.

Contents

Part III: The Regional Polity

Tables and Figures

Tables

Figures

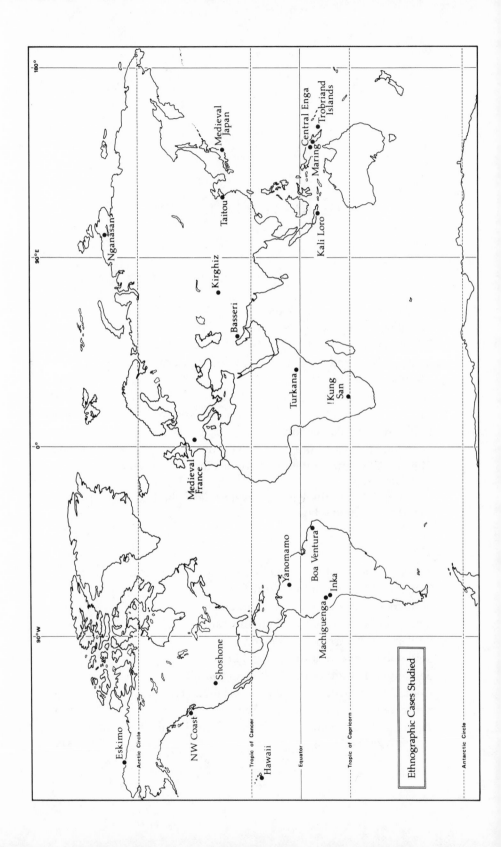

Ethnographic Cases Studied

THE EVOLUTION OF HUMAN SOCIETIES

ONE

Introduction

OUR PURPOSE IN THIS BOOK is to describe and explain the evolution of human societies from earliest times to roughly the present. Our emphasis is on the causes, mechanisms, and patterns of this evolution, which, despite taking many divergent paths, is explainable in terms of a single coherent theory. As teachers of cross-cultural economics and as field anthropologists—one an ethnographer, the other an archaeologist—we feel the need for a theoretical framework to help make sense both of the long-term prehistoric cultural sequences now available to us and of the diversity of present-day societies.

The San foragers of south Africa produce abundant foods with a few hours' work per day—are they the "original affluent society"? The Yanomamo of South America fight one another with peculiar ferocity—is this the unrestrained expression of innate human aggressiveness? In the striking North American Potlatch and Melanesian Kula Ring, "men of renown" publicly compete to gain personal prestige at others' expense—is this the human hunger for fame in its "primitive" manifestation? These are comparative questions of interest alike to the anthropologist, economist, geographer, historian, and political scientist, for they are ultimately questions about human nature—the common heritage of humankind as a species—and its expression in diverse environments, mediated by diverse cultural traditions. We hope in this book to provide a systematic theoretical approach for answering these and similar questions in a broad, cross-cultural frame of reference.

In one sense our approach is an outgrowth and elaboration of the work of such evolutionists and cultural ecologists as Childe (1951), Steward (1955), Leslie White (1959), Service (1962), Fried (1967), Netting (1977), and Harris (1977), who in turn depended on the work of nineteenth-century social evolutionists. To the extent that what we have to say here is new, its newness comes from two broad shifts in cultural evolutionism: first, a widespread and growing disillusionment with the idea that cultural evolution constitutes "progress" toward greater individual well-being; and, second, the emergence of a more complex, systemic, and "multilinear" sense of what cultural evolution is. To the extent that what we say is valid, its validity rests on an extensive documentation that has enabled us to avoid purely theoretical discourse and to develop our argument through detailed case studies. If the result is a longer book than we had hoped to write, it is because we have felt it necessary to do our utmost to avoid oversimplifying.

Cultural Evolution: A Brief History

The idea of a single pattern of cultural evolution was widely espoused by late-nineteenth-century social theorists, notably Morgan (1877), Tylor (1913 [1871]), Spencer (Carneiro 1967), and Maine (1870). All identified cultural evolution with progress from a worse condition to a better one; they saw new tools and new ideas as progressively liberating humans from an animal existence. Morgan—with his successive stages of Savagery, Barbarism, and Civilization—saw the biological chains of hunger and sex broken by cultural advances. Maine saw new public law ("Contract") liberating the individual from the tyranny of the family ("Status"). It would, however, be wrong to paint the evolutionists too simply. Engels (1972 [1884]) saw how individual greed could result in the suffering of slavery; Spencer saw how cultural developments based on competition and warfare resulted in oppression more often than liberation; and Malthus (1798) preceded evolutionary optimism with his "dismal" view that population growth led not to progress but to misery and death.

In the early twentieth century, reacting especially to the ethnocentric ideas of progress, Boas (1949 [1920]) and his students Lowie, Kroeber, and Benedict rejected cultural evolution. Each culture was unique and equally to be valued; if it changed, it changed in ways unique to itself, and no generalizations from culture to culture were possible.

This line of thought has in turn been discredited by the mounting archaeological and ethnographic evidence for cultural evolution. The "unilinear evolution" of Leslie White (1959; cf. Childe 1936, 1942, 1951) reinstated the nineteenth-century faith in progress as a cumulative growth in mastery over nature with increasing control of energy for human purposes. But White's theory was highly abstract. It lacked empirical confirmation and, when important economic institutions, such as the dramatic self-aggrandizing public displays of wealth found in the "prestige economies" (see Chapter 7), failed to support his utilitarian theory, he simply dismissed them as "social games" irrelevant to economic process (1959: 241).

The more empirically minded Steward (1955) questioned the idea of a single evolutionary typology, preferring instead to see cultural evolution as unfolding along several distinct lines depending on local history and ecology, a pattern to which he gave the label "multilinear evolution." Steward's approach stimulated an ecological orientation in much fieldwork carried out in the 1960's and 1970's. Ethnographers such as Rappaport (1967) and Netting (1968, 1977) sought to understand how specific aspects of culture, from subsistence practices to social organization to complex ritual cycles, served to solve critical problems of group and individual survival. Archaeologists such as Willey (1953) and Sanders, Parsons, and Santley (1979) documented the evolution of individual societies.

The development of a new evolutionary synthesis has been much slower to materialize. Service (1962) proposed his influential evolutionary typology of Bands, Tribes, Chiefdoms, and States, and Fried (1967) followed with a three-stage typology: Egalitarian Society, Rank Society, and Stratified Society. Both Service's emphasis on social organization and Fried's emphasis on social control have an important place in any theory of cultural evolution. Where both are limited is in their preoccupation with classification rather than with the actual causes and mechanisms of evolutionary change.

Harris's *Cannibals and Kings* (1977) was a crucial and innovative effort to understand these causes and mechanisms, to describe why and how a society evolves from a less complex stage to a more complex stage. We differ from Harris on some points of interpretation, as in our explanation of Yanomamo warfare in Chapter 5, and on many points of emphasis; but we embrace his general approach, with its focus on population growth as the primary determinant of social change. We have elaborated Harris's work in two ways. First, we have identified in considerable detail the key variables that change with cultural evolution and examined the systemic relations among them. Second, we have evaluated this model against archaeological and ethnographic material from every inhabited continent and every major type of human society.

In the simplest terms we recognize three basic components of the evolutionary scheme: environment, the individual, and culture. Individuals as active agents seek to meet their basic biological needs and those of their families. The environment presents the opportunities and the limitations: the ecological context within which individuals must find sustenance and avoid life-threatening hazards. The culture is the technology, the organization, and the knowledge that help individuals in their quest for survival; it includes broader demands on individuals, some of which may work against their perceived self-interest.

Issues in Cultural Evolution

Whether or not cultural evolution has taken place is no longer an issue. Recent archaeological work from all continents documents the basic development from early small-scale societies to later complex societies (Wenke 1980). Subsistence intensification, political integration, and social stratification are three interlocked processes observed again and again in historically unrelated cases. Foragers diversify and gradually adopt agriculture; villages form and integrate into regional polities; leaders come to dominate and transform social relationships.

The reality of cultural evolution is an accepted truth, but the reasons for this evolution are hotly debated. In part the debate continues because evolution inherently involves chicken-and-egg questions that have no ultimate answers. Does technological progress precede and generate population increases and changes in social organization, or do population increases and changes in social organization precede and generate technological progress? For many years theorists took it as axiomatic that technological progress was the cause, demographic and social change the result. Why did populations grow? Because new sources of food were discovered and made available by technological improvements. Why did village life replace mobile foraging? Because gardening is more secure and less arduous than constant moving about. Why did iron tools replace stone tools? Because iron is more malleable, can hold a sharper edge, and can sustain more rough use than stone. Why did paddy fields replace slash-and-burn cultivation of rice? Because irrigated paddy is more productive. Why did regional governments integrate politically autonomous villages? Because central government provides peace and prosperity beyond the means of a single village to provide for itself.

Even though we are skeptical today about the notion of progress, we often speak of technological and social change as making life "better." Indeed, if the changes were not for the better, why would people accept them? The theory of technological progress has the virtue of providing a direct and uncomplicated explanation for economic change: people come up with new inventions, some of which are found acceptable and learned and shared, remaining in use until yet more desirable inventions displace them. People accept changes in the way they do things because they recognize the benefits of doing so. In Childe's (1936) hopeful phrase, "man makes himself."

Yet it can also be argued that people invent or accept a new technology or a new form of social organization because they have no alternative: they must change if they are to survive. Something like this must have happened where the archaeological record documents the gradual broadening of the diet in prehistoric times to include costly and less-desirable foods (Mark Cohen 1977). As competition increases, people must live close together to defend themselves, their stored foods, and their lands. Leadership becomes a necessity for defense and alliance formation. In this light, population growth and a chain reaction of economic and social changes underlie cultural evolution.

That is the side of the argument that we emphasize in this book. To be sure, as with all chicken-and-egg questions, things work both ways: in an area of finite resources there can be no population growth beyond a certain limit without technological changes permitting more food to be provided per given unit of land. Population and technology have a feedback relationship: population growth provides the push, technological change the pull. But we shall argue later that it is fundamentally population growth (of which warfare, as in the example above, is one result) that propels the evolution of the economy.

Theories of Economic Behavior

A related question is whether the fundamental explanation of economic behavior is ecological or structural. In its simplest terms the question is an old-time conundrum of the social sciences. A woman goes shopping for food. How does she select food? Food that is nourishing to her family, or food that culture dictates as good even though it may not be the most nutritious? Her body and the health and growth of her family give her continuous feedback about nutrition (if she is able to perceive it), but her family, her friends, advertisers—in sum, her cultural group—also give her feedback concerning which foods are high-status, attractive, convenient, and so forth. Which will determine the outcome, the biological or the cultural influences?

The answer is clear: they are equally essential. Virtually nothing a human being does is ever either purely biological or purely cultural. Where one or the other must be emphasized as the more fundamental in a given situation, we tend to favor the biological explanation, perhaps because there were people before there were institutions. But neither explanation is sufficient without the other.

Let us now get away from conundrums and abstractions and examine the nature of these interacting explanations, and the major arguments put forward on behalf of each. Few serious students of economic behavior in cross-cultural perspective would deny the utility of any of the theoretical approaches we are about to review. The question is what weight to give each in relation to the others. Broadly the issues concern the role of biological processes, rational decision-making, and cultural constraints in determining economic behavior.

The Biological Bases of Economic Behavior

Evaluating the role of biology in human behavior always raises a dilemma for anthropologists. We know that human beings are animals and that the rules of biology apply to us as they do to other living things. But we also know that humans, with the acquisition of culture, have changed the rules of biology in ways never before seen. We begin here with a look at two versions of the biological approach to explaining economic behavior, and then turn to the critics of biological determinism.

Reproduction and Sociobiology. The definition of evolution as "the survival of the fit" is accurate only if it is understood to mean survival to reproduce. Recently scholars in biology and anthropology (Williams 1981) have sought to explain human behavior as serving to further the reproductive success of individuals. An apparent paradox exists where individuals sacrifice themselves on behalf of groups—for example, a young warrior's heroic death or a priest's vow of celibacy (cf. Emerson 1960). But "kin selection theory" (Hamilton 1963) has resolved this paradox by pointing out that in many cases of self-sacrifice the victim's genes do contribute differentially to the next generation by way of close kin: thus a man who dies in defense of his family is to some extent protecting the successful transmission of his genes. If we regret the individual's loss, his genes do not care. The individual was merely a vessel for carrying the genes across a generation; as long as other vessels remain safe, so do the genes.

Kin selection theory, then, clearly helps explain how individual sacrifice can still be "adaptive" in the strict sense. But how can we explain food sharing, defense, and information exchange between distant relatives or nonrelatives?

Fictive Kinship. Unlike other animals, human beings use reciprocity to create ties where kinship is weak or absent. In societies in which kinship relations are strong but economic relations exist between people who are not close kinsmen, such relations are usually cemented by modifying the institutions of kinship to make distant kin act like close kin and genealogical strangers act like biological kin. This is generally accomplished by a system of gift exchange.

Gift exchange is characteristically governed by cultural rules defining "the obligation to give," "the obligation to receive" (i.e. to accept a proffered relationship), and "the obligation to return" a gift (Mauss (1967 [1925]). Ties of reciprocity are strengthened when a gift is not repaid right away (showing that trust exists between the exchange partners), when over time a sense of "fair exchange" exists for both partners, and when a dissatisfied partner is free to end the relationship (Foster 1961). Culturally based reciprocity, structured in the language of kinship, engenders in exchange partners a sense of close kinship and interdependence. This fiction works because the relationships it engenders work: they produce behavior that simulates kinship and contributes in the same way to the welfare of the partners. The examples we shall encounter later include reciprocity in corporate kin groups and in patron-client relationships.

As we shall see, the central importance of fictive kinship is in making possible cooperation between people who would not cooperate on the strictly biological basis of kin selection. The limits of the biological capacity for forming ongoing human groups are generally reached at the level of a few families; larger groups tend to fission, and their component families may even become hostile to one another. In a real sense the social mechanisms that extend human cooperation beyond such minimal biological groups are peace-keeping mechanisms; they suppress family-centered individualism and allow the formation, at first uneasy, of multifamily communities. As we will argue later, the outbreak of warfare represents a "tragic failure" of the cultural mechanisms of social integration.

Ecology and Formalism. Let us now switch our focus back from genes to their vessels—that is, to flesh-and-blood people. Since healthy people are clearly in a better position to reproduce than unhealthy ones, people are motivated at the deepest biological level to strive for good health. To this need, it has been argued (cf. A. Johnson 1982), we may conjoin many less directly physiological needs such as the need for positive affect (Goldschmidt 1959) or a coherent world view (D'Aquili 1972). From the ecological point of view people are assumed to be motivated to produce a range of goods and services that will satisfy their basic needs to the fullest possible extent.

Closely linked to the ecological approach is the idea that a person "so disposes of his total resources as to obtain the maximum satisfaction" (Goodfellow 1968: 60). This maximizing or optimizing theory, called "formal economics" by anthropologists, assumes that all people have criteria by which they decide what to do at any given moment (Burling 1962; LeClair 1962; Homans 1967). Unlike the ecologist, however, the formalist does not assume that all individual behavior maximizes the satisfaction of biological needs. In formal economics it is "wants," not "needs," that are served; and a person may want something that he does not need or that is actually harmful from a biological standpoint (e.g. a diet drink sweetened with a carcinogenic substance). Neither the ecologist nor the formalist has all the answers. But where ecological logic and formalist logic are combined, as in "optimal foraging theory" (Pianka 1974), the result can be a powerful and productive theoretical synthesis.

The Cultural Bases of Economic Behavior

The diet drink in the above example is a cultural artifact in two senses: the chemical sweetener is a product of technology, and the desire for it is symbolically mediated through public images of youthfulness, slimness, and group acceptance. Thus to the constraints on the formalist's "want satisfaction" created by biological "needs" we may add the constraints created by cultural "values." In the famous phrase, "Man does not live by bread alone."

A dominant theme in economic anthropology has long been that mere biological need satisfaction cannot account for cultural behavior. Although the environment may limit cultural practices in some ways (there is no surfing in the Sudan), it does not determine the specific direction any culture will take (Kroeber 1939). Furthermore many cultural practices, such as the Trobriand chief's storing of huge quantities of yams until they rot (Malinowski 1922), are seemingly not self-serving in the material sense.

Substantivism. As developed by Polanyi (1957), this point of view became hardened into "substantive economics," which he saw as the antithesis of formal economics. Polanyi defined the economy as "instituted process"; his central concern was not with the process of "material want satisfaction" itself, but with how this process was "instituted" by society.

In particular Polanyi felt that the formalist idea of a rational, want-satisfying individual decision-maker was ethnocentric. However plausible it might seem in the West, people elsewhere did not make choices this way (Dalton 1961), but rather followed certain rules governing reciprocity and the redistribution of goods. Thus, for example, in some peasant societies people are required by the community on some occasions to underwrite lavish ceremonial feasts; and they have little choice but to comply, however much they may resent the expense.

We now recognize that the substantivist view is not contradictory to the formalist, but complementary. Choice, central to the formalist position, is ever-present: even the peasant has a "choice" whether to comply with the community's requirement or try to stand up against community pressure. On the other hand, all human choices are subject to institutional (cultural) constraints. Although both perspectives are necessary, theorists, as we shall see, still tend to favor one or the other.

Materialism: Vulgar and Effete. The long dispute between the formalist and substantivist explanations has resurfaced in recent years, with the two sides now calling themselves ecologists and structuralists. Leading an attack by structural Marxists, Friedman (1974) has tried to dismiss the ecological approach as "vulgar Marxism," a category in which he lumps cultural ecology (Steward 1955), cultural materialism (Harris 1979), and ecological anthropology (Rappaport 1971; Hardesty 1977). These authors believe that the central concern of a human population must be with making a living—meeting basic needs for energy and other nutrients, shelter, defense, freedom from disease, and reproduction. Apparently these materialists are so vulgar as to suggest that all bodies, all of us below our mouths, have irreducible visceral requirements.

In the effort to distance themselves from the vulgar materialists, Friedman (1974) and others (Godelier 1977; Meillassoux 1972; Legros 1977) focus on how social structure determines economic process. Like the substantivists these authors continuously draw our attention away from biology to culture. It is as if they wished to steer clear of any serious discussion of the body, with its insistent animal demands and frequently uncouth influences on behavior. They speak of a reified structure, acting superorganically to fulfill and "reproduce" itself (Godelier 1977: 4). Such a cool, intellectualized materialism, avoiding vulgarity by excluding flesh-and-blood organisms from it, might be termed "effete materialism."

Critically important to the structuralist viewpoint is how *ownership* of the means of production (land, labor, and capital) acts to channel the flow of goods and to support existing social arrangements. This approach is materialistic in that its advocates see the control of economic resources as central to maintaining a society, and we depend on this argument in our analysis of stratification. If we take social structure as our starting point, however, we are left with the nagging question of where it came from in the first place. Nor can we explain the worldwide occurrence of comparable social and political forms in societies that have had major differences in culture history.

This war of words is ultimately unenlightening. The truth is that people need food and people need rules; the interesting thing is how the two needs interact. Cultural rules stabilize the economy by sanctifying certain courses of action that, through time, have worked successfully in meeting the basic needs of the population (Rappaport 1971). Ceremonials of unity, for example, rarely attend economic interactions between the members of a small family or a cluster of close relatives; but they figure prominently in economic relations between "fictive kin," who may be needed to help defend a common territory or maintain an irrigation system. The cultural rules that "govern" economic behavior in the structural Marxist sense have evolved out of periods of experimentation. The successful solutions are encoded into rules (Chibnik 1981; Durham 1982). They take on a compelling power of their own, not simply because they are "traditional," but above all because they work.

The Origins of the Political Economy. In the last analysis, why are rules necessary—which is to say, why is culture necessary? Because in the crowded landscapes in which human populations live there is a constant potential for aggressive competition over the most desired resources. In families a certain amount of "family feeling," based on biological reinforcers but certainly strengthened by the myriad of small reciprocities among family members, helps minimize such competition; but when family feeling is not a factor, the difficulties of regulating destructive competition become massive. Where people who have no loyalty or feeling of common interest compete for scarce resources, we are in the realm of what Thomas Hobbes called "the war of all against all."

No one wins such a war, nor is there any winner in what Hardin (1968) called "the tragedy of the commons." Hardin's classic case of the problems that arise when "strangers" seek to exploit the same resources involves pastoralists exploiting a common pasture. If one herder conscientiously seeks to keep the pasture viable by restricting his herd's grazing time, the next herder may simply seize the opportunity for extra grazing for his own herd. The "good" herder's restraint thus operates to his disadvantage, and the "bad" herder's greed to his momentary advantage only; the pasture is degraded by overgrazing, and all herders who use it lose as a result. The only solution to this tragedy is for a group's members to observe a code of behavior that regulates all of them and protects the common resource.

There are rules that require the !Kung San (Chapter 2) to request permission to drink at another group's water hole, and rules that require the Eskimos (Chapter 6) to request permission to hunt in the home range of another, even though the host group in both cases is equally rule-bound to give permission. Like the rules necessary to prevent the tragedy of the commons, these rules coordinate the behavior of strangers to reduce the potential for resentment, violent aggression, and reprisals.

It is only through social structures and cultural rules that stable relations beyond small family groups can be maintained in a competitive environment. In this book we shall call any economy exhibiting such structures and rules a political economy. Although the political economy exists because it solves real economic problems of individual families—that is, for ecological reasons—it can exist only in a social matrix whose structure governs its workings. As we turn now to consider the general properties of the political economy and the patterns by which it evolves, we shall clearly see how impossible it is to single out the ecological or the structural aspect as uniquely fundamental.

The Subsistence Economy and the Political Economy

Consistent with our ecological approach, we define the economy as the provisioning of the material means of existence. It includes the production and distribution of food, technology, and other material goods necessary for the survival and reproduction of human beings and of the social institutions on which their survival depends. Whether we study the subsistence support of the household or the finance of the larger institution, the problem of material provisioning is basic.

Our definition of the economy is close to the ecologist's notion of niche, or the way a population derives its necessary matter and energy from the surrounding habitat (Odum 1971). It is also similar to the substantivist's notion of the economy as "man's . . . interchange with his natural and social environment, in so far as this results in supplying him with the means of material want satisfaction" (Polanyi 1957: 243). Unlike the substantivist, however, and in the tradition of the formalist, we see individual decisions as shaping the economy and economic change.

Analytically the economy can be subdivided into two sectors: the subsistence economy and the political economy. The basic dynamics of these two economic systems are different, and they contribute quite differently to the evolutionary development of societies.

The Subsistence Economy. The subsistence economy is the family economy. It is organized at the household level to meet basic needs, including food, clothing, housing, and procurement technology. The simplest form of the subsistence economy is the "domestic mode of production" (Sahlins 1972). Each household is ideally self-sufficient, producing all that it needs; it is also generalized, incorporating a division of labor by age and sex.

The nature of the subsistence economy is determined by the needs of the population and the cost of procuring different resources (cf. Earle 1980a). Theoretically no surplus is produced beyond what may be needed if things turn bad. The overriding goal is to fulfill the population's needs at the lowest possible cost that affords security.

To meet this goal families select from among potential procurement strategies those that seem best suited to obtaining food or other products from the environment. Each such strategy—for example, hunting large game animals like deer—has its own typical costs and productivity. Following the law of diminishing returns, the costs of producing food with any given strategy climb as output from that strategy increases; as more deer get killed, fewer deer remain available. Subsistence strategies differ in the initial cost of the strategy and in the slope of the cost curve. When a community first enters virgin territory, the strategies available for obtaining food differ in their initial costs; it may be cheaper to get a good diet by hunting deer than by collecting seeds and insects. Over time, as deer become less abundant, and hence more expensive to obtain, new strategies are added as their costs become comparable to the cost of deer. Thus, the number of strategies foragers use in obtaining food tends to increase the longer they inhabit a given area.

This logic is derived from formal economics (cf. Earle 1980a) and its application to optimal foraging in animal populations (Pianka 1974). The goal is not to maximize production but to minimize the effort expended in meeting household needs. A specific mix of strategies all exploited at the same cost level minimizes procurement costs to a region's population. This mix should remain stable except where upset by changes in population density, technology, or the environment.

Growth in the subsistence economy results from the positive feedback between population growth and technological development (cf. Wilkinson 1973). As population grows, total needs expand. In technologically simple societies population growth was often very slow, but over the centuries the overall growth rate has dramatically increased (Taagapera 1981). The availability of resources to support a population is determined by the environment and by the technology used. For our present purposes the environment can be considered relatively constant through time; it varies from locale to locale and fluctuates within locales, but for our time period it has been reasonably stable. Technology, by contrast, has changed dramatically under the pressure of population growth, making possible huge further increases in population and in food production from Neolithic times to the present day.

The Political Economy. The political economy involves the exchange of goods and services in an integrated society of interconnected families. All cultures have at least a rudimentary political economy, inasmuch as families can never be entirely self-sufficient but are linked by the need for security, mating, and trade. A true political economy, however, comes into being only at a certain stage of social evolution. Whereas the household economy is remarkably stable and enduring through time, the dynamics of the political economy make for dramatic changes in its nature. As it evolves, the political economy becomes geared to mobilizing a surplus from the subsistence economy. This surplus is used to finance social, political, and religious institutions that in their more elaborated forms are run by non-food-producing personnel; and these institutions in turn are used to support and justify ownership by the ruling elite of the region's productive resources, especially improved agricultural land.

Perhaps the most important difference between the political and subsistence economies is seen in their different rationalities and dynamics. The subsistence economy is geared to meeting household needs; if outside variables (population, technology, and environment) are held constant, it is inherently stable. By contrast, the political economy is geared to maximizing income for the ruling elite; it is growth-oriented in a highly competitive political domain, and thus inherently unstable.

An elite personage maintains his position and income through power—his ability to resist other elites' efforts to co-opt his sphere of economic control. His power in turn depends on maximizing his income by investing in income-producing projects; indeed, if he is to stay ahead of the game, much of the income from these new investments must itself go into further investments. The political economy grows by this process of positive feedback between investment and expanding income.

Such a system will grow unless checked by factors that cause declining yields. In most stratified societies we find a cyclical pattern in which the political economy expands to its limit, collapses by internal conflict, then begins to expand again. Elites recognize the limits to growth and try to overcome them by instituting major capital improvements. In Hawaii, for example (Chapter 10), where elite lines competed with each other for control of island populations, chiefs invested in such capital improvements as fish ponds, irrigation systems, and land reclamations in an effort to increase their income and with it their military power. The primary limit to the growth of the political economy appears to have been in the transport technology required to mount a successful attack on an island and resupply an invading army. Although many attempted conquests are recorded, none succeeded until the arrival of Western ships radically altered the possibilities. Money from the capital improvements was spent on such ships, and the newly acquired ships were used to fashion an interisland conquest state. With this radically new technology the limits to expansion were overcome, and the political economy was quickly centralized.

The Articulation of the Subsistence and Political Economies. As we have seen, the subsistence economy, designed to satisfy the basic needs of the family with minimum effort, tends toward conservatism, not growth and change; whereas the political economy, in the hands of leaders who seek continually to extend their sphere of control, tends toward growth. But why in fact does the political economy grow or fail to grow?

Since the political economy is financed by a surplus taken from the subsistence economy, the political economy cannot function, let alone grow, unless family participation is assured. Thus the limit of the political economy for integrating an economic community may be thought of as the point at which the costs of participation by the family exceed the benefits. Both the costs and the benefits of participation in the political economy change with its evolution. The impetus for this change comes in the first instance from population growth. Increasing population density puts pressure on existing resources and creates various problems in the subsistence economy that are typically beyond the family's capacity to solve on its own. Solving these problems requires group action and leadership, the very conditions that encourage economic control and expansion of the political economy.

Briefly, population growth leads to declining living standards and places the member families of a population at risk. By solving this problem in one way or another and managing the economy to the benefit of its constituent families, a ruling elite can gain effective control over production and thus guarantee the surplus needed for its support. As population increases, the benefits of participating in the political economy grow faster than the costs: more accurately, the costs of *not* participating in the political economy grow faster than the costs of participating, for in some ways the family in a developed political system is worse off than the household of family-level foragers.

We visualize the evolving political economy as expanding like a bubble. For the families inside the bubble the benefits of participating in the political economy exceed the costs; for those outside the costs exceed the benefits. Since it is extremely costly to control hostile populations that see no benefits to themselves from participating in the political economy, the political economy will not invest in the military control of outlying populations without compelling reasons to do so; and this attitude sets effective limits on the expansion of the political economy. By way of example, the Inka state succeeded in integrating settled agrarian communities over a strip of territory running some 2,000 miles north and south in the Andean highlands, yet failed to bring under its control the scattered hamlets of the adjacent Amazon rainforest a mere 50 miles to the east. As our discussion of the Machiguenga (Case 3) and the Inka (Case 16) will show, the costs and benefits of family participation in the Inka political economy were utterly different in the highlands and the rainforest.

In return for some measure of economic control the elites in a political economy provide services to the commoners, notably access to land that the elites own, often by right of conquest. As we shall see, the system of land tenure is the foundation of the political economy in agrarian chiefdoms and states. Elites and government officials also organize the construction and maintenance of technological improvements that benefit both the commoner and the landlord (Earle 1978), provide relief during periods of drought or crop failure, manage such military operations as may be necessary, and gain access to foreign goods through trade.

Perhaps equally important are the mediating services the elites provide. In a chiefdom or a state the local community is embedded in a regional system whose legal, religious, and military institutions are foreign to the commoners' daily lives and beyond their understanding. The elites act as patrons for their dependents in dealing with these overarching institutions. At this level the commoners' options are fully "circumscribed" (Carneiro 1970b), and there is no longer any possibility of remaining outside the political economy.

The Evolutionary Process

We see the evolutionary process as an upward spiral. At the lowest level the pressure of an increased population on resources evokes a set of economic and social responses that interact to create a higher level of economic effort capable of sustaining an increased population. The process repeats itself until eventually a growing population becomes possible only with the increasing involvement of leadership, with its concomitants of increasing dependence and political development.

Figure 1 presents a simplified scheme emphasizing the primary causal relationships that underlie the arguments presented in this book. The scheme shows the primary motor of population growth, a set of potential results of the intensification of production, and two sets of responses to those results, one leading to integration and the other to stratification. As we shall see in the case studies, however, the simplified scheme of Figure 1 becomes quickly complicated by feedback loops of all sorts. Not only do economic and social development become difficult to disentangle, but they may even reverse the causal arrows of the figure and directly affect the basic variables of population growth rate and environmental conditions or constraints.

The primary motor for cultural evolution is population growth. Human populations, like all other animal populations, have an inherent biological capability for growth that is necessary for a population to compete with other species for limited resources, to expand into unoccupied habitats, and to replenish itself after environmental disasters. Over the long course of human cultural development, the archaeological and historical record shows a consistent and eventually dramatic increase in human population worldwide (Coale 1974). The rates of this increase have been different in different regions, at different times, and at different stages of the evolutionary process. But the tendency toward increase is always there.

As a result of this population increase the subsistence economy must be intensified. In the process of intensification four problems become salient, their relative importance varying according to environmental conditions. The necessity of dealing with these gives rise to leadership and with it the opportunity for control.

The first problem is production risk. As a landscape fills in with people, the most desirable foods are soon depleted and less-desirable foods, those that once served as buffers against starvation in bad years, come to be part of the regular diet. With fewer buffers and less food generally per capita, the risk of starvation increases, and each household must provide a margin of security in food production against the possibility of lean seasons or years. Where production costs and risks are low this margin can be provided independently by each family, but as risks escalate families must rely increasingly on outside sources of security. The classic form of risk management is community food storage; another common form is reciprocal arrangements between communities for visiting in lean times. A region with community storage or visiting arrangements can sustain a larger population; but such arrangements produce a need for leadership and the potential for control.

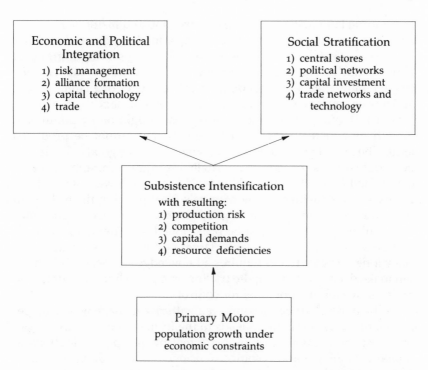

Fig. 1. A causal model of the evolution of the political economy

The second problem is resource competition. At all ethnographically known economic levels competition between families for prized resources occurs. At the family level, characterized by low population densities and dispersed resources, competition is relatively easy to manage. With intensification, however, locally rich resources such as fertile bottomlands become even more precious, and improvements to land such as long-yielding tree crops become more common. Both developments increase the benefits of the violent seizure of territory relative to the costs of violence. The general level of violence in a region accordingly rises, and small groups form alliances with other small groups for the more effective defense of their resources. Here again we encounter the beginnings of leadership and the potential for control.

The third problem, the demand for capital investment in technology, arises from the need to increase the range or the quantity of foods a population can produce in the resource area available to it. For example, the efficient use of marine resources may require the construction of huge canoes or whaling boats; the most efficient use of agricultural land may require the construction and maintenance of an irrigation system; the best way to assure a food supply in the winter may be to construct large-scale storage facilities serving a group of families. Such technological investments are beyond the capacity of a single family; they are dependent on the resources of a larger group, and in due course come under the control of a manager.

Fourth, the depletion of local resources brought on by population growth may increase the need for goods that cannot be produced locally but can be obtained in exchange for local goods. Trade can even out seasonal or annual shortfalls in production, and it can increase food production by making tools (e.g., axes) available in places poorly supplied with the raw materials for producing them. In both these ways trade in specialized goods increases the overall efficiency with which a population can be provisioned from limited resources, and thus the capacity to sustain a larger population. But trade, especially long-distance trade, requires a knowledgeable head trader who can make decisions binding the trading group: in short, a leader, who brings to his job the potential for control.

Production risk, then, is countered with risk management arrangements; resource competition leads to the formation of alliances to defend resources; capital demands are met by group contributions to capital technology; and resource deficiencies are made up by trade. None of these responses to the problems caused by subsistence intensification is within the capacity of the individual family. A larger, integrated group is needed, and comes into being; the problem is solved, population rises again, further intensification is needed, and the process is iterated up the spiral to the development of the modern state.

We will not dwell here on the secondary social effects of population growth, since the nature and workings of these effects are best understood from the study of detailed examples. Suffice it to say that what is most important for social evolution in the long run is the process by which economic intensification creates opportunities for economic control that in turn lead eventually to social stratification. We shall see this process at work in our case studies, and analyze it more fully in our concluding chapter.

The Evolutionary Typology

Nineteenth-century evolutionists tended to classify their evolutionary stages in technological terms: Stone Age, Bronze Age, Iron Age. As knowledge of the complexity of economic systems grew, these simplistic technological labels gave way to more generic terms such as hunter-gatherer, horticulturalist, and pastoralist that indicate broad economic systems rather than single features of technology. But anthropologists are no longer comfortable with typology that lumps together such divergent groups as the !Kung San and the Northwest Coast Indians as hunter-gatherers, the Machiguenga and the Mae Enga as horticulturalists, and the Turkana and the Basseri as pastoralists.

Following Service (1962) and Fried (1967), we have chosen more global designations based on the social organization of the economy. We have identified three critical levels of socioeconomic integration as a basis for organizing our discussion in this book: (1) the Family-Level Group, including the family/camp and the family/hamlet; (2) the Local Group, including the acephalous local group and the Big Man collectivity; and (3) the Regional Polity, including the chiefdom and the state.

The Family-Level Group. The family or hearth group is the primary subsistence group. It is capable of great self-sufficiency, but moves in and out of extended family camps or hamlets opportunistically as problems or opportunities arise.

The family/camp is characteristic of foraging societies of low density (less than 1 person per 10 square miles). Camp groups of 25-50 persons typically form when resources are highly localized or when a group larger than the individual family is required for risk management or for a particular subsistence activity. The group can then dissolve into small segments consisting of single families (5-8 persons) that independently exploit low-density, dispersed resources. These societies are characterized by a simple division of labor by sex. Suprafamily leadership is ephemeral and context-specific, relating to immediate organizational requirements such as a hunting expedition requiring the participation of numerous families. Although homicide is fairly common, organized aggression (warfare) is not. Ceremonialism is ad hoc and little developed. A camp characteristically has a home range, but does not claim exclusive access to this territory or strictly defend it against outsiders.

The family/hamlet is characteristic of somewhat higher-density societies (from 1 person per 10 square miles to 2 per square mile) in which families cluster into a settlement group or hamlet (25-35 persons) on a more permanent basis. The subsistence economy continues to rely heavily on wild foods, sometimes in conjunction with the beginnings of horticulture or herding. Storage is more prevalent. During the year individuals or families move out to exploit specific resources; from year to year, the hamlet re-forms and fragments as households change locations to minimize resource procurement costs.

The hamlet does not form a clearly demarcated political group, and leadership continues to be context-specific and minimal. Ceremonialism is little developed. As with the family/camp, the hamlet's territory consists of undefended home ranges and warfare is uncommon.

The Local Group. Local groups of many families, running to five or ten times the size of family-level groups, form around some common interest such as defense or food storage. They are usually subdivided along kinship lines into corporate lineages or clans. Depending on the extent of their common interests, these groups are either acephalous, village-sized units or larger groups integrated by regional networks of exchange headed by Big Men.

The acephalous local group is typically found in societies with densities greater than 1 person per square mile. The subsistence economy in most cases focuses on domesticated species, although in some cases wild resources, especially maritime resources, dominate. A frequent settlement pattern is a village of perhaps 100-200 people subdivided into clan or lineage segments of hamlet size (i.e., 25-35 persons). The local group forms a ritually integrated political group and may have a headman; but it typically fragments into its constituent kin groupings either seasonally or periodically as a result of internal disputes. Because of endemic warfare intercommunity relationships of various sorts are critically important for community security, but such relationships are contracted essentially on an individual, family-by-family basis. Ceremonialism is important for publicly defining groups and their interrelationships. Resources are held exclusively by kin groups, and territorial defense is common.

The Big Man and his managed intergroup collectivity are found at higher but variable population densities in areas in which warfare between territorial groups has traditionally been intense. Subsistence is focused heavily on agriculture, pastoralism, or extremely productive natural resources. The local community of perhaps 300-500 people is a territorial division, typically containing multiple clan or lineage segments that either live together in a village or are dispersed throughout the well-defined territory of the group. The local group is represented by a Big Man, a strong charismatic leader who is essential for maintaining internal group cohesion and for negotiating intergroup alliances. The Big Man is also important in risk management, trade, and internal dispute settlement, and represents his group in the major ceremonies that coordinate and formalize intergroup relationships. His power, however, is dependent on his personal initiative; if his support group deserts him for a competitor, little may be left of the reputation he has tried to build for himself and his local group, or of the alliances he has contracted.

The Regional Polity. Regional organizations arise out of formerly fragmented local groups under conditions we shall examine in detail. Depending on the scale of integration these are either chiefdoms or states.

Chiefdoms develop in societies in which warfare between groups is endemic but becomes directed toward conquest and incorporation rather than toward the exclusion of defeated groups from their land. The subsistence economy is similar to that of a Big Man collectivity and requires similar management. Economic strategies, however, notably irrigation agriculture and external trade, provide opportunities for elite investment and control, which are used to extract surplus production from the subsistence economy to finance the chiefdom's operations. As the regional integration of the polity proceeds, clearly defined offices of leadership emerge at the local and regional levels and are occupied by members of a hereditary elite.

Always in search of new sources of revenue, chiefs seek to expand their territorial control by conquest. Here a typical cyclical pattern is found, as local communities and thousands of people incorporate under the control of an effective chief only to fragment at his death into constituent communities. Competition is intense, both within a chiefdom for political office and between chiefdoms for the control of revenue-producing resources. Ceremonies legitimize the leadership and control of the ruling elite.

The development of states and empires involves the extension of political domination, usually by conquest, to a still larger area. States formed by conquest may incorporate vast populations, often in the millions, that are ethnically and economically diverse. As in chiefdoms, elites carefully manage the economy in order to maximize surplus production that may be translated into power and political survival. Elite ownership of resources and technology is typically formalized in a system of legal property. National and regional institutions are developed—an army, a bureaucracy, a law enforcement system—to handle the state's increasingly complex functions. Ceremonies mark significant phases in the annual economic round and legitimize unequal access to resources.

Quantity into Quality: The Emergence of New Social Forms. So far our emphasis has been on gradual, quantitative change, but in the chapters to follow we will address the difficult problem of qualitative change in the creation of new social institutions. In the evolution of social complexity a critical change occurs when it becomes necessary to integrate formerly autonomous or separate units (cf. Steward 1955). As Service (1962) has argued, larger sociopolitical units cannot be formed unless new integrating mechanisms arise that inhibit their segmentation into their component smaller units.

Mechanically, it appears that new integrative institutions such as the village or chiefdom are formed by "promotion" (Flannery 1972): from among the original autonomous units, one becomes dominant and subordinates the others. For example, in Polynesia a single local lineage may expand by conquest to form a regional chiefdom. The chiefdom is at first organized by the kinship principles formerly governing the local lineage, but its new regional functions inexorably lead to changes in this mode of organization. Kinship-based forms and institutions gradually give way to novel, more bureaucratic institutions designed to solve the problems of integrating society on a much larger scale.

Anthropologists in the past have insufficiently stressed the dynamic nature of evolutionary change, probably because the convenience of "stage" typologies has led them to ask simple questions of origin like, What caused the evolution of chiefdoms? As this book will demonstrate, chiefdoms are not suddenly created and cannot be explained as direct outcomes of some single condition. Rather, any such complex social form evolves gradually, responding to quantitative change in the variables of intensification, integration, and stratification. A new level of integration may not represent a significant qualitative change if it is not accompanied by changes in these underlying variables; it may be only weakly formed and subject to fragmentation like the Heian empire of medieval Japan (Chapter 11). In our view it is more important to understand how a new level of integration is achieved and stabilized than to answer any simple question of origins; and that will be our task in this book.

The Plan of the Book

This book is organized into three Parts corresponding to our three critical levels of sociocultural integration: the Family-Level Group, the Local Group, and the Regional Polity. Table 1 identifies the ethnographic cases we discuss and their level of integration. It is only by the careful examination of these cases, together with such archaeological evidence as we have for prehistoric times, that we can begin to understand the evolution of the political economy. It is here that a unilinear

TABLE 1
Cases Examined in the Book

Socioeconomic level	Chapter	Case	Area
FAMILY GROUP			
Without domestication	2	1. Shoshone, Great Basin	*North America*
		2. !Kung San, Kalahari Desert	*Africa*
With domestication	3	3. Machiguenga, Peruvian Amazon	*South America*
		4. Nganasan, Northern Siberia	*Asia*
LOCAL GROUP			
Acephalous local group	5	5. Yanomamo, Venezuelan Highlands	*South America*
	6	6. Eskimos, North Slope of Alaska	*North America*
		7. Tsembaga Maring, Central New Guinea	*Australasia*
		8. Turkana, Kenya	*Africa*
Big Man collectivity	7	9. Indians of the Northwest Coast	*North America*
		10. Central Enga, Highland New Guinea	*Australasia*
		11. Kirghiz, Northeastern Afghanistan	*Asia*
REGIONAL POLITY			
Chiefdom	9	12. Trobriand Islanders	*Oceania*
	10	13. Hawaii Islanders	*Oceania*
		14. Basseri, Iran	*Asia*
Early state	11	15. Medieval France and Japan	*Europe/Asia*
		16. Inka Empire	*South America*
Nation-state (peasant economy)	12	17. Boa Ventura sharecroppers, Northeast Brazil	*South America*
		18. Taitou villagers, Northeast China	*Asia*
		19. Kali Loro villagers, Central Java	*Asia*

theory of universal stages of development can be fruitfully combined with a multilinear theory of alternative lines of development arising from unique environmental and historical conditions.

In keeping with our basic focus on economic behavior in broad environmental and cultural contexts, we have concluded each of the three Parts with a short chapter examining economic behavior at the integration level in question from three perspectives that have often been viewed as conflicting theories of human behavior but that our integrated theory seeks to reconcile. These perspectives are, first, ecological, dealing with human behavior as environmental adaptation; second, structural, dealing with human behavior as governed by cultural rules; and, third, economic, dealing with human behavior as the purposeful and rational pursuit of the good life however defined.

The final chapter of the book summarizes the overall argument and examines some of its implications.

Part I

The Family-Level Group

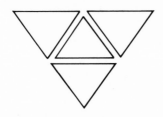

TWO

Family-Level Foragers

STARTING OVER TWO MILLION YEARS AGO, human foragers spread throughout the world to occupy a remarkable diversity of environmental zones. The very long growth and dispersion of human hunters and gatherers served as the context for our biological evolution and as the foundation for all later cultural development. Foraging economies have the simplest form of subsistence production, gathering wild plants and hunting wild animals. Although these economies are quite variable, they have in common certain elements of resource use, technology, ownership, and organization. These shared elements define what Lee (1979: 117-19) calls a forager mode of production.

This forager mode of production is predicated on a low population density, characteristically less than one person per square mile. At low population densities foraging is probably the most efficient mode of production; it has typically prevailed until higher population densities made it impractical. As we have seen, the efficiency of a subsistence strategy is inversely related to its intensity; the more people there are out looking for wild yams or wild boars, the harder it is to find one. Where population densities are low, efficiency is high and the relative attraction of domesticated agriculture or pastoralism is diminished.

At low densities foragers have been called "the original affluent society" (Sahlins 1968). Although this characterization downplays the seasonal and periodic hardships encountered by foragers, they do in fact live well in important ways. On the strength of data on the !Kung and on Australian aborigines, Sahlins argued that foragers' limited needs can be satisfied by a few days of work each week, leaving their remaining time free for noneconomic activities. A broad cross-cultural study by Hayden (1981a), which considers time spent processing food in addition to time spent procuring it, concludes that hunter-gatherers need expend only two to five hours per day in these activities.

In short, low-density foragers live a good life of sorts, and we feel that evolutionary change from this simple economy cannot be seen simply as a matter of developing improved technologies. Given that foraging efficiency depends on low-intensity resource use, why did population density remain very low for literally millions of years? Did people of those times not have a potential for rapid population growth and the technological capability to sustain this growth? The low growth rate in human populations during the foraging period must be explained if we are to understand the tempo and the causes of cultural evolution.

At least four biological and cultural factors associated with a foraging way of life combined to keep the population low. First, a chronic caloric deficiency lowers fertility rates; because of seasonal cycles in food availability and limited capabilities for storage, periods of food shortage were common. Second, a long nursing period delays renewed ovulation; since most wild foods are apparently not well suited to wean young infants, nursing among foragers typically remains a child's main food source for the first two or three years. Third, the intense physical exercise required for mobile foraging may lower female fertility (Frisch et al. 1980). Fourth, because closely spaced children are an economic hardship in a mobile society, infanticide may have been used to space births (Birdsell 1968a). Although these factors no doubt operate differently under different environmental conditions, the fertility of mobile groups is invariably low.

In addition, periodic disasters such as drought can cause famine in foraging populations, cutting them down to a fraction of their "potential density"; and at low growth rates such a population would be slow to regain its numbers. According to the important volume *Man the Hunter* (Lee and DeVore 1968), the population densities of foragers are characteristically only 20-30 percent of average carrying capacity. As Bartholomew and Birdsell (1953) point out, foragers must adapt to the worst conditions available seasonally and periodically, not the average conditions.

The efficiency of low-density foragers rests also on pragmatic decisions with regard to diet, technology, movement, and group affiliation. They are acutely cost-conscious, using only a portion of the available resources and varying their diet from place to place and from season to season to minimize their procurement costs and risks (cf. Reidhead 1980; Winterhalder and Smith 1981). The diet of many foragers, among them the Shoshone and the San, emphasizes plants over animals because plant foods are more abundant. When game is abundant, by contrast, hunting is more efficient than gathering, and meat sources dominate the diet, as among the Eskimo.

The technology used in food procurement is characteristically personal. It is small-scale, generally available to all families, multipurpose, and portable. The power of the technology to transform the ecosystem is limited, and the availability of resources is not generally much altered by human exploitation. (Exceptions of course exist, such as the overhunting of some large game species, the overharvesting of sessile shellfish, and various uses of fire.) The technology, however, is certainly not simple in the sense of being stupid. It is an appropriate and often ingenious solution to the problem of procuring resources at least cost.

Foragers follow a cyclical pattern of aggregation and dispersion that is responsive to the relative costs of procurement. When resources are uniformly distributed, the costs of exploiting them are uniform and maximum efficiency is gained with a dispersed population that minimizes competition among individual foragers. When resources are concentrated in one or two areas, the costs of exploiting them vary greatly according to the exploiter's distance from those areas; in such cases maximum efficiency is gained by aggregating population. Or, as we shall see in the Shoshone and San cases, resource availability may change through the year, with the population coming together at one season to exploit the concentrated resources of that season, such as the pine nuts of the Shoshone, only to break up again when food resources become more generally available.

Anthropologists have offered various explanations of forager social organization (Steward 1936, 1938; Service 1962; Lee and DeVore 1968; Williams 1974; Hayden 1981a). In this book we interpret the family level of low-density foragers as an adaptive response to particular environmental and economic conditions. The key economic conditions required for a family-level economy are often encountered. Technology is personal. The division of labor is elemental, and the labor required in a procurement activity rarely extends beyond the family. With little territoriality and comparatively free movement of population through a region, necessary resources are available more or less directly to all households. This elemental level of organization is, however, always part of a more complex social system that binds families together into camps and regional networks.

As we argue throughout the book, the primary causes of group formation are risk management, technology, warfare, and trade. Among foragers, risk management is of critical importance and results in the formation of informal and flexible social ties between families. The other three factors play an important role among some hunter-gatherers, such as the Eskimo (Case 6) and the Northwest Coast fishers (Case 9); but the social responses to these factors are of a different nature from the social response to risk management, reflecting, as we shall argue, higher population densities and more intensive subsistence economies.

The critical problem of risk stems from two somewhat different economic conditions. First, and more general, is the risk associated with plant gathering. On a daily basis gathering is quite predictable, since plants are sessile (immobile) and once located will continue to be available until taken up. On a year-to-year basis, by contrast, plant resources are unpredictable; an area that is good one year may fail utterly the next year. To compensate for this variability the population must be mobile, moving from one location to another to exploit the best available opportunities. But in order to do this, families must maintain broad regional networks of relationships, often with exchange and intermarriage, that give them access both to information on where food may be found and to the home ranges of groups with available food. A flexibility in group composition and a lack of territorial exclusiveness underlie the basic foraging economy and its use of fluctuating wild resources.

Second is the risk associated with hunting. Hunting, unlike gathering, is unpredictable on a daily basis: the game sought by the hunter cannot always be found and when found cannot always be killed. On a given day any hunter has a good chance of coming home empty-handed. The camp, consisting of a number of hunters, acts to average these high daily risks by sharing meat. Although the camp functions like the household in this regard, the sharing and cooperation is usually limited to meat and does not diminish the independence of the household, which can move from camp to camp.

By and large, the family level of organization is remarkably unstructured. Temporary social and economic rewards bring groups together only to have escalating procurement costs and social friction push them apart. Ceremonialism and leadership, two elements of group formation that we will track throughout the book, are ad hoc. They exist to resolve particular difficulties of group cohesion that occur only as long as the multifamily group is together. Both ceremonialism and leadership exist among foragers, but both are context-specific and comparatively unelaborated.

Where is the *band* of which so much has been made (Service 1962; Williams 1974)? In general, the band—a patrilocal group with exclusive rights to territory—appears often to be a construct of the anthropologist's search for structure in a simple society. The band in the sense of a camp certainly exists among foragers, especially where hunting requires a high degree of sharing. But the band as a territorially defined corporate group regulating marriages and resource use seems inappropriate to foragers because it would restrict the flexibility in movement on which their survival depends. The Owens Valley Shoshone come close to being a band in this sense, but as we shall see, they depend on relatively rich and dependable resources. Most low-density foragers, however, are not territorial because they cannot afford to be.

We now turn to the Shoshone and !Kung San cases to illustrate the similarities and differences among low-density foragers. Then we return to the more general issue of the place of foragers in the evolution of the political economy.

Case 1. The Shoshone of the Great Basin

The Shoshonean groups of the American Great Basin were historically low-density foragers. As we shall see, the Shoshone were in fact organized at different levels of complexity that represent the spectrum of hunter-gatherer "types" as outlined by L. Binford (1980; D. H. Thomas 1983a). But before evaluating this interesting example of evolutionary development, let us examine the family-level foragers as originally described by Steward (1938).

The organization of these foragers was minimally structured, with relations above the family being ad hoc, temporary, and minimal. Elemental family units of Shoshone came together and split apart according to the fluctuating availability of wild resources. Their organization of work and patterns of movement and association were adapted to exploiting sparse and unpredictable resources with a simple technology.

The Environment and the Economy

The Great Basin is dry, with rainfall at lower elevations typically less than 10 inches a year, falling seasonally in the winter months as snow; the vegetation is sparse and xerophytic. Especially during the hot, dry summers, water is restricted to small springs along the mountain bases and to the few permanent streams.

The topography of the Great Basin is broken, with elevations varying from valley floors at 3,900 feet up to towering mountains above 12,000 feet. Within the small home ranges of a local Shoshonean group, individuals had access to terrain with elevations varying by as much as 6,000 feet. Both rainfall and temperature change with elevation; for every thousand feet of elevation, mean rainfall increases about two inches and mean annual temperature drops about 3°F (D. H. Thomas 1972: 142).

This locally sharp variation in elevation and microclimate results in a vertical arrangement of microenvironments (Steward 1938: 14-18; Thompson 1983). Most important are the Basin Range Alpine Tundra (over 10,000 feet), the Limber and Bristlecone Pine zone (9,500-16,500 feet), the Sagebrush-Grass zone (7,500-10,000 feet), the Pinyon-Juniper zone (4,900-7,500 feet), the Sagebrush zone (4,900-5,900 feet), and the Shadscale zone (3,900-4,900 feet). Distinct plant and animal resources are to be found within these different microenvironments. In the high forested zones are the economically important pine nut trees, a number of plants producing useful berries, roots, and seeds, and several hunted species including deer, elk, and mountain sheep. In the drier lower elevations are seed-producing grasses, edible roots, jackrabbit and antelope, and fish in the permanent streams.

Seasonality is extreme in the Great Basin. Summers are hot and dry, with diurnal temperatures usually above 90°F (often above 100° F) and without significant rain. Winters are very cold and wet, with temperatures often below freezing all day (not unusually below 0°F) and with snow common, especially at higher elevations. These dry summers and wet winters make conditions difficult for a technologically simple society.

The Shoshone's home environment is harsh. Resources are scarce, unavailable in the wild through much of the year and unreliable from year to year. That foragers using a simple technology could survive here is a testament to their ingenuity. The population, subsistence economy, and social organization of the Shoshone are best understood as pragmatic solutions to the severe conditions of this environment.

Population density for the aboriginal foragers of the Great Basin was low, perhaps one person per 16 square miles (Steward 1938: 48), with variations from less than one person per 40 square miles to one person per two square miles (ibid., Fig. 6). D. H. Thomas (1972: 140-41) finds a modest correlation between the population density of the Shoshone and annual rainfall, much as Birdsell (1953) had found for Australian foragers. But the basic factor limiting the population density of foragers such as the Shoshone is not rainfall as such but the availability of food.

The Shoshone were broad-spectrum foragers. The bulk of their diet was provided by plant foods such as nuts, seeds, roots, tubers, and berries. Most important, when available, were pine nuts, which were harvested in large quantities during a brief period in the fall and stored for consumption during the winter, when they were the main food to be had. Late winter and early spring were times of hardship as stored foods ran out before new foods were available. Moreover, the pine nut crop is notoriously unreliable; the ripening cones are often damaged by wind, rain, and insect infestations, and harvests can be low. The severe seasonal famines recounted by Steward (1938) should caution us against any simple notion of forager affluence, especially where the availability of food varies seasonally and unpredictably.

Unlike most foragers, indeed perhaps unlike all others, the Shoshone made limited use of irrigation. In the Owens Valley, where population densities were unusually high, Steward (1930) reports that irrigation systems were developed to increase the yield and predictability of the grass seed harvest. As we shall see, the Owens Valley Shoshone illustrate certain aspects of intensification and social evolution in forager societies that foreshadow changes discussed in later chapters.

Hunting provided an important but clearly secondary addition to the Shoshonean diet (Steward 1938: 33-44). Included were such large game as deer, mountain sheep, antelope, elk, and bison, and such small game as jackrabbits, rodents, and reptiles, as well as fish and insect larvae. Although the range of animal species used appears quite extensive, hunting made up a small portion of the total diet, probably less than 20 percent.

The technology included simple, portable items, such as digging sticks, bags, and bows and arrows, that could be manufactured by each household. Most types of procurement, including all gathering of plant foods and grubs and some hunting of large and small game, required no cooperation beyond the individual household, which was organized internally by the sexual division of labor.

Warfare was rare or nonexistent, although raiding may have taken place in certain areas of higher population density such as the Owens Valley. Similarly, among the San individual acts of aggressive violence occur but intergroup aggression is rare.

Trade certainly existed among the Shoshone, as among other foragers. Most important appears to have been the exchange of food for raw materials, such as obsidian, for which local substitutes were limited or unsatisfactory. The extensive trade in obsidian has been well described for aboriginal California foragers (Ericson 1977).

To summarize, Shoshonean foragers had to solve six major problems of production and reproduction. They had to collect sufficient quantities of low-density plant foods, which they supplemented where possible with game. They had to cope with extreme seasonality and the extreme risk of food failure. They had to develop appropriate mating patterns, and to find reliable ways of obtaining needed raw materials. As we shall argue, family-level organization with ad hoc group formation, leadership, and ceremonialism was the effective way of getting all these things done.

Social Organization

By and large, gathering was an individual affair: although gatherers may work together for company, there is nothing inherent in the work that makes cooperation necessary. The risk is low on a daily basis. To be sure, resources such as the pine nut may vary from year to year; but within a year their availability is reasonably predictable once the status of the local crop has been ascertained.

Individual hunting was also common, but group hunting was perhaps more important in terms of its contribution to the food supply. Cooperative hunting of jackrabbit, antelope, and mudhen took place irregularly in the open lower valleys. Rabbit drives were impressive undertakings, requiring the coordination of fairly large groups. Huge nets, similar in height to a tennis net but hundreds of feet long, were placed end to end in a large semicircle. Then men, women, children, and dogs beat the brush over a wide area and drove the animals toward the nets. The jackrabbits caught in the nets were clubbed to death. "Rabbit bosses" provided the leadership needed in these drives— deciding when and where to hold a drive, where to place the nets, and what job to assign each of the participants.

Although much less frequent, perhaps only every twelve years, antelope drives were organized in a similar fashion. Animals were driven across a broad area into a funnel made of brush wings up to a half mile long that led into a circular corral where the herd was impounded and slaughtered. An "antelope shaman" thought capable of attracting the souls of the animals played a central role in coordinating the drive. These large-scale hunts sought to obliterate the local animal population in the interest of maximizing the immediate food supply; no attempt was made to save a breeding stock. A whole population was destroyed, and antelope were not hunted again until they reached numbers sufficient to justify another drive.

The most innovative part of Steward's (1938, 1955, 1977) work on the Shoshone was to show how the distribution and organization of groups were adapted to environmental patterns and corresponding resource procurement problems. The annual movement of population responded to the seasonal cycle of resource availability. In the fall families concentrated in the pine nut groves, where large harvests were prepared for storage. In the winter camps of some five to ten families were established near a spring for water and near the pine nut groves for ease in transporting nuts. In the spring, as the temperatures warmed, families departed from the monotonous life and monotonous diet of the winter camp and dispersed in search of new food sources. Nuclear families moved to higher and lower elevations and remained spread out through the summer. The environmental verticality and seasonality made for a pronounced patterned movement, often called a seasonal round.

Much of the year, then, the Shoshone moved as individual family units consisting of a father, a mother, children, and often a son-in-law, a grandparent, or some other closely related person. This unit, called a "kin clique" (Fowler 1966), corresponds to the elemental family of Steward (1977). Each family was a separate economic and decision-making unit.

During the fall and winter, camps of several family units formed around common resources such as water and the pine nut groves, but these camps of at most 50 people had neither a sense of communal integration nor a group leader (Steward 1977). Why did the winter camp form at all? Because of the water and pine nut resources, and because winter was a time of potential scarcity when it made sense to pool resources and average risks.

The weak development of the Shoshone camp as a suprafamily organization reflects the relatively minor importance of hunting, with its pressures for cooperation and sharing between families. The irregular rabbit and antelope drives were a different matter, occasioning a periodic shift toward a considerably more complex social organization. A large group, probably consisting of upwards of 15 families (75 or more people), gathered together for such a hunt. The group was led by a specialist, and its assembling was a time for gaiety and release.

On occasions of abundance, such as rabbit drives and unusually good nut harvests, many Shoshone families gathered for a fandango festival. As Steward (1938: 106-7) described for the Reese River Shoshone, the men from the assembled families hunted jackrabbits for five days, and at night all danced. The dancing was primarily for pleasure, and the festival was first and foremost a party when families who normally lived an isolated existence came together to enjoy each other's company, to dance, and to court. Although not a dominant element, ceremony was a part of this gathering; the round dance brought rain, and the recent dead were mourned.

The gaiety of the fandango marked the emergence of a temporary suprafamily group that, in addition to its recreational pleasures, had a number of important economic functions (D. H. Thomas 1983a: 86). First, the gathering pooled labor from many families without which the cooperative hunting of jackrabbits or antelope would have been impossible. Second, it made for the most effective possible use of the animals killed. Third, it facilitated the sharing of information about where food was to be found; that is, it radically reduced the costs of searching for food. Fourth, it served as an opportunity for trade in locally available raw materials, such as obsidian, and for building a network of friendships through exchange. Fifth, this was an excellent time to find a mate, not always an easy thing with small groups, low population densities, and infrequent encounters.

The ad hoc ceremonialism of the Shoshone illustrates an important characteristic of dispersed forager populations. Although normally spread out as families to make optimal use of dispersed resources, occasionally the population must come together in suprafamily activities that will benefit the collectivity. The ad hoc ceremonial, involving families from several winter camps, acts as a strong inducement to families to participate in such activities. As we shall see in the conclusion to this chapter and in later chapters, the development of ceremonialism becomes pronounced as territories become more defined and defended. The ceremony is an official invitation to neighbors to enter a group's territory without undue fear of attack.

These infrequent but economically and socially important gatherings highlight three points. First, population aggregation among foragers depends on locally dense resources that are frequently ephemeral and unpredictable. Second, leadership solves specific problems of organizing the activities of such a group, but like the large group itself this leadership is ephemeral and context-specific. Third, festival activities are very much tied to seasonal and irregular patterns of resource availability that encourage larger groups to form for economic reasons.

Among the Shoshone, there was also an apparent absence of strongly demarcated territories. Although families owned pine nut trees and facilities such as irrigation ditches, hunting blinds, and corrals, group territorial units were in most cases not demarcated (Steward 1977: 375-78). Rather, flexible and nonexclusive rights to use both plant and animal resources appear to have been characteristic. Warfare was of minor importance and not organized in precontact times.

Steward's description of the pragmatic and flexible Shoshone forms the basis for our model of a family-level society in which ceremonialism, leadership, warfare, and territoriality are of little importance. Service (1962), by contrast, argued that the family-level Shoshone were simply ethnographic remnants of a society of suprafamily "bands" who had been driven into marginal habitats by groups using horses and guns. Only archaeology can decide which model best fits the prehistoric Shoshone, and it has. In the 1960's and 1970's, D. H. Thomas (1972, 1973) studied the archaeological settlement pattern for the Reese River Basin, a Shoshonean area of relatively sparse and unpredictable resources. Finding that this pattern fits closely the predictions of site location, frequency, and type derived from Steward's model, Thomas (1983b) concluded that the prehistoric Shoshone of the Reese River Basin were a family-level society.

Recent work by D. H. Thomas (1983a) and by Bettinger (1978, 1982) shows that different groups of Shoshone across the Great Basin organized themselves in different ways, ways that can best be seen as local adaptations to specific resource conditions (Thomas 1983a). At one end of the spectrum were the Kawich Mountain Shoshone, living at very low densities (one person per 20 square miles) in a region with restricted water and sparse and unpredictable resources. Storage was uncommon because there was little to store; for the same reason the population was highly mobile, with a flexible family-level organization and no territoriality. Rules about who was eligible to marry whom were also flexible. Groups formed only irregularly for hunting drives and short fandangos. At the other end of the spectrum were the Owens Valley Shoshone, living at much higher population densities (one person per two square miles) in a well-watered environment that produced, with the help of irrigation, a comparatively rich and predictable resource base with a storable pine nut harvest. Populations were fairly sedentary, and some groups stayed in one centrally located camp through much of the year. These Shoshone were territorial and organized into local groups. Marriage rules were less flexible and became an important aspect of intergroup relations. The fandangos in Owens Valley provided the important function of permitting access across defended boundaries to food, trade, and mates (Bettinger 1982).

The Shoshonean case thus illustrates two kinds of forager organization. A population of low density, resulting from sparse and unpredictable resources, is organized at the family level, with suprafamily organization largely informal and ad hoc. A population of higher density, resulting from richer and more dependable resources, is organized at a higher level: as a local group with a defined territory. Why? We are not yet in a position to generalize. But it is clear that in the

Shoshone case the rich, predictable resource base of the Owens Valley both permitted a suprafamily group to form and may have required that group, if it was to survive, to defend its resources against encroachment by other populations living in less favorable environments.

We now turn to the San of the Kalahari, another family-level society but one in which camps are more enduring. The importance of hunting and camp organization will be explored.

Case 2. The San of the Kalahari

The !Kung San are our main ethnographic example of a forager society organized at the family level.* Until the mid-1960's, the San were relatively isolated from the outside, and they were one of the last societies to retain an independent hunting and gathering existence. Although their lifestyle is a specific adaptation to local environmental and economic conditions, we chose them for detailed analysis because so much has been written about them, especially with regard to the ecological and economic variables of central concern to our approach. Lee's (1979) excellent recent ethnography is our basic source; other valuable sources include Lee and DeVore (1976), L. Marshall (1976), Yellen (1977), Wiessner (1977), Howell (1979), Silberbauer (1981), and Leacock and Lee (1982). As we try to show, despite the absence of historical ties between the San and the Shoshone, they are similar in many elements of their culture cores, and differences between them can often be accounted for by contrasting environmental and economic conditions.

Like the Shoshone, the San are foragers dependent primarily on plant resources in a dry environment. Population densities are low, apparently limited by resource availability. Their family level of organization permits maximum flexibility in movement and in marriage, and suprafamily organization is informal and changeable. Territoriality, leadership, and ceremonialism are ad hoc and little developed, and warfare is nonexistent.

As in the Shoshone, we witness in the San the basic pragmatism of a family-level society. Decisions on what to eat, where to move, what group to join, and when to leave it are made by the family on the basis of straightforward evaluations of benefits and costs. As a correlate, the "affluence" of the forager even under severe conditions is evident, with some reservations.

*The ! in !Kung is a clicking sound that cannot be represented by a letter. Some other symbols of this sort are used in !Kung names in this chapter.

The San, however, do not live as isolated families but are organized into camps of several families and joined by personal networks of exchange that interconnect families and their camps across broad regions. The importance of these suprafamily organizations in handling the daily risks of hunting and the longer-range risks of an unpredictable resource base shows clearly the limitations of a family-level existence.

The Environment and the Economy

The !Kung San, the northern linguistic group of the San people, comprise over 15,000 people living in the modern nations of Botswana, Namibia, and Angola (Lee 1979: 34-38). Our information is particularly rich for the Dobe area that straddles the Botswana and Namibia border, where the traditional foragers have been studied in detail by the Kalahari Research Group (Lee and DeVore 1976).

The Kalahari desert is a large and dry basin 3,300-4,000 feet above sea level. The impression of this landscape is of an immense flatness (Lee 1979: 87). The underlying rock is covered by sand except for infrequent outcrops and downcut stream beds. The main topographic relief is provided by long, low dunes, separated by broad troughs, that run parallel to each other through the region. The dunes, stabilized by vegetation, create an undulating surface from the dune crests to the lower interdune trough, or *molapo*. Coarser, bleached sands are found along the dune crests, and finer sand and silt are deposited along the molapo.

The seasonal cycle in the Kalahari is characterized by a dry and cool winter and a rainy and hot summer. The !Kung San recognize five seasons based on differences in temperature and rainfall (Table 2). Bara is the time of resource plenty with rains and warm temperatures, and resource abundance continues into ≠Tobe as the landscape begins to dry out without rains. In !Gum days are a comfortable 75-81°F and without rains; but nights can become quite cool, with temperatures dropping close to or below freezing for about six weeks. Then in !Gaa

TABLE 2
!Kung San Seasons

Bara (summer) Dec.-Mar.	≠Tobe (autumn) Apr.-May	!Gum (winter) June-Aug.	!Gaa (early spring) Sept.-Oct.	!Huma (late spring) Oct.-Nov.
hot rainy	cooling drying	cool very dry	hot dry	hot showers

temperatures rise rapidly, with many days above 93°F, and the continued lack of rain parches the landscape. With the first showers in !Huma, the land quickly greens and plant resources again become more available.

A definite seasonality exists in the Kalahari, with times of resource scarcity, hot days, and cool nights. The extremes, however, are relatively mild, especially in comparison to the harshness of the Shoshone's environment. Wet does not combine with cold; and, except for a short period, heat is mitigated by rain.

From year to year, however, rainfall is variable, and dry years with poor plant yields are not infrequent. Lee (1979: 113) estimates mean annual rainfall at 18 inches, varying from a low of 8 inches to a high of almost 36. A drought (less than 15 inches) occurs two years in five, a severe drought (less than 13 inches) one year in four. Additionally, rainfall patterns, especially in the spring, are localized and can result in a marked local variation in food supplies.

The environment is fully natural, minimally altered by the San. The plant communities in the Kalahari are dominated by small trees, bushes, and grasses. Some regional differences in vegetation covary with rainfall and hydrologic patterns, but most correspond to different soil and water conditions (cf. Lee 1979: 97). In the loose and well-drained dune soils are sparse groves of broad-leaved trees like the mongongo (*Ricinodendron*); in the more compact and moister molapo soils are acacia trees and scrubs, with several important edible species.

Water is limited in the Kalahari. Permanent water sources are quite rare, restricted to rock fissures exposed in the dry stream beds. The Dobe area has nine permanent waterholes (Lee 1979: 306); some areas like Nyae Nyae are better supplied (L. Marshall 1976: 64), but others like ≠Kade are without permanent sources (Tanaka 1976: 100). Temporary water sources, existing for up to six months during the rainy summer, are found in the dry stream beds and molapo depressions. Additional water comes from small pools that collect in tree crotches following rain (Lee 1979: 94) and from plants (Tanaka 1976: 100-104, 114). For the San, water is a limiting factor.

There are about a hundred edible plant species in the Dobe area, including some forty species producing usable roots and bulbs and thirty producing berries and fruits. Fruits, melons, and berries are found in the summer and fall, and most roots, nuts, and bulbs are found in the winter and spring. Most important in the Dobe area is the productive mongongo tree, valuable for its fruit and nut. The mongongo fruit is seasonal, but its interior nut remains available on the ground throughout the year.

The !Kung rely on plant foods for about 70 percent of their caloric intake. In the diet recorded in Dobe for July and August, plants provided 71 percent of total calories and 64 percent of total protein (Lee 1979: 271). These figures are even higher in ≠Kade, where animals are scarcer (Tanaka 1976: 112).*

In Lee's (1979: 98-102, 158) summary description, the total range of plant resources eaten by the !Kung is impressive. This broad diet, however, shows considerable selectivity and flexibility to minimize procurement costs and to respond to environmental variability in space and time. The remarkable mongongo stands alone for its superabundance, year-round availability, and high nutritional value. In the July-August diet recorded by Lee (1979: 271), the mongongo tree provided 82 percent of the plant calories.

Both Lee (1979: 167-72) and Tanaka (1976: 105) record a hierarchy of preferred foods. Species are ranked on the basis of procurement costs (overall abundance, spatial distribution, seasonality, and collecting difficulties) and of desirability ("taste," perceived nutritional value, and side effects). For example, reflecting procurement costs, individuals preferred fruits to roots, and preferred roots found in soft shallow soils to roots that are harder to dig out. Interestingly, the ranking in foods varies markedly by region. In Marshall's area, mongongo is rarer than in Dobe and is of secondary importance behind Tsin bean (*Bauhinia esculenta*). In ≠Kade, where mongongo trees are absent, the nuts are not eaten (Tanaka 1976). Within the Dobe region the hierarchy of major and minor plants changes noticeably from waterhole to waterhole according to local species availability (Lee 1979: 176-80). In Lee's words (1979: 168), "plant foods are evaluated pragmatically and rationally; few species are restricted by magicoreligious taboos."

Animals are also important in the diet of the !Kung San. Over fifty mammalian species are recorded in the Dobe area, with various ungulates (notably kudu, wildebeest, and gemsbok) providing the most available biomass. The broken topography, however, restricts herd sizes to small groups or individual animals, and the scarcity of water limits animal populations. Animals are ranked by hunters on the basis of abundance and individual biomass (Lee 1979: 226-35); the most abundant ungulates are the most commonly killed. Other edible animals like lizards, mice, ostriches, African buffalo, and elephants are avoided because of low individual biomass, poor taste, high danger, or high procurement cost.

*It is important to remember that the /Gwi and //Gana of ≠Kade are different language groups from the !Kung. Most of the case's information is drawn from studies of the !Kung.

As might be expected, land-use intensity varies inversely with distance from a permanent water source (Yellen and Lee 1976: 44). Lee (1979: 175) describes the process of how a !Kung camp "eats its way out" from the living base:

The !Kung typically occupy a campsite for a period of weeks and eat their way out of it. For instance, at a camp in the mongongo forest the members exhaust the nuts within a 1.5-km radius the first week of occupation, within a 3-km radius the second week, and within a 4.5-km radius the third week. The longer a group lives at a camp, the farther it must travel each day to get food. This feature of daily subsistence characterizes both summer and winter camps. For example, at the Dobe winter camp in June 1964 the gatherers were making daily round trips of 9 to 14 km to reach the mongongo groves. By August the daily round trips had increased to 19 km.

This progressive increase in walking distance occurs because the !Kung are highly selective in their food habits. They do not eat all the food in a given area. They start by eating out the most desirable species, and when these are exhausted or depleted they turn to the less desirable species. Because plant food resources are both varied and abundant, in any situation where the desirable foods are scarce, the !Kung have two alternatives in food strategy: (1) they may walk further in order to eat the more desirable species or (2) they may remain closer to camp and exploit the less desirable species.

As the preferred mongongo continues to be exploited, it must be exploited at greater distances, which means increasing transport costs. As the unit cost of mongongo climbs, people shift to less preferred but now less costly alternatives.

Throughout the year the use of resources closely reflects seasonal availability and procurement costs. Mongongo, the least seasonal of all resources, is procured all year long, albeit with less frequency near the end of the dry season, when transport costs from the waterhole increase. Other resources are used in a more seasonal pattern (Lee 1979: 188-90). During the wet season, when plants are most available, emphasis in the diet is on such easily procured resources as fruits, berries, and melons. Waterfowl and some migratory ungulates are also hunted. Then, during the dry season, when food is less available, the diet is broadened (Yellen and Lee 1976: 44, 45) to include higher-cost foods like roots and bulbs. As indicated in Chapter 1, the broadening of the diet under resource stress is predicted by our model of the subsistence economy.

From year to year changing rainfall and other environmental factors also determine the availability of resources, and by extension affect the dietary pattern of the !Kung. Lee (1979: 174) records a major shift in plant food hierarchies between two years of differing rainfall. Good years for a favored species result in decreasing procurement costs for that species and an increase in its use. Bad years cause significant broadening of the diet.

The annual movement of the San population through the environment, as they gather around limited winter water and then disperse close to plant resources, is designed to minimize procurement costs (Fig. 2). In the Dobe region, as an example, water distribution creates a pulsating pattern in population that Lee (1976; 1979: 103-4) has called a "dialectic" of concentration and dispersion. During the winter, the dry season, standing water is limited to a few permanent waterholes around which the !Kung cluster. These base camps, such as the one at Dobe, can be quite large (perhaps 35 people in 12 huts) and are occupied for more than half the year. In the Kalahari there are many more camps than permanent water sources, so that several camps (from two to six) cluster around a single waterhole (Lee 1976: 79). When the spring rains begin in October and November, the camps quickly disperse to temporary camps in the mongongo groves, where they use the water that collects in the trees' hollows. These camps are smaller (perhaps only 10 people) and are occupied for only a few days each. As the seasonal pans fill up with summer rains, the population becomes maximally dispersed while remaining close to both water and resources. As autumn comes on, the pans begin to recede; the population falls back on the larger pans, and eventually on the permanent sources.

The goal is to maintain the maximum possible dispersal of camps consistent with the availability of water. This goal corresponds to the strategy of minimizing procurement costs in terms of movement to and from the camp. In addition to this annual pattern of movement, the distribution of the San population responds to unpredictable changes in water sources under the drought conditions common to the Kalahari. The "permanent" waterholes do not always retain water in dry years, and these sources can be ranked according to the severity of a drought in which they will dry out. Under drought conditions, San camps progressively aggregate around the more amply supplied sources. During one bad drought, for example, J. Marshall (1957: 36) found seven camps at one waterhole. During such droughts whole areas in the Kalahari may be abandoned as the San emigrate in search of suitable water and food (Hitchcock 1978). This flexibility in movement is essential to the San economy, which relies on social ties rather than storage for handling risk (Wiessner 1982).

The !Kung's diet and movement patterns illustrate the importance of cost considerations in deciding what resources to exploit and how to

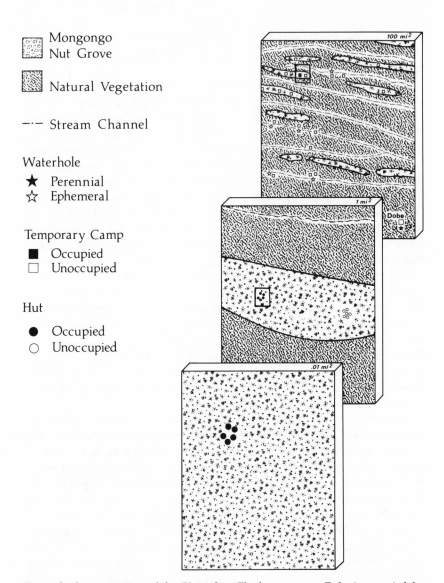

Mongongo
Nut Grove

Natural Vegetation

—·— Stream Channel

Waterhole
★ Perennial
☆ Ephemeral

Temporary Camp
■ Occupied
□ Unoccupied

Hut

● Occupied
○ Unoccupied

Fig. 2. Settlement pattern of the !Kung San. The base camp at Dobe is occupied for much of the dry season, but during the wet season camps scatter and sites are occupied for only a few days at a time.

exploit them. Their high selectivity and flexibility in space and time correspond to changing resource availability and procurement costs. Environmental conditions, because of their direct effect on procurement, determine in large measure the nature of the subsistence economy and related social and cultural characteristics. This close interrelationship between ecology, economy, and society, which is summed up in Steward's (1955) notion of "culture core," is central to our concerns in this book.

Sahlins (1968, 1972) has called the !Kung "an original affluent society" because they are able to meet their limited wants without major difficulty and thus have ample leisure time to gossip and gamble. The limited wants of a foraging society are linked to the mobility of groups and the costs of carting around any excess baggage. Equally important for hunter-gatherer affluence is a low population density; for it is this low density, with its corresponding low demand on resources, that keeps costs low in the subsistence economy.

The regional population density of the San averages one person per 10 square miles, varying from one per 13 square miles in ≠Kade to one per 8 square miles in Dobe and one per 1.4 square miles in Nyae Nyae (Tanaka 1976: 1100; Lee 1979: 41; L. Marshall 1976: 18-19).* As might be expected, the population density order roughly corresponds to the availability of permanent water sources for these areas (none in ≠Kade, 9 in Dobe, 16 in Nyae Nyae). Population is not evenly distributed in a given area, but is highest within one day's travel of the permanent water sources. In Dobe this "economic density," as Lee calls it, is about one person per square mile (Lee 1979: 306).

What limits population to such low densities in a foraging society like the San? The obvious explanation is the scarcity of resources, and the correlation of population density with water availability seems to bear this out. Yet Sahlins sees the San as affluent, and Lee makes it clear that they rarely have trouble getting enough to eat. In a one-month study Lee (1979: 271) estimated an individual's average daily intake of 2,355 calories and an average daily expenditure of 1,975. On the strength of these figures, obesity might be a bigger problem than starvation!

Maybe, instead of *average* scarcity, we should focus on *periodic* scarcity. The San do not eat that much or expend that little every month of the year; if the year is taken as our unit, Wilmsen (1978) postulates a cycle of weight gain and loss of roughly 5-10 pounds.† The San diet,

---

*The relative order of these three San regions is probably more accurate than the absolute figures, since there is no standard way of calculating population density.

†Lee disagrees (1979: 440-41). See also the discussion in Konner 1982: 372-73.

generally well balanced, may at times be calorie-deficient (Truswell and Hansen 1977), with the result, according to Howell (1979), that fertility rates in San women are low. This argument is based on more general work suggesting that a minimum level of body fat in females is necessary for fertile ovarian cycles (Frisch 1978). In addition, the periodic droughts cut the population back to levels well below average carrying capacity (cf. Hitchcock 1978).

Another factor limiting population growth in foragers such as the San is long spacing between births. Lee (1979: 324) suggests that births were traditionally spaced about four years apart; with a relatively short reproductive period, population growth in these circumstances would thus have approached zero. Why was spacing so long? Perhaps it was based on the mother's potential productivity as a gatherer. Lee speculates that because a San woman carries her children under four years old with her as she moves about in her gathering tasks, her workload is greatly affected by the number and weight of children that she must carry. With a birth spacing of four years, a woman does not need to carry more than one child at a time. As spacing increases from two to three to four years, the maximum weight of children that must be carried decreases from 47 to 41 to 26 pounds.

How is this apparently desirable spacing maintained? One possibility is that women choose to limit births as one way of lowering their food procurement costs. Birdsell (1968a: 243) has suggested that infanticide may be used by foragers to space births; but San women rarely practice infanticide—Howell (1979) records 6 cases out of 495 births. A more likely explanation, we think, is nutritional deficiencies as discussed above. Another is the long nursing required by a lack of suitable weaning food (Konner and Worthman 1980; Lee 1979: 328). Both appear to inhibit ovulation and thus provide a biological mechanism to limit growth rates.

Briefly, then, we think some combination of biological and economic factors, along with the occasional disastrous year, acted to keep the !Kung San population low, and that this low population density perpetuated foraging as a way of life. In other words, the San preferred the foraging economy to its local alternative of pastoralism because of its cost advantage over domestication *on the average*, and this cost advantage remained in place because of the biological and ecological limits to population growth.

!Kung technology consists of a few multipurpose tools made from locally available materials (Lee 1979: 119). Included are the women's kaross (a treated animal skin used to carry foods and other materials), the digging stick used for procuring roots and bulbs, the man's bow and arrow used for stalking the local mammals, the general-purpose knife used for all cutting jobs, and the ostrich eggshell canteen. The tools typically are made from natural materials that require little formal alteration. Today scavenged metal is cold-hammered to the desired form for arrowheads and knives, and trade goods such as pottery and Western metal pans are becoming increasingly important. But the traditional tools of the !Kung were manufactured individually from local materials for the maker's own use.

Unlike the Shoshone, the !Kung San have no storage facilities for vegetable foods, presumably because unstored food is available in adequate quantities throughout the year. (Although the !Kung also do not store water for extended periods, the /Gwi San are reported to bury several hundred filled ostrich eggshell canteens in preparation for the dry season [Lee 1979: 123].)

For all its simplicity, !Kung technology is effective and often ingenious. For example, the light bow (20-pound pull) is deadly effective against even large game thanks to the use of arrows tipped with a poison derived from the pupal form of chrysomelid beetles (Lee 1979: 133-34). The arrow itself is an ingenious composite tool made from the stem of a large perennial grass, a bone linking shaft, and a metal point hammered into shape from a piece of fencing wire. Other effective implements include an iron-tipped hunting spear to finish off wounded animals, a springhare probe to hook the sleeping animal deep in its burrow, and baited rope snares for smaller mammals and game birds.

E. M. Thomas (1959) titled her book about the San *The Harmless People*. Warfare, in the sense of organized intergroup aggression, is not present among them, and outward signs of violence are discouraged. To be sure, homicide, especially between men in conflicts over women, is not uncommon: angers swell, and poison arrows fly. Such conflicts, however, are seen as disruptive, and the aggressors are not supported. Lee (1984: 96) records the following dramatic scene:

/Twi had killed three other people, when the community, in a rare move of unanimity, ambushed and fatally wounded him in full daylight. As he lay dying, all the men fired at him with poisoned arrows until, in the words of one informant, "he looked like a porcupine." Then, after he was dead, all the women as well as the men approached his body and stabbed him with spears, symbolically sharing the responsibility for his death.

Personal violence is not allowed to expand into intergroup conflict because of the overriding importance of intergroup ties; rather, disputes are settled by separation.

Trade in special craft products, and now especially in Western goods, exists among the San, and, as in other foraging societies, probably existed on a small scale in prehistory (Lee 1979: 76). As with the Shoshone, frequent trading would not have been necessary because the range of objects used was limited and generally of long life. Nothing like economic specialization existed.

To summarize briefly, the major problems of production and reproduction facing the San are remarkably similar to those facing the Shoshone. Like the Shoshone, they had to collect an adequate supply of low-density plant foods, which they supplemented by hunting dispersed and unpredictable game. Their problems of seasonality and possible food failure, if less extreme than those confronting the Shoshone, were broadly analogous, as were their needs for a reliable system of finding mates and a way of obtaining special craft goods by trade. Not surprisingly, the overall pattern of San organization is also very similar to the Shoshone pattern. The different role of hunting and its implication for camp organization is the main difference.

Social Organization

Among the !Kung San, as in other simple subsistence economies, the family with its own shelter and hearth is the elemental economic and social unit. The individual or the family makes all basic economic decisions: what to collect, how to collect it, when to move, what group to join (cf. Yellen 1977). Goods are brought into the family by its members, who are involved in different procurement activities according to the sexual division of labor. Within the household, resources are pooled and shared out freely. Most plant foods eaten by the family are gathered by family members.

The organization of work among the !Kung is, like the technology, a simple and direct response to procurement problems. Most subsistence activities can be performed by individuals working separately. Women are the gatherers; they also do some manufacturing (e.g. of canteens), most food preparation, and all child care. Men are the hunters; they also do some gathering and considerable manufacturing, especially of materials used in hunting.

Gathering is usually done individually or in small groups (Lee 1979: 192-93; L. Marshall 1976: 98). Groups involve parallel work, with no division of labor and no obvious gain in efficiency over lone procurement. In the mongongo nut harvest, for example, individuals go out in a group but each works a separate tree, and the roasting and cracking of nuts are done individually. Tasks are performed together but are not coordinated, except to establish a rhythm for the work.

In hunting, men also work singly or in small groups. Because there are no herds in the Kalahari, large hunting parties are impractical. Among the San a hunting party for large game consists of one to four men (Lee 1979: 211; L. Marshall 1976: 132). When an animal is spotted, a single hunter stalks it so as to minimize the chance of alarming it (Lee 1979: 217). The group of three or four is, however, important at several points. As an animal is pursued, its tracks often become obscured and the hunting party fans out to pick up the trail. When the animal is spotted and a single hunter begins to stalk it, others position themselves at likely escape points to get a second shot. Once an animal has been mortally wounded, three to six people butcher it and carry the meat back to camp. This activity requires a cooperative work effort, because a single hunter cannot carry a large kill himself and without assistance would have to abandon edible meat to scavengers.

For the San, even more so than for the Shoshone, an organization above the family level is essential for the family's survival. The two levels of suprafamily organization are the camp and the regional interfamily and intercamp network. Although these levels are highly flexible and informal, they are essential to handle problems of subsistence risk.

The camp is the basic local group, a noncorporate, bilaterally organized set of people who live together for at least part of the year. A camp settlement commonly has five or six small grass huts (about six feet across) that face onto a central space (Fig. 2). A hut houses a nuclear family, and a camp consists of several closely related families (Yellen 1976: Fig. 2.1). In Lee's (1979: 56-57) study, camp size ranged from 9 to 30 persons. Camp groups, adding and losing members, move through the environment to position themselves close to critical resources. Sometimes, especially in the winter, camps are close to each other and the staccato sound of mongongo nut pounding (called !Gi speech by the San) drifts from camp to camp. At other times the separate camps are spread out across the vast and empty landscape.

The upper limit on camp size is apparently set by internal disputes that fragment the camp, and by the higher resource costs associated with larger groups. (Larger groups eat up the resources in the immediate camp area more rapidly, resulting in rising transport costs and more frequent moves.) The lower limit is established by the desire to maintain a ratio of producers to dependents of about 3 to 2 (Lee 1979: 67) and by the requirements of hunting.

Sharing, an important cultural value among the San seen most clearly in the distribution of meat from a large game kill (L. Marshall 1976), binds the camp together economically. According to Marshall's account, for example, the carcass belongs to the owner of the first arrow to wound the beast. Because of reciprocal exchanges of arrows among camp hunters, however, the owner of the arrow is often not the successful hunter (Lee 1979: 247). Meat from the kill is distributed by the owner to close and more distant relations within camp until everyone in camp has a share.

This broad sharing of meat handles two problems. First, it distributes food that could not possibly be eaten by a single family without storage; second, it averages the risks of unpredictable hunting so that all families get a share regardless of an individual hunter's success. Meat exchange eliminates what might otherwise be moments of intense envy and friction when one hunter's success is contrasted with another's failure. Hunting creates the need for an exchange-based group larger than the nuclear family and socializes it through generalized reciprocity.

Although the group is of great economic importance to the San, its membership is not rigidly defined. People may affiliate with a camp through bilateral descent or marriage, so that a household can join any of several camps. Marriage rules are very flexible and help create a web of kin relations between camps. Visiting, which incurs a reciprocal obligation, is so common that the number of people in a camp varies from day to day. Individuals form broad networks of exchange (*hxaro*) that bind families and give access to a partner's camp and territory (Wiessner 1977, 1982). These regional networks, which permit a family and its camp to move comparatively freely in space, also permit rapid adjustments to changing economic opportunities across the San landscape. They are central to the San's adaptation to changes in resource availability (Lee 1976).

Regional networks are created when camps cluster together around the permanent waterholes during the dry season. This is the time for ceremonies and intercamp activities. The dry season is a social time, and the whole tempo of life changes as population gathers. Lee (1979: 446-47) argues that this concentration of population offers strong social rewards in addition to economic ones, so strong in fact as to override the consideration that such aggregates may not be optimal for resource procurement in the short term. When several camps gather for all-night trance dancing and curing ceremonies, marriage brokering, socializing, and exchange make and strengthen ties within and between camps.

The concentration and dispersion dialectic described by Lee represents a very real and basic human ambivalence toward life in groups.

The self-sufficient individual and his family enjoy independence and an ability to control their own destiny. The group offers social rewards and critical economic assistance, but also limitations, frustrations, and personal conflict. The small social group of the camp has probably been necessary to the individual since earliest hominid times, satisfying as it does what Goldschmidt (1959) calls the "need for positive affect." The tension between family and group persists, but is secondary to the group's manifest economic and social advantages. The group eventually fragments when resources are broadly distributed and predictable, only to come together again when resources are localized and uncertain.

The regional mobility of the San requires a deemphasis on territoriality. Steward (1936: 334-35) described the San as organized territorially into patrilineal bands, but the reinterpretation presented by Lee and his collaborators emphasizes nonexclusive access. A group's "home range" is simply the area it uses most frequently, an area that is not sharply bounded, is nonexclusive, and is not actively defended. (See DeVore and Hall's [1965] description of a baboon group's home range.) Territoriality is not based on recognized boundaries but is focused on a key resource, which for the !Kung is the waterhole.

The !Kung recognize a locality (*n!ore*), of roughly 100-200 square miles, that is associated with a core group with long residence in the area (Lee 1979: 334):

Within a n!ore, the water hole itself and the area immediately around it is clearly owned by the Kausi group [a camp], and this ownership is passed from one generation to another as long as the descendants continue to live there. But this core area is surrounded by a broad belt of land that is shared with adjacent groups. In walking from one n!ore to another, I would often ask my companions, "Are we still in n!ore X or have we crossed over into n!ore Y?" They usually had a good deal of difficulty specifying which n!ore they were in, and two informants would often disagree.

Access to resources within the n!ore is said to be unrestricted for members of and visitors to the associated camp (Lee 1979: 335-36). A different camp must ask permission of the core group to use a n!ore's resources, especially its permanent water. This right apparently may be refused; if it is accepted its acceptance imposes a reciprocal obligation on the visiting camp. The general impression, however, is that access to resources is only minimally encumbered, and that individuals may gain access to resources either as visitors or as members of requesting camp.

Yellen and Harpending (1972) have emphasized the lack of territ

riality among the San, which they see as inevitable in an unstable environment in which population must continually distribute itself according to variable resource yields. Although group ownership of land tracts is recognized, as is individual ownership of tools, harvested foods, and perhaps some natural resources, the low population density and minimal capital improvements that characterize San existence appear not to require restricted access to land. In addition, to defend such a low-density resource as a San land tract would almost surely cost more than the tract is worth.

To summarize, the San camp has a fluid composition and no clearly demarcated corporate nature. Although the pervasive exchange of meat among camp members may give the camp the appearance of a clearly demarcated group, in many other aspects it is an opportunistic aggregation of independent households.

As this assessment suggests, the camp is largely without established leadership. Leadership is minimal and informal. Lee (1979: 343-44) summarizes the situation as follows:

> In egalitarian societies such as the !Kung's, group activities unfold, plans are made, and decisions are arrived at—all apparently without a clear focus of authority or influence. Closer examination, however, reveals that patterns of leadership do exist. When a water hole is mentioned, a group living there is often referred to by the !Kung by a single man's or woman's name: for example, Bon!a's camp at !Kangwa or Kxarun!a's camp at Bate. These individuals are often older people who have lived there the longest or who have married into the owner group, and who have some personal qualities worthy of note as a speaker, an arguer, a ritual specialist, or a hunter. In group discussions these people may speak out more than others, may be deferred to by others, and one gets the feeling that their opinions hold a bit more weight than the opinions of other discussants. Whatever their skills, !Kung leaders have no formal authority. They can only persuade, but never enforce their will on others.

To be sure, age and special skill confer respect, and a respected person's opinion carries weight when a decision must be made, e.g. a decision to move camp. L. Marshall (1976: 133) notes that in a hunting party a recognized good hunter acts as an informal leader. However, when asking a !Kung elder about local leaders ("headmen"), Lee (1979: 348) was told: "Of course we have headmen! . . . In fact, we are all headmen. . . . Each one of us is headman over himself!" A repeatedly successful hunter is respected but can also be envied, and he will often stop hunting for a while rather than try to assert strong leadership over the group. Leadership appears to be largely specific to a context, such as an individual hunting party, and does not extend

generally to camp affairs. Most decisions made by the group are by consensus; they are largely informal and are reached through long discussion by all concerned (Silberbauer 1981).

The Shoshone and San cases are remarkably similar for cultures without historical ties and at opposite ends of the earth. In both cases aridity and environmental variability make the region marginal for agriculture or pastoralism, and as a result foraging has continued as the basic subsistence mode down to the recent past. In both cases population densities are very low and highly dispersed, and both subsistence economies are essentially pragmatic, selecting among possible food resources so as to meet the group's needs. The resulting diets derive the bulk of their calories from gathered plant resources. Hunted meat, although highly desired, is of secondary importance. (This deemphasis on hunting is very important; as we shall see, it may not apply to all forager groups.)

Technology is simple, portable, and general-purpose. Social organization is equally simple, with built-in flexibilities that permit populations to adjust quickly to changing resource needs. The elemental unit is the nuclear family, organized especially by a sexual division of labor to be a self-sufficient economic unit. Families, although free to move on their own in search of food, unite at least during some seasons into camps with several families. These camps are not highly structured politically, and membership fluctuates as families join and leave the group. Throughout the year there is a pattern of concentration and dispersion of population in response to seasonal shifts in resource availability. Camps come together around resources, and this becomes a time for ceremonies and exchange. Leadership within the camp is informal, and decisions are reached through consensus. Territorially, camps are identified with specific home ranges, perhaps focused on a particularly important resource such as the San waterhole, but these territories are not defended and are generally open to people from other camps through reciprocal agreement.

Certain differences, however, do exist between the Shoshone and San, differences that reflect specific contrasts in their environments. For example, the marked seasonal differences in the Shoshone region require the use of food storage to avert starvation in the late winter. In winter camps the sharing of stored pine nuts is an important element in group cohesion. The San pool their risks in a different way, though one that finds an echo in the Shoshone's occasional rabbit or antelope drive. Hunting contributes daily to the San diet, and reciprocal arrangements within the camp are ideal for distributing the more risky yield of meat in comparison to the more predictable yield of plants.

The Shoshone represent a dichotomized society, spread out in nuclear families for gathering plant resources and then briefly concentrated for group hunting. The San represent a more stable intermediary position, with a balanced gathering-hunting economy and less variation in organizational makeup.

Prehistoric Foraging Societies

What do these close looks at the Shoshone and the !Kung San tell us in a general way about foraging, especially in prehistoric times when it was the universal mode of human existence? To review the evolution of foraging societies prior to agriculture, we shall consider briefly three periods that witnessed major changes for early human populations: the Lower and Middle Paleolithic, the Upper Paleolithic, and the post-Pleistocene. These periods saw three progressive "revolutions" in human society, resulting finally in a foraging mode that we believe was analogous to that of the Shoshone and the San.

First and by far the longest period was the Lower and Middle Paleolithic (very roughly 2,000,000-35,000 B.P.), the time of human origins both as a biological species and as tool users. During the Pleistocene or Ice Age, our erect hominid ancestors developed into modern *Homo sapiens*, and members of this species, with greatly enlarged cranial capacities (increasing from perhaps 650 cc to 1,450 cc), began using tools as their basic means of adaptation. A slow but consistent population growth resulted in some increase in population density, but more significant was the spatial expansion from an initially restricted distribution in Africa to a very broad distribution throughout Africa, Europe, and Asia. This unprecedented expansion resulted in part from population growth in the core areas, and in part from the opening of unexploited environments with the help of new technologies. Critical technological inventions included fire and clothing for survival in harsh European and Asian winters close to the glacial ice masses, and efficient strategies for hunting large game. Early in this period, the diet was apparently quite eclectic, including small and large game either scavenged or clubbed to death at close range. By the Middle Paleolithic hunting appears to have become more important thanks to the development of an effective hunting technology, including well-crafted stone projectile points that must have been hafted to throwing spears.

What archaeological evidence we have suggests that these hominids were organized into small mobile groups not unlike a San camp. East African sites as early as DKI from Olduvai Gorge (1,750,000 B.P.) and Olorgesailie (480,000 B.P.) have been interpreted as base camps for groups of 25-30 members (Isaac 1978; Wenke 1980). Isaac believes that concentrations of stone tools with butchered animal bones document home base camps, and that these early hominids were already organized as family groups with sharing and a sexual division of labor. In a recent article, Potts (1984) suggests instead that the observed concentrations were tool caches distributed in the environment for use in butchering. Although these were not base camps, which would have required defense against carnivores interested in the fresh meat, he suggests that they were repeatedly reused locations associated with a small group of early humans. Certainly Middle Paleolithic sites (100,000-35,000 B.P.), including the important caves of the Dordogne region in France, were repeatedly occupied base camps.

Artifact assemblages from the end of the Middle Paleolithic have been grouped into "tool kits" associated with different economic activities (Binford and Binford 1966). For example, Tool Kit I had twelve tool types, including borers, end-scrapers, and knives, that were apparently used to work bone, wood, and skins; Tool Kit V had six tool types, including projectile points, discs, scrapers, and blades, that were apparently used in hunting and butchering. Sites differed systematically in the tool kits found, suggesting to the Binfords that certain sites emphasizing a broad range of activities including manufacturing were probably base camps and other sites emphasizing food procurement were short-term special-activity sites. The implied settlement pattern fits the general aggregation-dispersion model outlined in our forager cases.

The basic family-level organization described for the Shoshone and San fits well with our still-sparse information on the earliest humans. Group size was apparently small, and the lack of stylistic differences in the artifacts suggests that membership may have been flexible. The importance of protection from predation in the Lower Paleolithic and the importance of hunting in the Middle Paleolithic would have necessitated a camp group integrated by generalized reciprocity.

Much has been made of the Upper Paleolithic transition that took place at the end of the Pleistocene between 35,000 and 12,000 B.P., especially in Europe (Conkey 1978; Gilman 1984; Hayden 1981b). The dramatic changes of this time in the economy and social organization of humans were apparently impelled by a continued growth in population; the spread into the New World took place during this period, and a sharp increase in the number of recorded sites argues strongly for higher population densities.

Coupled to this population growth must have been a significant intensification in resource use. New technologies included spear throwers (*atlatl;* said to increase the range of a throw from 200 feet by hand to 500 feet by atlatl), barbed harpoons, and fish gorges (Wenke 1980). In many economies the staple food seems to have been large migratory game animals such as reindeer or wild cattle. It is not clear why, since intensification characteristically results in a broadening of the diet (Earle 1980a). Perhaps because intensification is most difficult in the leanest season (winter), when few if any additional food sources are available, people in some areas solved their food problems in the Shoshone manner, not by broadening their diet but by increasing the exploitation of a rich resource available in fall that can be stored for use in the winter. Whatever its origin, this focus on a rich, storable resource appears to have had a profound effect on human society.

The settlement pattern of the Upper Paleolithic probably continued to include base camps and special activity sites. The main change is in the size of settlements at the base camps. Settlements such as Solvieux in southern France could be quite large (over seven acres) and probably represented a group of several hundred people (Sackett 1984). At Dolni Vestonice, a palisaded camp in Czechoslovakia, five huts were occupied. One large hut (45 feet long) contained several hearths, suggesting that it housed several nuclear families. In general, the size of settlements in the Upper Paleolithic implies a local group larger than is found commonly among foragers and much closer to our expectations for village communities like the Yanomamo (see Case 5).

As we shall see in Part II, with the formation of more or less long-term groups of a hundred or more people comes considerable institutional elaboration that includes ceremonialism and group leadership. In the Upper Paleolithic, cave art from sites such as Altamira in Spain and Lascaux in France, and carved "venus" figures from eastern Europe, offer unambiguous evidence of group ceremonial activities. Various Upper Paleolithic artifacts, such as the large and carefully flaked Solutrean spear points and the bone batons with engraved animals, are almost surely status markers of leadership.

Can this be? Can hunters be organized well beyond a family level? As we argue in the Eskimo and Northwest Coast cases, we believe that foragers develop higher levels of integration under particular economic and political conditions. It seems plausible that local groups and Big Man systems existed for the Upper Paleolithic. Three conditions potentially important for this development were risk management, large-scale hunting, and territorial defense.

The need for risk management in hunter-gatherer populations like the Shoshone and the San is commonly identified as leading to social relationship beyond the nuclear family. Since there is no reason to suppose that the nature of risk changed significantly from earlier periods to recent times, it would seem that the camp arrangements and flexible regional exchange network that characterize foragers like the San are little different from what they were 30,000 years ago.

The problems of hunting large migrating game have been singled out by both S. Binford (1968) and Wobst (1976) as causing the cultural elaboration seen in the Upper Paleolithic. To simplify their arguments, hunting migratory species like reindeer requires far more hunters than the local camp can provide and therefore leads to ceremonial elaboration as a way of integrating normally dispersed camp groups. As the Shoshone case shows, such a tie between aggregation for hunting and ceremonial elaboration is plausible enough, but its importance for the Paleolithic has recently been questioned by Gilman (1984). Essentially, cases of dependence on migratory herd animals vary considerably and do not correspond closely to cases of cultural elaboration. In particular, the Upper Paleolithic populations in Spain that produced some of the most sophisticated art were dependent on deer, which were apparently nonmigratory and would probably not have been hunted in large group drives.

Territorial defense may also have been critical for some Upper Paleolithic societies that depended on large game. Gilman (1984) argues cogently that the local style groups that characterized the Upper Paleolithic can be seen not as mechanisms for including greater numbers of people for cooperative hunting, but as mechanisms for *excluding* people by defining a bounded social group. In essence, Pleistocene societies, with their high population densities (for hunters), were dependent on exclusive access to favorable locations for hunting. In this context the stylistic differences that differentiated one local group from another in the Upper Paleolithic may in fact represent attempts to limit the extent of social obligations and thus restrict access to critical resources by neighboring groups.

Specifically, it can be argued that linked technological improvements and population growth focused hunting on highly productive and storable animal foods. To hunt such species efficiently, hunters must control their migratory paths, such as mountain passes and river ponds for reindeer (S. Binford 1968) and river bends for salmon fisheries (Jochim 1984), or, for nonmigratory species like deer and mammoth, their optimally productive home ranges. An intensified use of animal resources would tend to emphasize differences in the costs of hunting from locale to locale, and would thus increase the benefits to be obtained from group defense of the best hunting areas.

The development of local groups in the Upper Paleolithic is getting ahead of our story and into issues described in Chapters 5-8. Let us return to our family-level foragers. During the immediate post-Pleistocene period (12,000-7,000 B.P.), referred to as the Mesolithic in Europe and as the early Archaic in the New World, the diet of human populations in many areas changed radically to include a large number of new species (L. Binford 1968; Mark Cohen 1977). Changes in the environment helped make this change necessary, but its main cause was the growth of human populations. In many locations, such as the Desert Cultural areas of western North America, the subsistence economy for the first time incorporated plant resources. This process of intensification, which has been called the "broad-spectrum revolution" (Flannery 1969), appears to have taken place on a worldwide basis (Mark Cohen 1977; Christenson 1980). With the expansion into virgin territory largely complete, further population growth required intensification.

The most common outcome of this broadening diet was a concentration on plant foods that created a subsistence economy generally analogous to that of the Shoshone and the San. It was during this period that the basic family-level society spread throughout the world, and it is from this base that we trace the evolutionary development described in this book.

Even from this broadly shared foundation considerable diversity existed. In some areas intensive hunting continued along with a territorially organized society: for example, the camelid hunters and early pastoralists studied by Rick (1978, 1984) in the Central Andean *puna* (see Chapter 11), whose local groups were apparently sedentary and distinguished by stylistically diagnostic stone projectile points. In other areas intensification concentrated on rich resources that could be stored to carry the population over lean periods: for example, the pre-agricultural Natufian villages of the Levant (Flannery 1972), where sedentary local populations harvested and stored abundant wild grains. But overriding this diversity was a common pressure that was to result gradually in a shift to domesticates and fundamental changes for human society.

Conclusions

To understand the overall evolution of hunter-gatherers, we must consider the three main evolutionary trends—intensification, integration, and stratification—as they relate to the changing economy and society.

The intensification of food procurement activity in a given area is necessitated by an increasing population and/or a deteriorating environment. In the Pleistocene and immediate post-Pleistocene periods, a slow growth in population spread humans throughout the world and gradually raised population densities in areas capable of sustaining more people. Intensification in food procurement was the result (Mark Cohen 1977). First came the gradual occupation of new habitats with suboptimal resources, such as low-density large game, small game, and plants that require more costly procurement strategies. Next came the diversification of diets as increasingly costly species were added to support a larger population. Both trends increased the amount of labor devoted to procuring food. Logically and historically, the next step was domestication.

Integration, our second trend, occurs only in human groups of a certain size and complexity, but the degree of integration of a society does not correspond in any simple way to the degree of intensification of its economic activity. In some environmental conditions intensification brings integration in its wake; in others it does not.

Three levels of social integration can be seen in all hunter-gatherer societies, but their relative importance varies significantly with resource availability, with the specific form of resource intensification, and with technological development. The *family* as the basic subsistence unit was nearly universal, although its importance was temporarily diminished when the camp took over some of its economic functions. The *camp*, of four to six families, was also nearly universal. As we have seen, however, its importance and its degree of institutionalization varied widely, being least among plant gatherers like the Shoshone and greatest among the hunters of large game. The intensification of hunting, by creating a need for territorial exclusion, may have caused the camp in some areas to become a basic defensive group with stronger ceremonial integration. The *region*, a collectivity of some 10 to 20 camps, was organized to handle problems of security and defense. Regional networks of plant gatherers made it possible for camps or individual families to learn of and obtain food elsewhere when their own home range came up short. With intensive animal hunting, the regional network may have provided the alliance system used in the defense of hunting territories. Where population densities were unusually high, the result was a local community composed of several co-residential sections, each analytically equivalent to a camp.

The importance of territoriality is undeniable but variable. In the original discussion of patrilocal bands, Steward (1936) and Service (1962) identified a territorial group of camp size as typical of hunter-gatherers. More recent work, however, has tended to refute the corporate and territorial aspects of forager organization, and shown instead that minimal territoriality made possible a flexibility of movement in search of food that was essential to a forager's survival. Without territorial boundaries restricting regional movements, populations could easily concentrate on the most favorable resource available at any moment. Often enough there was no other.

Territoriality in hunter-gatherers should thus be associated with more stable resources, as with the Owens Valley Shoshone. Territoriality also restricts access to critical resources that either are naturally circumscribed, such as waterholes or pine nuts, or have been improved by technical means such as local irrigation in the Owens Valley or fish weirs on the Northwest Coast (see Chapter 7). Where resources are concentrated, access to them can be restricted more easily. With the increasingly widespread and increasingly successful efforts to restrict access to critical resources, we encounter the beginnings of warfare.

Associated with territorialism and warfare is a suggestive change in the importance of ceremonialism. In comparatively low-density groups that lack territoriality, ceremonialism is closely tied to periods of aggregation, as when the San aggregate around the winter waterhole or the Shoshone come together for an animal drive. The ceremonialism is particular to the larger group rather than to its component families, and acts to offset tendencies in the group to fragment from internal disputes.

Among territorial hunters and gatherers ad hoc ceremonies may play a different role, both to define a social group with rights of access and then to override such social divisions as part of broader alliance formations. According to Yengoyan (1972), initiation rites in Australia were timed to take advantage of abundant yields of unpredictable wild resources. When such a good yield was noted, a local territorial group would invite neighboring groups to join in its initiation ceremony and at the same time to join in harvesting the bountiful resource. Similarly, among the territorial Pomo of northern California, unusually good yields in seeds and fish were the occasion for a major ceremony (Vayda 1967). Neighbors of the sponsoring group acquired the seeds or fish in exchange for shell money, which was in due course exchanged for food during lean periods. Some such mechanism to compensate for regional differences in the availability of wild foods appears essential. In some cases, e.g. the Northwest Coast fishers (Chapter 7), such mechanisms are one sign of a developing political economy.

Stratification is not seen in the two cases analyzed in this chapter. In general, foragers are characterized by minimal social differentiation and a strong ethos of equality and sharing. Their goal is the subsistence of each and all, not the differential economic advantage of one or some. Stratification depends on differential access to resources, which in turn must be predicated on a strong notion of land tenure not seen in the San and Shoshone cases. Elsewhere, however, certain economic and ecological conditions tied to intensification have produced hunter-gatherer societies with differential access to resources and thus with stratification. As we have seen, some Upper Paleolithic cultures may have been of this description. And as we shall see in Chapter 7, social elites are a prominent feature of the high-density, territorial hunter-gatherer societies of Northwest Coast fishers.

THREE

Families with Domestication

WE HAVE ARGUED that the family is a natural unit of human social and economic organization, rooted in biological capacities and tendencies that evolved over the millions of years when hominids lived by foraging. Our prototypes for the family-level economy were the Shoshone and the !Kung San, classical forager groups. In this chapter we generalize our argument to show that domesticated food production as such does not necessarily lead to a more complex social and economic system. In the two cases we examine, the Machiguenga and the Nganasan, the technology of domesticated food production is available and contributes significantly to the economy, yet the family remains the dominant unit of economic integration.

To be sure, we now encounter somewhat more stable settlements, which we refer to as hamlets. But these merely reflect the existence of such stable resources as gardens among the Machiguenga and winter fishing spots among the Nganasan, and do not signal the emergence of a significantly more complex integration of the economy. Apart from the formation of hamlets, we find little to distinguish our cases here from those of the previous chapter: the family remains opportunistic, aggregating and dispersing as the availability of resources dictates, maximizing flexibility and minimizing structural constraints such as territoriality and leadership.

Although interpersonal violence and homicide are known in these groups, organized raiding and warfare are rare, except on the part of more highly organized and more powerful neighboring groups. Multi-family groups cooperate in food production or food sharing only on particular occasions, and household autonomy is repeatedly affirmed in the seasonal or permanent dissolution of hamlets into their constituent households.

In both cases domesticates serve as a dietary supplement to wild foods, which remain very important. The Machiguenga of the Peruvian Amazon, who live in semi-sedentary family or hamlet settlements, grow most of their food, but also prize a diversity of wild foods. Although they have abundant unused land suitable for horticulture, they prefer to scatter in family units for easy access to wild foods. The Nganasan, reindeer hunters of the Siberian tundra, keep small herds of domesticated reindeer not for food but for transportation and for use in hunting. They employ the technology of domestication but remain essentially foragers.

Why did these groups not take advantage of their technology of domestication to complete the expected evolutionary transition to more densely settled, internally differentiated societies? As we shall argue in Chapter 4, this poses the question backwards. After all, as we have seen, in the right circumstances settlement in small, scattered groups offers cost-efficient solutions to basic economic problems. The more interesting question is, What leads people to forsake their family autonomy for larger, more concentrated settlements in which food procurement is less efficient and social tensions are greater?

The archaeological evidence is clear that agriculture by itself is not responsible for revolutionary changes in societal organization. So far as the archaeological record shows, sedentary village life first occurred in societies dependent on hunting and gathering; the Northwest Coast fishers (Case 9) illustrate ethnographically how this is possible. In this chapter we shall argue that the family-level organization that characterized most hunter-gatherer societies after the end of the Pleistocene persisted in at least some instances well after the beginnings of agriculture.

In both the Middle East and Mesoamerica agriculture and pastoralism appear not as economic revolutions permitting a sedentary lifestyle, but as long and gradual transitions not directly linked to villages. Indeed, in the Middle East, sedentary villages predate the beginnings of agriculture; the villagers stored wild cereals to be eaten during lean periods (Flannery 1969). The village of Ain Mallaha, occupied between 11,000 and 10,000 years ago in what is now Israel, contained about 50 circular, semi-subterranean houses that suggest the kind of horticultural village described in Chapters 5 and 6. The first evidence of plant and animal domestication, however, is found only at the end of this period, around 10,000 years ago. At the archaeologically important village of Ali Kosh in southwestern Iran, Flannery (1969) documents a slow adoption of domesticates into the diet. For more than a thousand years after the earliest use of domesticated cereals (emmer wheat and two-row barley) and animals (goats and

TABLE 3
Developmental Trends in the Tehuacan Valley

Phase	Population	Percent domesticates	Largest settlement
Late Ajuereado (?-7400 B.C.)	25	0%	camp
El Riego (7400-5800 B.C.)	50	4	base camp
Coxcatlan (5800-4150 B.C.)	150	14	base camp
Abejas (4150-2850 B.C.)	300	22	hamlet?
Purron and Ajalpan (2850-1000 B.C.)	insufficient data		hamlet
Santa Maria (1000-150 B.C.)	4,000	58	village

SOURCE: Christenson (1980).

sheep), hunting and gathering continued to provide the majority of the diet. In the Middle East as elsewhere, the subsistence economy shifted to domesticates over several thousand years as human populations gradually grew.

The best documented long-term sequence for a growing population, a changing subsistence economy, and a changing social organization comes from MacNeish's seminal research in the Tehuacan Valley of Mexico (MacNeish 1964, 1970; Byers 1967; Christenson 1980). Table 3 presents the basic data showing the relationship among these three key variables. As we interpret this sequence, the long-term development was propelled by a growth in human population and an intensification of the subsistence economy. Initially there was a shift from a mixed hunting and gathering economy in Late Ajuereado and El Riego to a broad-spectrum economy, relying on plants for about 65 percent of the diet, by Coxcatlan. Domesticated food products (maize, beans, squash, etc.) were picked up starting in El Riego, and gradually provided an ever higher percentage of the diet.

Linked to population growth and subsistence intensification was a slowly changing settlement pattern. The foragers of the El Riego and Coxcatlan periods were organized at a family level like the Shoshone or the San, with a characteristic pattern of base camps and smaller short-term camps. Probably by Abejas, base camps had increased in size to perhaps 50 people and become more sedentary, indicating a transition to hamlets. Not until Santa Maria, some five thousand years after the initial use of domesticated plants, are true sedentary villages found.

In sum, we see archaeologically not a technological revolution but a slow increase in farming in a family-level society much like that now to be described for the Machiguenga and the Nganasan. Evolution beyond the family level to more complex forms cannot be explained by domestication as such.

Case 3. The Machiguenga of the Peruvian Amazon

The Machiguenga are tropical horticulturalists who live at considerably higher population densities than the !Kung San or the Shoshone, but closely resemble those foragers in social and economic organization. Like the classic foragers, the Machiguenga are pragmatic in their search for food, aggregating and dispersing frequently as the situation may dictate. Although their multifamily groups are more permanent than San camps, the Machiguenga eschew village-level integrated groups and clearly value the economic autonomy of the household. In the extent of their social organization of production, they fall somewhere between the San and the Shoshone.

The Machiguenga environment and technology would appear to make a decent life possible for a much larger population. From the air one's first impression is of an endless and empty natural forest. Occasional small gardens and clearings with one to five houses dot the landscape (Fig. 3). For the forager groups discussed in Chapter 2 we found limiting factors—water for the San, water and winter food for the Shoshone—that kept population densities low. For the Machiguenga, however, no single obvious scarcity limits population growth. Food production is ample to meet basic needs and is secure enough to ward off starvation under most environmental conditions.

We shall argue that what keeps population low is the economic desirability, if not necessity, of keeping settlements small and widely scattered. Unlike the Yanomamo (Chapter 5), the Machiguenga do not have food-producing resources that are sufficiently dense to be worth fighting over. Moreover, despite their reasonably abundant food, they are less amply supplied with other life essentials and must scatter and move frequently to keep subsistence costs down and to preserve their health. All aspects of their adaptation reinforce their family-level economy.

The Environment and the Economy

The Machiguenga (A. Johnson 1983; O. Johnson 1978) reside in the western fringe of the Amazon rainforest, along the slopes of the Andes mountains in southeastern Peru. The high Andean altiplano

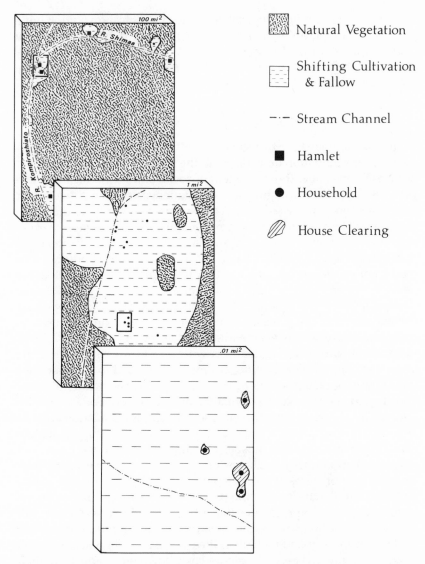

Natural Vegetation

Shifting Cultivation
& Fallow

—·— Stream Channel

■ Hamlet

● Household

▨ House Clearing

Fig. 3. Settlement pattern of the Machiguenga. People settle in single households or small hamlets that move every few years when local resources become scarce. Small gardens, both in production and abandoned, are close to the settlements, forming islands in a sea of tropical rainforest.

supported politically complex societies on an intensive agricultural base long before the European conquest of the New World. By the time of the Inkas (ca. 1400 A.D.), an empire state had managed to integrate an area running along the Andean chain more than two thousand miles north and south from the administrative center at Cuzco (Chapter 11). Yet the Inkas, despite numerous incursions into the tropical forest, were never able to dominate politically more than a few miles to the east into the *montaña,* a rainforest region of rugged mountains between 1,000 and 6,000 feet above sea level. The montaña was inhabited by extensive horticulturalists like the Machiguenga, living in small scattered hamlets. Fierce as these people could be, they could never have resisted the Inka armies in a direct confrontation; yet the Inkas feared them, calling them *antis,* "savages."

In the Machiguenga environment it is not necessary to band together for defensive purposes or hunting. Machiguenga settlements fluctuate between individual households, isolated from others by expanses of virgin forest, and hamlets of three to five cooperating and related households (Fig. 3). The choice and its timing are determined primarily by the local scarcity or abundance of basic resources.

Machiguenga households are semi-sedentary. Their main residence is in sturdy houses built to last the three to five years that they will normally reside in a given location. Indeed, an older house at an abandoned site may be left standing to serve as a hunting lodge or a temporary stopover while families visit old gardens, where crops can still be harvested and where small herds of peccary and other game, attracted by the availability of untended root crops, can be hunted.

During certain times of the year, when wild foods are abundant, the Machiguenga leave their houses to live in temporary huts along the rivers or in distant gardens. People value these times as opportunities to get away from their "crowded" hamlets, where the social costs of sharing and cooperation are high and where wild foods have been locally depleted. Population density is 0.8 persons per square mile, high for family-level societies but low enough for overall resources to be adequate for a healthy existence. Although we see the forest of the Machiguenga as unsettled, they often find it "crowded." Why this should be is a question to which we will turn shortly.

Of the labor invested by the Machiguenga in food production, nearly two-thirds goes into their highly productive gardens; the other third goes into procuring wild food, notably game, fish, and insects. Although wild food makes up only about 10 percent of what they consume, they consider it essential to their diet. Judging from people's recollections, as recently as 1965 wild foods made up a much larger percentage of the diet. Increased outside contact accounts for the change, partly by making steel tools for gardening easy to obtain, partly by increasing the population density and thus decreasing the availability of wild foods. Later we shall examine some of the implications of these recent changes. For now we note that garden foods provide the bulk of food energy in the diet and are also the main basis of Machiguenga food security, which is achieved by "overproducing" root crops that can be stored in the ground until needed.

Given the Machiguenga's ability to produce a large surplus of starchy staples above subsistence needs, it is puzzling why their population densities remain low and the family-level organization persists. Early observers assumed that the "limited potential" of tropical soils acts as a brake on population growth in the "green desert" of the tropical forest, just as extremes of drought or cold limit population among foraging groups (Meggers 1954). Tropical soils are often more fragile than soils of the temperate zones. The luxuriant vegetation of a tropical rainforest rests on a delicate balance of nutrients cycling rapidly from forest to soil and back again. A constant rain of detritus—leaves, branches, fruits, animal feces, etc.—falls to the ground, where it is rapidly decomposed into nutrients by insects and bacteria working in the warm and humid topsoil. These nutrients are picked up by the shallow root systems of the forest and quickly used to support new growth. Strip the land of the forest cover and the sun and rain beat down unimpeded, destroying the crumb structure of the thin topsoil. Nutrients percolate down beyond the reach of new roots, occasionally leaving behind laterites (oxides of iron and aluminum) that can solidify into "hard pans" in which nothing will grow. More commonly, the erosion or depletion of soil nutrients by continuous cropping decreases fertility and in extreme cases destroys it.

Not all observers agree, however, on the poverty of tropical soils. In some parts of the Amazon lands have been cropped continuously for decades without apparent loss of productivity, whereas in other parts intensive cultivation has left large areas useless for farming. We do not yet understand why these different outcomes occur (see Sanchez 1976; Moran 1979: 248-90).

If tropical soils are in fact poor, slash-and-burn, or shifting, agriculture might well be appropriate to the Amazon rainforest. Slash-and-burn agriculture, as among the Machiguenga, requires cutting and clearing small gardens from the forest. After one or two years of farming, the field is allowed to revert to natural vegetation as new plots are cleared. Periods of fallow, while the fields are unfarmed, are essential to regain soil fertility. Such agriculture was once assumed to be a backward and inefficient technology. Observers familiar with the clear, plowed plots of intensive agriculture in temperate zones were dismayed by the sight of fields littered with half-burnt logs and the mingling of diverse crops seemingly without plan or form. Long fallowing was viewed as a wasteful practice because it kept so much land out of production, and yields of gardens were assumed to be low. Our growing knowledge of the vulnerability of many tropical soils to degradation has led to a more sympathetic view of slash-and-burn agriculture. What concerns most current critics of the system (e.g. C. Webster and Wilson 1966: 87) is the shortening of the fallow period (to get more gardens into production) in areas where population is growing. They complain that this practice inhibits the restoration of the soils that longer fallowing provides.

In a common type of slash-and-burn garden often found in less intensive horticultural systems, where wild foods still play an important part in the diet, several distinct species of food crops are intercropped. As Geertz (1963) remarks, intercropped gardens "imitate" the tropical forest and go far toward protecting the integrity of the soils. Ground-hugging crops like pumpkin and squash lay down a bottom cover; above these an intercropped matrix of staples like maize, manioc, and yam fill the middle zone; and above these, tree crops like banana, cashew, and guava form a canopy.

The diversity of crops is also some protection against pests or diseases, which are most devastating when they strike a field of identical plants. The Machiguenga, for example, not only plant from six to ten different crops in the same field,* but plant several varieties of each because, as they say, "we like the differences." They name 15 varieties of their basic staple, manioc, and 10 varieties of maize, their second most important crop. It would take a highly unlikely combination of hazards to prevent all these varieties from producing.

In general, according to Beckerman (1983: 3), the advantages of intercropped fields for subsistence farmers are as follows:

1. Lower losses due to pests and/or plant disease
2. Increased erosion protection
3. Lower risk of total crop failure, attributable in part to 1 and 2 but also to a spreading of risk over several crops that are unlikely to fail all at once

---

*The Machiguenga recognize at least 80 different species of cultigens, but most of them are grown in small quantities in house gardens and serve as condiments, medicines, building materials, etc. In house gardens they also experiment with new crops.

4. More efficient use of light, moisture, and nutrients

5. Production in a single garden of many commodities needed for household economic self-sufficiency

6. The spreading of labor demands more evenly throughout the year

7. Diminished storage problems

Nonetheless, monocrop fields of such staples as manioc or plantain are found in tropical regions under some conditions. Beckerman (1983) explains this practice as a form of intensification, reflecting a community's increased dependence on horticulture for food. Even in these cases, however, many of the advantages of intercropping are maintained by planting several varieties of the staple crop.

It is not true that slash-and-burn gardens are relatively unproductive. They typically return 100 to 200 times as much seed as planted, compared to the returns of less than 100:1 reported in annual plow cropping in Mesoamerica, for example, and less than 10:1 reported in European grain farming before the modern era. Returns to labor are also high: 20 calories for every calorie of labor invested, making it possible to produce a substantial "surplus" above ordinary subsistence needs. With less than four hours of combined labor per day, the members of an average household produce more than twice the food energy they need to maintain themselves.

Even the long fallows are efficient. Boserup (1965) has shown that the length of a fallow is a fundamental feature of an agricultural system, and that it is closely related to population pressure on resources. In systems with a significant fallow period, she distinguishes three types: forest fallow, where a year or two of cultivation is followed by a long fallow period allowing the forest to regrow; bush fallow, where several years of cultivation are followed by less than 10 years of fallow, so that only "bush," not true forest, regrows; and short fallow, where a few years of cultivation is followed by an equal number of fallow years, then more cultivation, so that not even bush regrows. According to Boserup, a shorter fallow period requires more labor for the same yield on the land, that is, a loss of labor efficiency. Two lines of argument support this view.

The first argument is that shorter fallows lower the fertility of the soil. Forest growth restores the crumb structure and nutrients to the soil that were lost during cultivation. We do not know exactly how long it takes for the ecosystem to restore itself fully following cultivation. Soil constituents are substantially restored after 10 years, but full restoration of the forest complex may take from 25 to 50 years.

TABLE 4
Machiguenga Soil Constituents by Age of Garden

Age of garden	Number of observations	Percent organic matter	Percent nitrogen
Primary forest	4	6.8%	0.32%
First-year	12	6.7	0.32
Second-year	2	6.2	0.30
Third-year	3	4.6	0.21
Fourth-year	4	3.6	0.16
Fallow (2 years)	2	3.2	0.14

Boserup could not prove conclusively that shorter fallows lower soil fertility, but evidence from the Machiguenga tends to bear her out. In Table 4 we see that the fertility of Machiguenga soils, as measured by organic matter and nitrogen, declines steadily with the number of years in cultivation. Primary forest and first-year gardens are of virtually identical fertility, with fertility dropping off most sharply after the second year of cultivation. (This, incidentally, is the point at which Machiguengas are likely to begin abandoning their gardens.) The two gardens listed as fallow in Table 4 had been abandoned for only two years, and we see no evidence that their fertility has as yet been significantly restored.

The data in Table 4 support the argument that tropical soils rapidly lose fertility under continuous cultivation and require long fallows to restore fertility. Other soil constituents may be equally important: for example, soil acidity, which rises dramatically with age of garden (Baksh 1984). But whatever the chemistry, it is clear that shorter fallows will not restore fertility fully, and this means yields will be lower. Since labor input will not be less, lower yields mean a lower return to labor, or a loss in labor efficiency, as Boserup argued.

The second argument against shorter fallows turns on increased weeding costs. Excessive weeds are the main reason Machiguengas give for abandoning their gardens. In a new garden weeds are a problem, but weeding is needed only every six weeks or so; moreover, it proceeds rapidly, since young weeds are delicate and easy to pull by hand. But as the garden ages, weeds become entrenched until hand pulling becomes impossible and a machete must be used. Ultimately, tough grasses, nettles, and other durable species begin to predominate and the cultivator must give up. With fallowing, which allows the original complex of plants to return, tough weeds diminish to their original small proportion of the whole.

Here again Boserup's evidence is not conclusive. But Bergman (1974: 191) reports that the Shipibo Indians of the Peruvian montaña need invest only 260 hours of labor per hectare in maize gardens cleared on virgin land, whereas they require over 480 hours per hectare in land cleared after short fallows. He attributes the difference almost entirely to the need for extra weeding in the short-fallowed fields.

Boserup's reasoning and the evidence we have just presented support Meggers's contention that there are limits on the potential of tropical rainforests for agricultural intensification. So do the spectacular failures of modern agribusiness technology in such Amazon enterprises as the Ford Motor Company rubber plantations at Fordlandia (Wagley 1976: 89-90) and Daniel Ludwig's paper pulp mill at Jari (*Veja* 1982). More catastrophic failures of subsistence are recorded prehistorically for tropical forest environments on Pacific islands (Chapter 8).

But the situation is more complicated. In an influential paper Carneiro (1960) showed that the Kuikuru Indians, of the upper Xingu region of Brazil, had enough land to support ten times their present population and still afford the luxury of 25-year fallow periods. The Machiguenga also have an apparent abundance of cultivable land. Hence land shortage cannot be the only limiting factor in the human ecology of the tropical rainforest.

It is, to be sure, *one* limiting factor, since the Machiguenga distinguish carefully between potential garden sites and regard most land as inferior. They want soils that are soft and free of rocks, fertile, well drained, not too steep, and not too distant. They are constantly on the lookout for good land, and a family may lay claim to an attractive piece more than a year in advance of actual clearing. Good land produces more and requires less labor than inferior land, thereby keeping overall labor costs low. Since increased population density decreases the availability of desired land, it has (and is perceived to have) the effect of increasing production costs.

Finally, even tropical rainforests are unpredictably subject to years of excessively dry or wet weather that can reduce garden productivity drastically. Crop pests or disabilities from injury or illness to family members can also interfere with normal garden productivity. Thus the large food surpluses that the Machiguenga produce in normal years have an important security function, and cannot be taken simply as proof that their abundant land could support a much larger population.

Given the fundamental importance to any society of producing sufficient food energy, it is understandable that cultural ecologists like

Meggers should think first of agricultural potential as the factor limiting population growth among extensive horticulturalists. But of course
more than calories are at stake. Especially in tropical regions the most
common staple crops (manioc, bananas, sweet potatoes, etc.) are high
in calories but low in other essential nutrients. In densely settled regions overreliance on crops like these can result in chronic nutritional
deficiencies (Jones 1959). Since protein is second only to calories in
nutritional importance, the next stop in the search for limiting factors
was protein. Gross (1975) argued that because of the scarcity of protein foods in the Amazon, horticulturalists there need large territories
in which to hunt, fish, and collect protein-rich grubs and nuts.

Problems soon developed with this explanation as well (Beckerman
1979, 1980), notably that protein cannot be shown to be scarce in the
Amazonian diet. Recent careful studies in native Amazonian communities show people obtaining twice as much protein as nutritionists
recommend for good health (Berlin and Markell 1977), and the Machiguenga are no exception. Indeed, the Machiguenga normally exceed the recommended levels of intake of virtually all essential nutrients (Johnson and Behrens 1982).

Yet the protein explanation cannot be simply dismissed. The Machiguenga do not consider themselves to be rich in protein foods.
They treasure the nuts, seeds, insects, fish, and game they obtain
from their forest and rivers; and they willingly spend far more work
energy in procuring such foods than in producing the equivalent
weight of garden foods (A. Johnson 1980). Wild foods are sources of
high-quality protein, and also of diverse nutrients in addition to protein, such as vitamins and fatty acids.

Horticulture does provide some vegetable protein, but tropical root
crops are notoriously poor sources of protein. For example, although
it costs the Machiguenga ten times as much effort to produce a kilogram of fish as to produce a kilogram of garden foods, fish have about
ten times as much protein per kilo, and so the cost of protein in each
case is roughly equal. Hence extensive horticulturalists like the Machiguenga compensate for the deficiencies of their garden foods by
foraging for wild foods, and their continued nutritional well-being depends on maintaining the small, scattered settlements and low population density necessary for access to an adequate supply of wild foods.

Though they get enough protein, then, they think of protein-rich
foods as scarce and they work hard to obtain them from the wild.
Their diet is also low in fats and oils (Baksh 1984: 389-93), of which
they consume hardly more than the minimum levels recommended by
nutritionists. This may help account for their practice of identifying

the quantity of fat (*igeka*) in a food with its good taste (*poshin*). In their terms, poshin foods like meat, fish, and peanuts are delicious because of their igeka.

The Machiguenga frequently complain of other scarcities as well, notably the perennial scarcity of palm leaves for roofing. After residence in a fixed location for a few years, even a small hamlet will exhaust the local supply of fish, game, palms, and firewood. A favorite topic of conversation is who went where and saw what palms, fruit trees, fish, game, or spoor. Such matters are recounted and discussed most enthusiastically.

Given this perceived scarcity of good agricultural land and other natural resources, it is perhaps surprising that the Machiguenga have no history of warfare. Tales of homicide are occasionally told, but one hears more often of suicide. Much like the San, the Machiguenga emphasize peaceful relations among themselves, whom they contrast to their "wild" and violent neighbors living at lower elevations. When disputes flare, families move apart to live until hostilities cool. Belligerent persons are shunned.

One reason for this peaceable way of life is the marginality of the Machiguenga's forest environment with respect to alluvial land for farming, most game, and especially river fish. Land of this sort is unattractive to populations used to the comparatively rich riverine environments at lower elevation or dependent on intensive agriculture in the highlands. As among the San, the scarcity of resources apparently favored low family sizes and a dispersed population. But why did this scarcity not result in interfamily competition for pockets of good agricultural land and natural resources? Apparently because such resources were not sufficiently dense and dependable to make territorial defense cost effective. The group aggregation required for defense would soon deplete the resources, and the rapidly escalating cost of procuring food would cause the group to disperse.

To summarize, the fundamental problem faced by the Machiguenga is the scarcity and occasional unpredictability of wild resources in their forest environment. The low population density that results from this scarcity has benefits, especially an absence of warfare. The response to scarcity has been to maintain the flexibility of the family-level society that we will now describe.

Social Organization

The Machiguenga keep production costs low and maintain a healthy and comfortable standard of living by keeping their social groups small and widely scattered. Fully self-sufficient at the household level, they live as isolated households for up to several years at a time, residing at other times in hamlets of several households.

At least 90 percent of the food consumed in a household is produced by its members. Table 5 shows how men and women spend time in an average household. Although both men and women do a variety of tasks in gardening, foraging, manufacture, and other work, the division of labor between the sexes is so strict that their actual labor rarely overlaps. The starred items show statistically significant differences between the sexes, but even in nonstarred categories such as fishing or harvesting, the sexes work with different species of fish or crops and rarely duplicate each other's work. Together, a husband and wife have all the skills needed to meet a household's normal subsistence requirements.

Being semi-sedentary rather than nomadic, the Machiguenga build more elaborate houses and acquire more goods than their nomadic counterparts. Still, because they are seasonally nomadic in search of wild foods, and because they must move settlements every five years or so, they do not acquire cumbersome stores of goods; and at a moment's notice they can travel light and live as foragers off the forest. Goods obtained by trade with outsiders are few, being confined chiefly to axes until very recent times.

The Machiguenga are good craftsmen whose products are typically more serviceable than beautiful. Men build houses, spin fibers for the nets and bags they knit, and fashion bows and arrows from cane and hard palmwood. Women spin cotton thread, weave cloth for their gownlike *cushmas*, make face paint and other dyes, and weave nets and baskets. A Machiguenga house, made from hardwoods and palms, strikes the Western observer at first as flimsy and crude, but it soon gains respect as a secure, durable, and comfortable structure.

The Machiguenga do not have economic specialists, but as is true everywhere, some people do higher-quality work than others. A man is known for making better bows, a woman for her fine weaving. Criticizing and admiring the handiwork of others is a popular pastime. Men may have their bows made by a better craftsman and repay him with favors, though not in a strictly calculated sense. Some young women who do not yet weave are looked down on and considered lazy for their dependence on more experienced women. But these differences are not institutionalized in any sense into occupations or classes.

There is also a division of labor by age. Infants are warmly and indulgently treated, but toddlers are expected to be increasingly self-reliant and useful until the age of five or six, when they become

TABLE 5
Machiguenga Time Allocation
(Hours per day)[a]

Activity	Men		Women	
Food production				
Hunting*	0.7		0.0	
Fishing	0.7		0.3	
Collecting	0.4		0.4	
Agriculture*	2.5		0.8	
Livestock	0.0		0.1	
		4.3		1.6
Food preparation		0.2		2.2
Food consumption		1.1		1.0
Commercial activities		0.0		0.0
Housework				
Housekeeping	0.1		0.4	
Water and fuel	0.1		0.1	
Pet care	0.0		0.1	
		0.2		0.6
Manufacture				
Acquiring materials*	0.7		0.1	
Manufacturing artifacts	1.2		1.3	
Manufacturing clothes*	0.0		0.8	
Manufacturing facilities	0.5		0.1	
		2.4		2.3
Social				
Socializing, visiting	1.2		1.0	
Child care*	0.1		1.0	
Caring for others	0.2		0.1	
Public ceremony	0.1		0.0	
Public recreation	0.2		0.1	
Education	0.1		0.0	
		1.9		2.2
Individual				
Hygiene	0.2		0.1	
Sleeping	0.2		0.1	
"Nothing"	1.6		2.1	
Ill	0.4		0.4	
		2.4		2.7
Other		0.5		0.5
TOTAL		13.0		13.1

SOURCE: Johnson 1975a.
NOTE: Asterisks indicate significant differences between men and women ($p < .05$, t-test).
[a] Daylight hours only (6:00 A.M.-7:00 P.M.).

responsible contributors to the household economy. Children's tasks include fetching water, carrying seed during planting, relaying messages, and, for girls, the care of younger siblings. After age six, children's work becomes more sex-specific. One finds boys hunting for sparrows and lizards with small bows and arrows, and girls spinning uneven but useful thread on small spindles. By age twelve, boys and girls are able to carry out most of the adult tasks of their sex. At this age they show little initiative and are inclined to avoid work when they can, but their attitude changes as they mature and begin to take on family responsibilities of their own.

In polygynous households there is also a division of labor between wives (O. Johnson 1978). Younger wives are more often involved in work outside the home, in gardens or foraging. Older wives are likely to be found in the home, organizing the productive labor of their children and concentrating on manufacture. Thus younger women are away more often with their husbands, and this evokes the jealousy of older wives. On the other hand, older wives are more productive and earn respect from their husbands and from other women. Older wives also have much larger social networks and increase the flow of exchange with other households.

Each wife in a polygynous household maintains a separate hearth, symbolizing her control over her own food production and the independence of her contribution to the household economy. She prepares her own foods as well as the staple foods brought in from her husband's garden, and distributes them to household members. Mothers in polygynous households interact primarily with their own children and uncommonly with the children of co-wives. Co-wives also tend to interact and share food with their husband more than with each other, especially where relations between co-wives are tense.

But most Machiguenga households run smoothly as units of generalized reciprocity. Food is constantly circulated among members. A woman passes an ear of roasted corn to her husband, who breaks it in half and returns half to her. He then breaks his half and gives part to his young daughter, who shares it with other children. Mother's half of the corn likewise divided, the children begin passing bits of corn back to their parents. Food is seemingly enjoyed as much in the sharing as in the eating.

Every food item has an "owner": the one who procured it or planted it. In fact, all possessions are owned by individuals, and one must ask to borrow them before using them. If a child should refuse to share a possession, the parent will not force the issue, but teasing and chiding will make the child uncomfortable and generosity becomes instilled over time. The one who shares is made to feel proud that he or she has been able to give something of value to others in the household.

Isolated households may go for weeks at a time with very little social contact or exchange with other households. They accept isolation because it gives them free access to the natural resources around them. But there are also advantages in living in hamlets of three to five households of related kinsmen, commonly married brothers and sisters. The ties of affection and nurturance established in childhood prepare the way for friendly, cooperative relations as adults.

Machiguenga hamlets typically are cleared areas on a flat or promontory well away from the river. Three to five households will be built in various locations near one another, so that a common clearing can serve them all for work and company. Sometimes one or two households in the hamlet will be built 300 feet or more away from the others, so that fruit trees or stands of sugarcane between the two clearings can provide a measure of privacy. The houses remain close enough for easy visiting, food sharing, and mutual aid in child care and food preparation, but each household maintains its own racks and sheds for smoking, drying, or storing food and its own pens for muscovy ducks or chickens, if any. Nothing is owned communally by the hamlet members. Even when brothers cooperate to clear a garden, they usually divide it into two parts to be cultivated individually. A brother rarely helps himself to products from his brother's garden without asking permission first.

From each garden located fairly close to the hamlet the food is brought home, prepared, and eaten by each household apart, but households often come together for a common meal when wild foods are available. Fish, game, and grubs, being both scarce and greatly desired, are occasions for sharing from a common pot; indeed, such sharing of wild foods is the main economic basis of the hamlet. Women from each household arrive with pots of manioc; they may also bring greens or other foods they have harvested and prepared, and some manioc beer for their husbands. Such meals are unexpectedly complex and structured, and give insight into the balance of individual and group interests the Machiguenga seek to achieve.

On one occasion the three households of a hamlet came together to share a fish caught by one of the men. The three household heads were related: the first was the older brother of the second, and the third was their sister's husband. The younger brother was married to the older brother's wife's daughter (by a previous marriage). The older brother ranked highest socially, the brother-in-law lowest. When special foods like fish were available for a communal feast, it was usually held at the older brother's house.

On this occasion, as on most others, each married couple sat together as people chatted while the fish soup was cooking. Then they separated into male and female groups: one group the three men and the brothers' twelve-year-old nephew, the other the women and younger children. The older brother's wife dished out a large plate of fish soup and set it before the men along with a bowl of manioc. The men began to eat the manioc but ignored the soup until the older brother took a spoonful. After a moment the younger brother took a spoonful, then the brother-in-law, and finally the nephew. They continued eating manioc until the older brother took another spoonful of soup; then again, in the same sequence, the others did. This orderly cycle continued until the soup was gone; then another pause ensued until the older brother broke off a small piece of fish and ate it. Then the others followed suit in the same order until the fish was gone. All this was done in a matter-of-fact way, without discussion.

Meanwhile the women and children were sharing from a common pot. As among the men, individuals helped themselves to manioc without restraint. But the fish soup was carefully doled out by the women, ensuring a fair distribution. When the eating was done, husbands and wives turned back toward one another and soon shifted position so that spatially the nuclear families had reasserted themselves.

This small episode teaches us two important facts about the social organization of a Machiguenga hamlet. First, despite the fundamental individual freedom of the separate households, they accept a measure of hierarchy and control so that precious resources like fish can be distributed with minimum resentment or dispute. An abundant food like manioc does not occasion such care. Second, the "social" nature of the fish—that it ultimately belongs to the group and not to the person who caught it—is apparent in the dissolution of the nuclear families into hamlet-level men's and women's eating groups. As soon as the "socialized" fish has been consumed, the nuclear family units reconstitute themselves, for they remain the primary units of Machiguenga society.

When households collaborate, it is usually to obtain or distribute special foods. A single family may do all its own hunting, fishing, foraging, and growing of pineapples, papayas, and other favorite foods. But these are often available only sporadically, and then in unwieldy quantities. Sharing can not only reduce a sudden windfall to manageable proportions, but ensure that similar windfalls in other households will be shared, thus making special foods available more frequently to more people. The good feelings surrounding such exchanges help alleviate the constant small frictions that arise from daily competition over scarce resources, and are the primary social glue holding a Machiguenga hamlet together.

As we have seen, evidence of hierarchy appears in the distribution of the fish soup. But there is no paradox. The Machiguenga household itself is hierarchically ordered, primarily on the basis of age although occasionally an especially productive member may outrank an older but less productive member. Complex cooperative tasks are managed smoothly because a clear chain of command and compliance exists. When children grow up and form separate households, these lines of authority persist in memory and tend to reassert themselves when group cooperation is required.

The three households in the fish soup example express their hierarchical structure in many areas of behavior. For example, a higher-status household receives far more visits from a lower-status household than it pays in return: the older brother's household receives about six visits from the younger brother's household, and nine from the brother-in-law's, for every visit it pays them. Likewise, the brother-in-law's household pays a great many visits to the younger brother's household, which almost never visits it in return (A. Johnson 1978: 106-9).

Lines of authority and prestige between households materialize in cooperative ventures. The most cooperative task in the Machiguenga case is fish poisoning, which may involve from two to as many as ten households. Here a leader always coordinates the activities: men build dams to slow the water, and women construct weirs to catch the drugged fish as they float downstream. Each of these activities involves a complex division of labor, and timing is important. The level of water in the river, the number of workers needed, the provision and preparation of the poison, the exact moment it is to be introduced into the water, all require coordination by senior men and women who are used to authority and toward whom others are compliant.

In the early phases of the work, as in the multi-household meal, husbands and wives separate into same-sex groups and work apart. Once the poison has been introduced into the water, however, husbands and wives rejoin one another at preselected points and gather fish for their own households. Sometimes a household that provided much labor will find few fish in its stretch of river. In the later exchange of fish these differences will be evened out to some extent, but no authority or institution exists to allocate the catch fairly, or even to define what "fairly" might mean.

If disputes arise within a household or hamlet, they are resolved locally by a senior family member. For example, a man was attempting to catch a stunned *segori*, a troutlike fish of excellent flavor whose roe are especially valued. It eluded him and disappeared in a pool. A minute later his seven-year-old nephew caught the fish, a look of pure pleasure lighting the boy's face. But the uncle saw and said, "Here, that's my fish. I was chasing it!" The boy refused to hand over his prize until his other uncle, a highly respected man, ordered him to. Later the boy caught his own segori, and his happiness was restored. But had he not, his disappointment would have been seen as an unavoidable outcome of the need to acknowledge seniority and keep peace between households.

Periodically, particularly during a full moon, members of a hamlet will plan a beer feast. The women spend several days making manioc beer while the men devote themselves to hunting and fishing. Members of more than one hamlet may attend if invited by a respected man or woman. With feelings and tongues loosened by the abundance of beer and meat, many "political" issues, such as forming cooperative fishing groups, claiming garden land, or teasing rulebreakers, are aired. A man who is organizing a fishing project will seek cooperation and in this capacity may set the tone for some of the discussions. Or a man of wit may become a center of attention as he directs caustic gibes at some unfortunate who has offended him. But no clear leader is to be found, and the conversation ebbs and flows as this topic or that is picked up, passed around, and dropped.

For extensive horticulturalists like the Machiguenga cooperation between households will always have both costs and benefits. Sociability, security, and the distribution of windfalls all make cooperation attractive, but at the expense of a certain autonomy in deciding how to serve one's own best interests. The tensions that arise may grow into resentments, but standards of courtesy and respect keep these from being freely expressed. During the drunkenness of beer feasts, hostilities break out in intense, humiliating joking and in verbal and physical fights. These may ventilate feelings and restore equilibrium, but often they lead to a sense of injustice and a decision to leave the hamlet. On the whole, the Machiguenga fear aggression and much prefer disengagement. In most cases, one who feels intense anger simply flees (*ishiganaka*). He may later return to stay, or he may gather his family and move away. Slowly, over time, troubled relations will be soothed, and in a later phase of the cycle—especially when wild food is again abundant—the same households, and perhaps some new ones, will reestablish their hamlet and enjoy cooperative living again.

Hamlet groups do not own corporate property, nor are they as groups validated by the ceremonial occasions that we will discuss at length in Chapters 5-7. Except for a loose sense of a home range, as described for the San (Case 2), no territoriality can be said to exist. Individual families own garden plots that they have carved out from the natural forest, but only as long as they cultivate them; the plots revert to common lands during fallow. All natural resources of the forests and rivers are open to all Machiguenga, although a foraging group generally keeps its distance from another's home range.

In summary, the Machiguenga illustrate the conditions under which horticulturalists may maintain a family-level economy and social organization. In an area where competition from other groups is low and where valued wild foods are scarce and widely scattered, the Machiguenga function very effectively in small, scattered households or hamlets. By the simple device of "overproducing" and storing certain edible roots, they can live for years at a time as independent, self-sufficient households.

On the other hand, they find advantages in cooperating with other households in poison fishing and sharing windfalls of wild food. Within and between households, natural hierarchies exist that establish chains of command for coordinating labor and food distribution. But such leadership, and the splitting of married couples into separate men's and women's groups that sometimes occurs when wild foods are procured or consumed, is always temporary. The autonomous household regains command when the specific event is over.

As an interesting footnote to this case, Baksh (1984) has studied a Machiguenga village formed recently under a program of the Peruvian government. About 200 people agreed to live together under the direction of a charismatic leader who focused their wish to have access to modern technology, especially medicines and steel tools. They had the unusual opportunity of settling in an area rich in wild resources that had become depopulated as a result of Western contact. Although families lived in separate households and in hamlet-sized neighborhoods within the village, they agreed to cooperate in planting cash crops to earn money and "advance ourselves."

As long as wild foods remained abundant, the village succeeded in handling the numerous tensions of group life. Villagers would remain in the community four days a week, mostly working in the communal gardens; on the other three days, commonly referred to as "vacations," individual families would scatter to fishing places where yields were abundant. As local fishing areas became depleted, however, and travel time to ever more distant streams reduced the efficiency of fishing, tensions rose. People complained of being "hungry" even though garden foods were abundant, and disputes erupted into open conflicts that threatened to splinter the community. In the end the leader resolved matters by moving the whole village to a new site, a day's hard walk downriver, where fish were still abundant.

The Machiguenga case we described earlier is the usual one: families are scattered, but no sizable patch of territory is uninhabited. Thus wild resources everywhere are at a low level, for no good fishing spot or hunting trail is abandoned for long. Any unit larger than a family simply depletes local resources that much faster, requiring more rapid abandonment of a site or else disrupting the community with increasingly frequent disputes.

We may anticipate later chapters by noting that when there is nowhere to run to, when the environment is too full of competing families, some other means of resolving disputes becomes necessary, and the most likely at this level of political development is warfare (Carneiro 1970b). The seeming abundance of the Machiguenga economy, therefore, does not imply underpopulation. Indeed, the speed with which even a small local increase in population can lead to depletion and hardship indicates that the Machiguenga are living closer to environmental limits than at first appears.

Case 4. The Nganasan of Northern Siberia

We now examine briefly a family-level forager society in which domesticated animals play a significant economic role. Here again domestication as such—in this case animal domestication—is not a sufficient condition for socioeconomic development beyond the family level. Among the Nganasan, small family herds of tame reindeer served almost exclusively as a way of facilitating the foraging way of life. With pressure from an expanding European population, however, new conditions arose that encouraged the Nganasan and similar groups to increase their herds of tame reindeer at the expense of wild reindeer. And it was this process, a reaction to population pressures rather than to the attractions of domestication, that eventually led the Nganasan to form larger, more economically complex social units and exert a tighter political control over resources.

The Environment and the Economy

The Nganasan (Popov 1966) inhabit the frozen, windswept tundra of the Taymyr Peninsula in the northern extremity of Central Siberia. They are found from the northern limits of the forested tundra north across a hilly plain to the Arctic Sea. The landscape is locally variable, with dry, rocky hills, grassy slopes, swampy lowlands, and numerous lakes. Trees are rare; shrubs, lichens, and sedges are the main vegetation. The fauna of greatest importance to the Nganasan are reindeer, polar fox, fish, and various species of geese and ducks.

On the Taymyr Peninsula the temperature falls below freezing on 263 days of the year. The summer is short, and frosts are likely to occur in late spring and early fall. Owing to the intense summer sun, however, the tundra blossoms in July and August, when it is visited by great flocks of birds and swarms of biting insects. At this latitude (75° N), well north of the Arctic Circle, there is about a month in summer when the sun never sets, and another month in winter when it never rises.

Reindeer, or caribou, are the central focus of the Nganasan economy. For most of the year reindeer are scattered in small groups, but they aggregate in large herds in the fall for the migration south, and again in the spring for the return north. In the summer and fall reindeer grow fat on grass, sedges, leaves, and fungi. Over the winter, however, they depend on lichen and their own stored fat for survival. The availability of lichen limits the population of reindeer, probably more so than predation by wolves or, under traditional methods of exploitation, by humans (Ingold 1980: 20, 35).

Human settlements are widely scattered, with population densities below one person per 50 square miles, and population movements are responsive to reindeer movements. Unlike wolves, which can follow the reindeer herd at its usual speed (from 10 to 40-odd miles per day), the slower humans must use stratagems to ambush the reindeer or lure them to destruction. The most popular and productive of these are communal hunts in the spring and fall at places reindeer are known to visit. During these hunts large numbers of reindeer are trapped, killed, processed, and in the fall, stored for winter consumption.

Migrating reindeer stop at certain lakes and cross rivers at favorite crossings; old hunters, who know these places and the time for best hunting, take charge of organizing the hunt. Men are prohibited from hunting at such sites except during the communal hunt, so that the game will not be frightened away by excessive contact. In October 1936 Popov (1966: 20) observed a migrating herd so large that it took several days for the "dense mass" of reindeer to cross the frozen Pyasina River.

Reindeer prefer to gather near lakes or rivers into which they can flee for safety when attacked by wolves. The Nganasan take advantage of this by using dogs to drive reindeer into the water, where hunters in dugout canoes spear them. Another strategy uses "flags" made of poles from which strips of leather are flapping. Since the flapping intimidates reindeer, simply by placing poles every 15 feet or so the Nganasan construct long funnel-shaped "fences" along which they can drive the reindeer into corrals where hunters lie in wait.

Particularly in the fall, the take from these communal hunts can be prodigious. People gorge themselves on food in the late summer and fall, and process the surplus for storage. Meat is dried, fat is rendered and stored in containers made from skins and internal organs, and skins are prepared for tents and clothing. Often groups of men leave the seasonal camps for days at a time, returning with quantities of animals for the already busy women to process. To a greater degree even than among the Shoshone, the Nganasan must store large amounts of meat and fat in order to survive the long winter ahead. They have two simple rules for eating: in the spring "eat as little as possible," and in the fall "eat as much as possible."

Other foods are important in some seasons. The annual cycle of food production is approximately as follows. With the spring thaws Nganasan families disperse and move north, away from their winter hamlets, to hunt reindeer, partridge, and ducks. With the advent of the summer fishing season in June and July, small, scattered family groups enjoy a "relatively settled life" until late July and August, when they come together for communal hunts of geese, which are molting and can be taken in large quantities with nets. Like the reindeer these are rendered, and their fat is stored for winter consumption.

In late August the move back south begins, interrupted periodically by reindeer drives until November, when the Nganasan settle once more in winter hamlets. Over the winter some hunting of scattered individual reindeer and polar foxes continues, along with ice fishing. In this period clothing is made, tools and sleds are repaired, and other sedentary activities are pursued. Early spring, before the thaws have come and when stored foods have run out, is a period of scarcity and hunger both for humans and for the animals they eat.

Routes of human migration are fairly well established. A hunter often leaves the frozen carcass of his catch lying alongside a trail that he knows his family will be passing a month or two later. In order to protect the carcass from wolves and polar bears, he may cover it with rocks and pour water over them. The water quickly freezes, making a secure "icebox" in which food remains stored until needed.

The Nganasan's movements reflect the whereabouts of their quarry. For most of the year reindeer, fowl, and other game are widely scattered, and the Nganasan follow them in groups of one or two families. At other times, when reindeer or geese are locally available in large quantities, families congregate to exploit that opportunity. Periods of stable settlement—in summer near favorite fishing spots, in winter near ice-fishing spots and (more important) near pasture for domestic reindeer—alternate with periods of movement in pursuit of migrating reindeer.

Tame domestic reindeer are used chiefly for transportation. The Nganasan family, although mobile, is not light on its feet. In the fall and the long winter, a family requires several reindeer to pull large sleds piled high with the heavy tents, clothing and skins, food stores, and firewood that are essential to surviving the harsh Siberian winter. Tame reindeer also pull the light, fast sleds on which hunters pursue small herds of wild reindeer in the winter, and can be trained to act as decoys to lure wild reindeer within range of camouflaged hunters. Additionally, though only when the alternative is starvation, a household may slaughter its domestic reindeer. So reluctant are the Nganasan to slaughter a tame reindeer that they consider shedding its blood a sin; thus they kill the animal by strangulation, a difficult task.

It takes labor to pasture domestic reindeer and protect them from wolves, and in the winter a family may have to move when nearby pastures of lichen have been exhausted. Traditionally Nganasan households kept fewer than ten reindeer, enough for winter transportation and hunting but not so many as to require frequent moves to new pastures.

Warfare is not reported. Stories of hunger and famine do occur, however, in which men came to blows over caches of food during the hungry spring season. Men nowadays do not know of such cases and admire those fierce ancestors who would fight over food. But it appears the common response to the spring scarcity is to join a hamlet group and share stored foods until the scatter to summer resources can begin. As with the San (Case 2) and the Machiguenga (Case 3), scarcity may provoke personal violence, but organized intergroup aggression must have been discouraged because of the importance of broad interpersonal and intergroup ties in spreading risks.

The scarcity of resources in the far north might suggest the likelihood of intergroup trade, but what scant information we have indicates that until recent times the Nganasan were essentially self-sufficient. As we shall see, however, the situation changed; an extensive trade of animal products for technological items developed historically as part of a general intensification of resource use.

To summarize, the critical problems faced by the Nganasan are the extreme scarcity and unpredictability of resources in the Arctic environment. Because of these problems population density remained very low until historic times, and a family-level existence could be maintained. Cooperation between families was needed only for large-scale hunting and for the sharing of stored foods.

Social Organization

The autonomy of nuclear family groups and multifamily clusters is a strong ideal among the Nganasan. Nuclear families often live separately in their own small tents. Possessions for individual use are treated as private property: as with the Shoshone, only very large items like the nets used in reindeer drives are owned by a group. Families share with each other, to be sure, but each keeps careful track of its contributions. Popov (1966: 108) comments on "their extraordinary frugality with food products. In the spring when the people who are short of food go for help to their better supplied neighbors, the latter give them a meager amount: two or three cracknels or a small piece of meat the size of a fist. However, no one is offended by this, for food at this time of year is of great value and precious to everyone."

When two or more families share the same tent, one man and his wife are accepted as leaders of the tent and occupy the place of honor at the right of the entrance. Other residents of the tent inform the leaders about their own economic activities. Popov does not mention whether separate families keep separate larders, but it does seem that sharing a tent implies at least a degree of communal food supply. In winter a large tent (up to 30 feet in diameter) may house as many as five families. Clusters of tents are common, as are clusters of stone and earthen huts. Still larger aggregates occur temporarily when geese or reindeer are abundant.

Where several families share a tent, each occupies its own portion of the tent, within which men, women, and children have their assigned places according to commonly accepted principles (e.g., the men are nearest the central hearth). The parking place for each member's sleds is also established, indicating the degree to which individual behavior must be structured in the multifamily co-residential group.

In the larger groups the unequal distribution of skills may lead to a division of labor. A good fisherman may be a poor sled maker and vice versa; hence exchanges between them are natural, if by no means free and easy. Popov (1966: 57) writes:

Collective consumption does not in any way . . . mean that food products, tools of production, or everyday objects are freely lent; on the contrary, strict accounts are kept of everything. A hunter's family, for example, will share the meat of a killed wild reindeer with neighbors. . . . But the hunter's neighbors who receive a potful must give help to the hunter's household, by means of their own labor or reindeer. They are obliged to tend his domesticated reindeer, clean his fishnets, lend him their sledge reindeer, and even occasionally supply him with a gun and ammunition. If a hunter did not receive help from his neighbors, he considered it his right not to share a potful with them.

For all this emphasis on individual ownership, concessions must be made to the needs of the group. For example, as we have seen, experienced hunters are allowed to regulate communal reindeer drives, and individual hunters agree not to hunt in ways that would threaten the group's success. Also, when camps or hamlets break up in the spring, families reach an agreement about which trails, streams, lakes, and so forth each will exploit, in order to avoid unnecessary overlap and competition.

With these exceptions we find no evidence of political activities beyond the household level. There is no group territorial control over resources except in the sense of a "home range" that a group occupies by tradition or mutual agreement; understandings concerning winter fishing spots are perhaps the strongest forms of resource control. A dominant man may attract hangers-on who will work under his direction, but they do not depend on him for access to resources and may strike out on their own at any time. The Nganasan operate on the !Kung San principle that "we are all headmen."

In sum, the Nganasan reveal the basic pattern of a family-level economy. Living in an environment of scattered resources, they pursue food sources opportunistically, moving much of the year in single-family households in pursuit of reindeer and other wild foods. Then periodically they congregate to harvest seasonally abundant foods such as migrating reindeer herds and flocks of molting geese. Stored meat and fat from these harvests are essential to their survival through the arduous winter and spring. Families remain obdurately independent even in their winter camps and hamlets, and are always free to separate from the group to follow an independent course. The family herds of reindeer are small and facilitate the foraging way of life: tame reindeer provide transportation, help in the hunt, and serve as insurance against starvation.

Recent history has seen significant changes in the economy of the Nganasan, leading to their transformation from reindeer hunters to true reindeer pastoralists. Essentially, as population expanded northward, consuming more and more of the temperate forests, the demand for the animal products of the far north increased drastically. The Nganasan found that they could sell reindeer and furs in an ever-growing market, and with the proceeds they could afford to buy guns, canoes, nets, traps, iron pots, tea, tobacco, and supplementary foods.

But as the demand for reindeer meat grew it became advantageous to manage reindeer production by increasing the sizes of domestic herds, which could be pastured on lands from which wild reindeer were being depleted by overhunting. Tame reindeer are clearly marked by coded notches carved in their ears, and will not be hunted by other Nganasan. The consequences of this transformation to true pastoralism have been several. The costs of production rose since private herds had to be protected from wolves and poachers. Many more animals were herded: whereas formerly eight or nine reindeer was a large number for a family herd, now 50 animals is considered a small herd. In the winter a family with a larger herd must move frequently in search of pasture. Hence the semipermanent hamlets of sod houses built near fishing spots have now been abandoned in favor of large, heavy tents that must be cumbersomely dismantled, transported, and erected in a new location every few weeks. It is even necessary to seek out and transport fodder in winter for the domestic herds.

Camp size has increased and kin relationships have become formalized around property in herds. Dowry payments and patron-client relations have emerged as important features of social life. With larger camps has come increased capital investment in such technology as large nets for the communal hunts of reindeer and geese. And reindeer hunting and sale of reindeer (both domestic and wild) are now monitored by a community extending far beyond the limits of the family group (see Ingold 1980).

The transformation of the Nganasan from reindeer hunters with small domestic herds into large-scale reindeer pastoralists was a response to a great increase in the demand for reindeer meat in an expanding market. Once reindeer were being hunted for sale rather than household consumption it became necessary to own more of them, since the numbers of unowned (i.e., wild) reindeer were being rapidly depleted. With changes in the provisioning of winter fodder the numbers of reindeer that could be maintained in this environment, especially in winter, increased, a form of intensification of production. The resulting increases in the scale and complexity of social organization are clearly responses to the underlying economic change.

Conclusions

Although they possessed the technology of food domestication, neither the Machiguenga nor the Nganasan made use of it before recent times to evolve beyond the family-level economy. The clear tendency is for small social units to be scattered uniformly across the landscape as long as the wild foods on which they depend are widely scattered. Aggregation is temporary, for the purpose of cooperation in getting food, as in the Machiguenga poison fishing or the Nganasan reindeer hunt, or sharing food, as in the seasonal hamlets of both groups.

Family autonomy is evident in a number of ways. Productive capital such as tools, weapons, herds, and gardens is individually owned, and its use by others is regulated and carefully reckoned. Similarly, a family keeps its own food supply, sharing food only reluctantly with families of the same hamlet. The ultimate autonomy of the family, of course, is its freedom to move, to detach from other families and pursue its own interests with minimal interference.

Evidently it is pressure on resources that brings about a greater dependence on domesticated foods and an increase in community size and economic integration. Internal population growth, encroaching populations from outside, access to new technology (e.g., shotguns) that facilitate intensification, and the opportunity to earn cash by intensifying production all make for a greater dependence on domestication. With this change comes larger community size and a new level of social stratification involving tighter control over resources on behalf of the larger group (as distinct from the perceived self-interest of the separate families that make up the group). All indications are that families as such are not particularly happy about this development, but accept it because they have no alternative.

FOUR

The Family Economy

ANTHROPOLOGISTS HAVE BEEN SLOW in acknowledging the family-level society as a distinct, theoretically significant type of human community. We have taken it for granted that families, households, and kindred groups are fundamental economic units, but usually think of them as subordinate to larger social institutions such as lineages, villages, or tribes. Moreover, we have tended to focus on societies with more developed social structure, such as corporate kin groups, hierarchical political systems, and ceremonial associations, and this focus has too often caused us to describe family-level societies in terms of what they "lack," as though they exhibited some reprehensible failure to achieve a respectable degree of size and organization.

Even Steward, who did the most to clarify the concept of a family-level society with his accounts of the Shoshone, considered them to be "typologically unique" (1955: 120) and denied their theoretical significance in prehistory. Service (1962: 64-66) went further and denied their very existence except in isolated instances of modern contact.

Yet Steward's analysis remains valid today, not only for the Shoshone but for many other forager peoples, including ones known to us only through prehistoric records. He was right to concentrate on a scattered population seeking locally variable resources consisting primarily of wild vegetable foods and small animals. He was right that in a family-level society, by which he meant a nuclear family augmented only by a few relatives (1955: 102), the family is self-sufficient and not permanently tied to a stable multifamily group. Moreover, he made it clear that the family makes sense as the elemental economic unit when the mode of food getting is "competitive." Unless there is some economic basis for cooperation, two or more families living together simply get in each other's way, and the natural tendency is to move apart far enough to minimize competitive interference in the food quest.

As we have seen in the two previous chapters, the existence of a family-level society depends on the family's ability to retain access to the three main factors of production: land, labor, and capital technology. At the family level, access to land is essentially unrestricted, labor is largely provided by the nuclear family, and the simple technology is produced within the household for its own use. In future chapters we shall see how erosion of the family's unrestrained access to the land, labor, and technology necessary for subsistence production underlies the formation of more complex social institutions.

Let us now summarize the main elements of the family-level society:

1. Population density is very low, characteristically well below one person per square mile. Factors responsible for this low density include recent settlement (as happened throughout the world in prehistory) and the scarce resources of a marginal environment.

2. The environment is marginal for human occupation and especially for intensive procurement strategies. Resources are characteristically scattered, unproductive, and/or highly variable. We surmise that family-level societies once occupied the full range of environments, but that in due course the pressure of population growth transformed all societies except those in the most marginal environments.

3. The technology consists of personal tools, such as the ubiquitous digging stick and bow and arrow, that are used individually to procure and process foods and raw materials. The most common food sources are wild plants and animals. Domesticates are adopted reluctantly to solve particular problems and come to dominate the diet only when natural foods are depleted.

4. Settlement patterns fluctuate between aggregation and dispersion. Camps and hamlets form at certain seasons when resources are concentrated or in short supply, only to break down into small family units when resources are abundant but dispersed. Settlements are generally located close to resources, but tend to change frequently to avoid depleting resources.

5. Social organization is familistic and informal. The main units are the biological family and the extended or "grown up" family living together as a camp or hamlet. Kinship is bilateral and flexible, allowing small groups to form and disband opportunistically. Production is organized largely according to the sexual division of labor within the family. Sharing based on reciprocity among families exists to solve problems of daily unpredictability. In general, the family unit is sufficient in most instances when plant resources dominate, but the camp becomes critical to handle the uncertainties of hunting.

6. Territoriality is weakly defined and access to resources is rarely defended. The scarcity and unpredictability of resources favor a nonterritorial system such that families and camps can move to a more promising area whenever they need to. The defense of a given area is impractical because resources are not concentrated or are unpredictable.

7. Warfare is virtually nonexistent. Personal hostilities among men, sometimes resulting in impulsive homicides, flare up over women or for other reasons. Aggression as such, however, is discouraged so that an extensive network of social relations may be maintained to minimize risk.

8. Ceremonialism is largely family ritual, as discussed in the Machiguenga case. Ad hoc ceremonies, as among the Shoshone and San, serve to bring people together when this seems necessary or desirable; but they remain informal and do not involve any kind of ceremonial hierarchy.

9. Leadership is similarly ad hoc, with leaders such as the Shoshone rabbit bosses providing direction when needed for short periods and then reverting to normal status. The authority of skill or experience does not confer superior status in any formal or permanent way.

10. Considerations of cost and risk minimization are paramount. Food is selected, people move, groups form, and ceremonies are performed according to a pragmatic evaluation of costs and benefits. The first and foremost consideration is the need to minimize risk; this is responsible for a broadly eclectic diet, an extensive network of kin and nonkin relationships, and the periodic formation of camps or hamlets.

This description constitutes a basic model for the subsistence economy and social organization of most low-density hunter-gatherers. An important conclusion of Chapter 3 is that this basic organizational pattern can continue when some resources are domesticated. Gardening or the use of domestic animals may permit somewhat higher population densities and bigger, more permanent hamlet groups; but the radical change from family-level organization to a more complex level does not necessarily follow the adoption of domesticates and may lag thousands of years behind it.

We now recast the commonalities evident in the case materials by looking at the systematic relationships involved. We approach them from the standpoint of the three theoretical orientations introduced in Chapter 1: the ecological, the structural, and the economic.

The Ecological Perspective

The diet, use of storage, and social organization of family-level societies, and even their ceremonial practices, can be understood as
a direct adaptation to the use of natural resources with a simple
technology. The diet of such a society is highly variable, changing
through the year and from locale to locale according to where food is
available. Perhaps most interesting is the importance of natural foods
in the diet and the extreme reluctance to shift to domestic foods,
which are characteristically more laborious to procure and less varied. The shift to domestication seems to allow higher population density with a minimal sacrifice of the forager lifestyle. Diet reflects the
natural availability of wild foods and the need to augment these with
domestic foods.

We also underscore the extent to which surplus food production
and storage are required by seasonal availability and unpredictability.
The mongongo nut stores itself, and in a sense the root crops of tropical horticulturalists do the same, though the gardens must be maintained to keep a surplus available. The storage of pine nuts by the
Shoshone and of dried meats and rendered fat by the Nganasan are
unequivocal cases of more than casual surplus production for storage.
In three of our four cases—the exception being the well-provisioned
!Kung San, who may "store" food seasonally as body fat—stored
foods are critical for survival during the scarce seasons.

Indeed, we have to understand such surplus production not as a
sign of affluence (Sahlins 1972), but as a form of insurance against bad
weather, health problems, or other contingencies that can cut production drastically and threaten families with starvation. Social organization and demographic distribution are also closely linked to resource
availability. The changing availability of wild foods seasonally and between good and bad years largely explains the pattern Lee (1979) calls
"the concentration-dispersion dialectic." In the !Kung San wet season, the Shoshone and Nganasan summers, and the Machiguenga
dry season, when resources are quite uniformly available and abundant, families scatter competitively in search of wild foods and are
little involved in larger groups. In other seasons, however, they come
together and form camps and hamlets for the most efficient procurement and distribution of food. These larger groups work together in
such large-scale activities as rabbit drives, reindeer ambushes, and
fish-poisoning expeditions, and share food when famine threatens.

The Structural Perspective

Turning now to the "structural" or ideological elements of family-level society, we find that many structural principles also clearly reflect ecological and economic necessity. For example, the sharing-hoarding dialectic of the camp's ideology mirrors the concentration-dispersion dialectic that we have interpreted as basic to resource exploitation in these societies. Similarly the principles of ownership and division of labor have clear pragmatic value for the flexible family pattern of resource use.

A striking characteristic of family-level societies is their comparatively simple and flexible social organization. As noted, anthropologists (e.g., Holmberg 1969: 124-60) have had a tendency to stress what is "lacking" when discussing family-level social organization. By contrast, we view it as an expectable economic form under specifiable circumstances, and its simple, flexible social rules as a natural consequence of that form.

To be sure, family-level societies abound in structures, albeit restricted in scope for the most part to family-sized units, that regulate access to resources, patterns of production, the distribution of food, and economic relations beyond the family unit. Typical of these are the rules governing the division of labor by sex. These are not so much formal principles as common understandings concerning proper spheres of activity for men and women. A violation of these rules is not a crime but an embarrassment; the violator is not punished but teased and ridiculed. The structural basis for these rules is so deep and enduring that men and women rarely perform the same tasks: even in a common pursuit like gathering they tend to divide the tasks into male and female activities rather than perform identical tasks. Thus the social organization of the economy, flexible and individualistic though it is, nonetheless constrains behavior powerfully and pervasively through cultural understandings of what is courteous, respectful, or proper.

The Economic Perspective

The structural and economic aspects of the family-level economy are closely linked. For example, although ad hoc ceremonies are structured by rules of proper comportment that strengthen mutual respect and reinforce the rights and duties of exchange partners, the social relations that remain after the ceremony ends are personal networks of "friends," not formal groupings dictated by rules of kinship or clanship. The individual in a family-level economy is constantly negotiating his social relations as he or she makes demands or responds to the demands of others. Generally speaking, individuals and nuclear families must strike a balance between the assertive pursuit of self-interest, as seen in the hoarding of resources and the frequent escape of families from camps or hamlets, and the group's requirement of generosity and self-effacement.

What surfaces repeatedly in ethnographic accounts of family-level societies, Sahlins (1972) and others to the contrary notwithstanding, is the importance of economizing behavior in all aspects of life from production to social and political organization. Perhaps the most dramatic examples of such behavior from an evolutionary perspective concern the reluctance with which family-level economies adopt such seemingly advanced cultural elements as domesticated foods, capital investments in production, and complex political controls on access to resources. The Nganasan kept only small herds of domesticated reindeer, and the Machiguenga persisted in a seminomadic forager life, not because they failed to grasp that intensified domestication could support a larger population, but because a more modest technology was more efficient for meeting their individual needs. Indeed, in these two cases the recent shifts in economic focus toward greater reliance on domesticated crops and animals reflect outside pressures that have changed the relative values of alternative procurement strategies, for the first time making the more "advanced" technology attractive to the individual family.

We are now ready to consider the forces and lines of development that lead to a more complex form of social organization. We shall find that the principles we have already discovered remain in effect in more complex economic systems, where they are overlain and modified by higher-level processes. We shall also find that more than one path leads to suprafamily organization, and that several such paths may coincide to produce greater degrees of organization in some cases than in others. In most cases, however, the factors sustaining development beyond the family level are extensions of processes already evident, though not yet dominant, in the family-level societies we have examined.

People in family-level societies are closely attentive to matters of possession and ownership. As a rule, every item produced has an individual owner who decides how it is to be used. Items change hands only through deliberate acts of giving and receiving that have significance within and between communities. Between families, and to an extent within families, there is little or no communal ownership of products in the sense of freely making one's goods available to the community on demand.

Just as the acknowledgment of individual ownership is a structural reflection of the day-to-day economic autonomy of the family unit, so such "territoriality" as exists at the family level reflects the family's right to forage close to home without interference or undue competition. It is rarely necessary to use violence to defend territory because of the consensus that no family or camp should intrude too deeply into another's home range. But a needy person may ask permission to intrude and, according to the rule of generosity, will receive it, having formally acknowledged that the resources he will procure there are not really his but the giver's. Often, as among the !Kung San, the formal request reciprocally binds the family receiving access to resources to open up its own home range in times of need.

Such mediation and indeed all suprafamily relations occur ad hoc. We have noted the absence of formal, calendrical ceremonials at the family level. Families and even individuals do celebrate or conduct rituals, but only on the spur of the moment, when windfalls or personal considerations dictate. Seasons of unusual scarcity or abundance make pooling and cooperation more likely, but these are trends, not regulated features of a ceremonial cycle.

When suprafamily ceremonies do occur, however, they are of real economic importance. They draw people from distant settlements, exploiting dense temporary resources while resting or fallowing their home territories. Further, ceremonies are opportunities for mating and gift exchanges that create and reinforce personal ties throughout a large region.

The social networks engendered and maintained by ceremonial exchanges are economically essential, even if they are only occasionally activated. The success of a family, camp, or hamlet group depends on an adequate year-round supply of resources in the home range, but no home range provides this ideal over the long term. The personal ties established during multifamily feasts and dances act as a security net by which families who are suffering local disasters, such as droughts or shifts in the migration routes of game, may reliably gain access to the home ranges of other families.

Part II

The Local Group

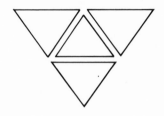

FIVE

The Family and the Village

OUR CASES IN THIS CHAPTER and the next are in many ways similar to those of Chapters 2 and 3. An emphasis on wild foods in the diet persists, seasonal fluctuations in food supply still generate an alternation between feast and famine, and population patterns constantly oscillate between aggregation and dispersion. But now we also encounter local groups of several hundred people with strong ties of economic and sociopolitical cooperation and exchange—ties whose breakdown means serious economic loss and even disaster. These groups, usually taking the form of nucleated villages or property-owning clans, are of a higher order of economic integration than family-level societies. They arise in response to specific problems encountered in the natural and social environment, which in turn pose new organizational problems for their members.

In Chapter 4 we argued that the benefits of a larger community must outweigh the costs before people will join it, or even form one. At the level of the local group, two broad kinds of benefits arise from larger size: sharing of food (and other resources) and defense. Where food sharing is primary, larger communities form either when it is necessary to pool resources during periods of food insecurity or when windfalls must be distributed before they spoil. We have explored this phenomenon at length in Chapters 2 and 3.

As we shall see in this chapter and the next, the necessity of defending resources against raiding by enemies also places intense pressure on local groups to increase their size, both by living in closer proximity and by entering alliances. Because people who share defensive alliances are likely to share food, and vice versa, it is difficult to disentangle and measure the relative importance of defense and pooling in increasing group size; indeed, reliance on one for security is likely to lead eventually to reliance on the other, mutually enhancing the value of the local group to its constituent families. This pattern is especially evident among the Yanomamo, the subject of this chapter.

What is clear is that both strategies entail costs, notably the depletion of resources within easy distance of the local group and the bickering, theft, adultery, begging, and other unpleasant consequences of multifamily group life. Like family-level camps, local groups disperse whenever possible, coming together for a time, then scattering, only to come together again (though usually with some changes of membership) once conditions warrant. Our next four cases show considerable variation in this regard, based largely on economic factors: they range from the seasonal aggregates of the hunter-gatherer Eskimo to the stable, sedentary local groups of the higher-density Tsembaga horticulturalists. The more permanent village group seems closely tied in many cases to warfare.

Of particular interest in these cases is the extent to which ceremonial life reflects underlying social and economic variables. Whenever the local group comes together, ceremonial life is intense. Some ceremonial activities seem designed to dissipate factional tensions, as in the Eskimo foot races and the Yanomamo club fights. Others reinforce exchange partnerships and alliances, as in the Tsembaga *kaiko*, the Eskimo Messenger Feast, and the Yanomamo feast. Yet despite the economic advantages of the local group, and despite intense ceremonial efforts to hold the group together, local groups are constantly in danger of fissioning. This tendency is strongest in the Yanomamo feast, whose participants are often uncertain whether the feast will end in friendship or in treachery and death.

Finally, with the local group comes an increased need for leadership and centralized control of resources, productive capital, labor, and storage areas or facilities. Exceptionally skilled and admired men gain wealth and prestige above the common run, but are obligated to share their wealth with their followers. In return for good management of the group economy, they enjoy fine clothing, special foods, additional wives, or other marks of high status. As in simpler societies, leadership is typically context-specific, without offices or hereditary titles. Fortunes rise and fall over time. But such leaders do represent a meaningful departure from the emphasis on modesty and conviviality found in family-level societies, and most househeads' willingness to accept the lower status accorded followers argues that they believe they obtain value from their leader.

Case 5. The Yanomamo of the Venezuelan Highlands

The Yanomamo have become a test case for materialist theory, primarily because of the difficulty of explaining their peculiar form of warfare (Harris 1974; Chagnon and Hames 1979). The central issue has been and remains this: do the Yanomamo, who appear to fight frequently, impulsively, and with extraordinarily high rates of mortality, fight over scarce material resources or for other, nonmaterial reasons?

In his original descriptions of the Yanomamo, Chagnon (1968a, 1968b) emphasized that they fight for many reasons: for women, who they say are scarce; for revenge of suspected sorcery and of real past injury; and because the political system is too weak to prevent warfare. Harris (1974: 102; 1979) correctly pointed out that such an eclectic view did not provide a satisfactory *explanation* for Yanomamo warfare. In his opinion the Yanomamo were competing over hunting lands, and in particular over access to scarce supplies of dietary protein. Chagnon (1983) replied that whereas the Yanomamo do indeed view meat as both highly desirable and rather scarce, his data showed them to be adequately supplied with protein in the diet.

Recently Chagnon has espoused the bioevolutionary concept of "inclusive fitness," marshaling evidence to show that success in war, intimidation, and political maneuvering correlates with success in reproduction: in the simplest terms, Yanomamo men do not just fight over women, but over the "means of reproduction." Aggressive men, within limits, succeed in leaving more offspring than men who allow themselves to be intimidated and dominated. Thus as the debate between Harris and Chagnon over the causes of Yanomamo warfare has matured, it has tended to coincide with the debate, discussed in Chapter 1, between theories focusing on reproduction and natural selection (Chagnon, sociobiology) and those focusing on production and adaptation (Harris, cultural materialism).

In this chapter we shall argue that it is indeed competition over scarce resources that ultimately explains Yanomamo warfare. Chagnon's important conclusions concerning competition over reproductive rights in women seem entirely correct to us, within the limits of his pioneering data, and we incorporate them in our explanation. But the critical issue in Yanomamo warfare is control of vital resources: hunting territories, to be sure, but also improved agricultural land and territories rich in other resources. This critical issue explains not only the fact of Yanomamo warfare but the peculiar form it takes, a form part way along the evolutionary spectrum from the relative absence of warfare among peaceful family-level groups like the Machiguenga (Case 3) to the more highly organized and routinized forms of warfare that we shall encounter in later chapters. Although like the peaceable Machiguenga in much of their adaptation to a humid tropical forest, the Yanomamo have crossed the fateful threshold from a cultural emphasis on control *over* aggression to an emphasis on control *by means of* aggression. We shall now examine why this should be so.

The Environment and the Economy

The Yanomamo (Chagnon 1983), also known as Yanoama (Biocca 1971; Smole 1976) and Waika (Zerries and Schuster 1974), traditionally inhabit the highlands of the headwaters of the Rio Orinoco and Rio Negro in Venezuela and Brazil. Although this region is generally considered part of the tropical forest lowlands of South America and its peoples are often considered "Amazonian," these highlands are different from the tropical lowlands that surround them. Most studies of the Yanomamo describe communities that have recently migrated from the highlands into "virgin" lowland territories (Smole 1976: 226), communities significantly different from the highland Yanomamo on whom we focus here (see Map B in Migliazza 1972: 17).

The Rio Orinoco originates in the Guiana Highlands, a landscape of rocky promontories and ridges with uplands of ancient granites and metamorphic rocks heavily weathered and eroded into an intricate and irregular sequence of hills and valleys. Except for a spotty distribution of alluvial deposits, the soils are poor: "None of these uplands of sedimentary rocks is suitable for agriculture. Throughout the Guiana Highlands, it appears that agricultural advantage is limited to the valley bottoms and that these are not of remarkable fertility." (Sauer 1948: 320; see also Lathrap 1970: 42.)

The region traditionally occupied by the Yanomamo ranges in altitude from about 1,000 to 4,000 feet. The higher elevations (up to 6,000 feet) are used less frequently for foraging and tend to be "uninhabited, covered by scrub and brush and very rocky" (Smole 1976: 32-33). We may visualize the region as an island rising above the sea of tropical rainforest that spreads outward along the rivers of the Orinoco and northern Amazon drainages. The highlands are cooler and drier than the lowland rainforest, and support a unique biota (Anduze 1960: 186-87; Chagnon 1983: 55).

Anduze (1960: 173) noted that whereas game abounded along the river at lower altitudes, it became increasingly scarce higher up, a fact he attributed partly to intensive hunting by the highland Yanomamo.

Smole (1976: 41, 131) also notes that wild foods in general and game in particular are less abundant in the highlands. Fish are also scarce and add little to the diet of highland Yanomamo, as compared to lowlanders (Chagnon 1983: 102). Thus, the Guiana Highlands are not merely an island in the physical sense, but a distinctive ecological zone characterized by poor soils and scarce game.

The Yanomamo have long inhabited the Guiana Highlands and may even be descendants of original "foot nomads" of those parts, who inhabited a much larger region until they were displaced by expanding Cariban and Arawakan groups (Smole 1976: 17-18; Wilbert 1966: 237-46; Atlas 1979: 320-21). The Yanomamo language is unrelated to Arawakan and Cariban, and the culture is quite distinct. The Arawakan and Cariban groups were canoe peoples of remarkable strength and ferocity who dominated the navigable rivers with an economy centered on bitter manioc, fish, and game hunted in the comparatively rich lowland forests. Most often, where the Yanomamo were in touch with such groups, it was the Yanomamo who were either subservient, if contact was peaceful, or killed and driven off if contact was hostile (Smole 1976: 228, 230; Chagnon 1983: 61).

Unlike lowland groups, the highland Yanomamo do not use canoes, consume little fish, and generally avoid water whenever possible. Their economy is also distinct in its emphasis on plantains and sweet manioc. According to Smole (1976: 13-14), the Parima region of the Guiana Highlands is "one of the last great cultural redoubts of the South American continent. . . . Most of the traditional Yanoama territory is inaccessible by water navigation, effectively protecting its inhabitants from outsiders." When we examine Yanomamo warfare and "fierceness" later, we should remember that the Yanomamo primarily occupy a difficult refuge zone historically surrounded by powerful antagonists.

Following the European colonization of the Americas, riverine groups like the Arawakans and Caribans were highly vulnerable because they occupied rich environments easily accessible by boat. They were enslaved, decimated by disease, and finally incorporated within the expanding frontier of Western civilization. By contrast, the Yanomamo seemed to draw into their highland shell even more completely, perhaps chiefly as a way of avoiding disease, which they equate with white men (Biocca 1971: 213; Chagnon 1983: 200).

At least by the nineteenth century, however, the Yanomamo had acquired steel tools and plantains, after which their population rapidly increased (Chagnon 1983: 61). Since the 1940's, a number of Yanomamo have ventured out of their highland redoubt and begun to colonize the larger rivers at lower elevations, where population pressure was comparatively low owing to the collapse of the Arawakan and Cariban populations. As a result of Yanomamo reticence and lowland depopulation, an unpopulated no-man's-land or buffer zone had grown up around the Yanomamo. Good garden land was available in this zone, and game animals were both abundant and virtually unafraid of men since they were so seldom hunted (Steinvorth-Goetz 1969: 195). Chagnon reported abundant game in this region in 1968-71. Ten years later, however, the game had been hunted out by the Yanomamo and others, and Chagnon likened the region to a desert (1983: 157, 202).

The costs of moving into the lowlands were also great. Malaria is endemic to the lowlands and undoubtedly acted as a barrier to Yanomamo migrations in the past (Smole 1976: 228). Malaria and yellow fever are major causes of death among contemporary Yanomamo (Smole 1976: 23), and the threat of other diseases being spread by contact with whites, such as measles (Chagnon 1983: 199), is much greater in the lowlands.

Partly for these reasons, the population density of the lowland Yanomamo is comparatively low, but there is considerable evidence for higher densities in the traditional highland environment. Smole (1976: 48) summarizes the situation as follows: "Mean population density over their entire territory is approximately 0.5 person per square mile. Since such a calculation is based upon heavily populated highlands as well as virtually empty lowlands and uninhabited high highlands, the effective density is much higher locally. Observation of the extent to which portions of the high Parima have been transformed into savanna leads to the suspicion that only a few decades ago, if not centuries, population densities there were considerably greater than they are now."

Like the family-level economies we reviewed previously, the Yanomamo economy provides a sufficient livelihood at comparatively low cost. In particular, the diet of the highland Yanomamo contains ample supplies of protein (Chagnon and Hames 1979). Yet the resources they depend on for a high-quality diet and for what they see as the basic necessities of life are scarce, and as a result they experience "crowding." To a limited extent the crowding is a result of warfare, which forces the Yanomamo into large villages for defense and hastens the degradation of the environment within reasonable traveling time from the village. This is paradoxical because, as we shall argue, the warfare itself is an outcome of scarcity and competition over resources.

Wild Foods and Hunting. The Yanomamo are dependent on wild foods for nutritional diversity and for the spice such foods add to meals that are primarily made up of garden produce. They have eclectic tastes and relatively few restrictions on what may be eaten (see Taylor 1974). Their main "taboo" is against killing anything needlessly, for animals are scarce and nothing should be wasted (Smole 1976: 183). Otherwise "everything from the mosquito on up is fair game" (Anduze 1960: 203). Their foods include crabs, shrimps, and occasional small fish from mountain streams, frogs, ants, termites, insect larvae, hearts and fruits of palm, other fruits, and various roots. Indeed, although large game is the meat preferred by highland Yanomamo, because of its scarcity they probably depend as much on insects as on game (Smole 1976: 163). Some fruits are preserved by drying and are then stored in caves (Biocca 1971: 76; cf. Smole 1976: 237), and some groups prepare special "breeding grounds" in which large numbers of frogs will reproduce to be harvested (Smole 1976: 247).

Owing to the virtual absence of fish and the scarcity of large game, the highland Yanomamo are better described as gatherers than as hunters. While living in villages they have a hard time satisfying their desire for wild foods. So they frequently leave their villages and form hamlet-sized extended family groups to forage in less densely populated areas of their territory and even in the remoter sections of neighboring village territories.

During some periods these small foraging groups depend heavily on wild foods, but they always carry along garden produce, and especially plantains, as supplementary food. Frequently, they set up camp in an old garden that is still producing food, and then "eat their way out of camp" !Kung-style, returning to camp each day to make a meal of wild foods and the remaining crops in the garden. When they come across a windfall of wild fruit too large for their needs, they send word to friends and relatives in other groups to come and share the good fortune. Thus it happens that frequently throughout the year a village is either completely empty or else inhabited by a small fraction of its total population.

Yanomamo groups vary in how much wild food they eat. The inhabitants of large, sedentary villages may be militarily secure but regret the comparative absence of wild foods in their diet, whereas small, mobile groups enjoy access to wild foods but are vulnerable to raids and may be driven from their territories. In one case reported by Helena Valero, a Brazilian girl captured and brought up by the Yanomamo, a powerful group of villagers captured the women of a small foraging group. When the captives fled, the village women called after them angrily: "Go on, go on! Go back to eating wild fruit and bad fruit. Stupid women for running away! If you had stayed with us, you would have eaten *pupugnas* [peach palm fruits] and bananas from our *rocas* [gardens]. Now you'll have to work hard enough to find wild fruit in the woods!" But the forager women were not impressed, calling back: "We haven't come to ask you for fruit and bananas" (Biocca 1971: 34-36). Indeed one group, the *Gnaminaweteri* ("solitary people"), received its very name because of its preference for a peaceful, mobile, small-group lifestyle.

Certain species of game are the only animals the Yanomamo designate as "real food"; the major garden foods, plantain and peach palm fruit, are also "real food." Only real foods can form the basis of meals during intervillage feasts and ceremonies. Thus when a feast is anticipated large groups of men leave the village for about eight days (the length of time required for freshly harvested green plantains to ripen in the village), returning only after they obtain sufficient supplies of game. Among the most prized species are the tapir and the peccary (both very meaty animals), the agouti, the armadillo, and secondarily certain monkeys and birds (Smole 1976: 182). In these expeditions the men range far from their village and often into hunting lands of adjacent friendly villages. The preferred hunting areas are usually at higher elevations, where neither gardens nor villages are located and hence where game is not so frequently hunted. But even when hunting is done for household use and not for a feast, hunters expect to be gone for several days. By contrast, the typical hunting trip among the family-level Machiguenga (Chapter 3) lasts five to seven hours.

In other respects the highland Yanomamo and the higher-elevation Machiguenga have much in common. Neither group gets as much wild food as it wants, and no highland area exists in either region that is not systematically and continuously foraged. Even in the only zone both groups avoided, the recently settled lowland fringe where once-dreaded canoe Indians and recently the more dreaded whites lived, game is scarce. Chagnon (1983: 157) reports a hunt for a feast held by the Bisaasi-teri in 1965, 14 years after they had moved into the lowlands. Although a large group of men spent a week hunting, they returned with only 17 monkeys, seven wild turkeys, and three large armadillos, scarcely enough to feed a hundred guests for several days and supply them with meat to carry home after the feast. The Yanomamo understandably have a special term for "meat hunger" as distinct from other hunger (Smole 1976: 175), and even in the lowlands "meat is always the most desirable food and is always considered to be in short supply" (Chagnon 1983: 119; cf. Harris 1974: 102-3).

Coupled to this pervasive sense of scarcity is perhaps a more important sense of imbalance. Some locales are perceived as better for game than others, and most of these probably *are* better. These locations are coveted and, one can assume, defended actively from outside hunters. According to Chagnon (1983: 70), "The Yanomamo prefer to remain in one *general* area a long time, especially one that has a reliable source of game within a reasonable walk from the village. My research has revealed many cases of the same village remaining in one area for 30 to 50 years, leaving it only when the military pressures on them are overwhelming."

Gardens. Near the villages, plots for slash-and-burn agriculture are cleared from the forest. The centrifugal attraction of wild foods is counterbalanced by the centripetal pull of the gardens, which are as highly productive as the Machiguenga gardens (Smole 1976: 150-51). The highland Yanomamo make effective use of those scarce lands that are suitable for horticulture, growing a number of foods and other materials without which they simply could not exist in anything like their present numbers.

The soils the Yanomamo clear for gardens are "fertile, friable loams" (Smole 1976: 24), often capable of supporting a village population in the same general location for many years. But large areas of the highlands are unsuited to agriculture: in these areas "soils tend to be sandy and at least mildly leached" (Smole 1976: 37). Even the "friable loams" are often very acidic (pH 4.5), which limits the production of some crops. In addition, many areas have slopes too steep for gardening. The higher altitudes (above 3,500 feet) are marginal for agriculture, and the lower altitudes are avoided for health reasons. Finally, extensive areas of the highlands, though at a good altitude, are sterile savannas of absolutely no economic utility (Smole 1976: 37).

As among the Machiguenga, the selection of good garden land is a major concern. Potential sites for new gardens are a popular topic of conversation between the men during hunting trips (Chagnon 1983: 60). The best lands (*ishabena*) will be covered by a forest of large trees; be at the proper altitude; have sufficient slope for good drainage and for different crop associations at different elevations in the same garden, but not an excessively steep slope; have heavy, dark soil; be sufficient in extent to support a village-sized population; and be near pure drinking water (Smole 1976: 26, 107-10, 116, 132, 239). Such ideal conditions are rare. As a consequence the Yanomamo are very unevenly distributed throughout the highlands, with great concentrations in areas with the best garden sites and few if any settlements in other areas. Once located in an area of good soils, the Yanomamo tend to stay, clearing garden after garden until whole regions are checkered with old and new gardens, blending into open savannas that appear to be man-made.

Altitude, as we have seen, is an important factor in Yanomamo horticulture, and gardens at a given altitude will not often fill all a household's needs. Crops like plantains, peach palms, and tobacco prefer low, humid soils, whereas crops like arrow cane do better at higher elevations. Since all these crops are essential to the Yanomamo, we find not only frequent trade between villages at different altitudes, but the cultivation by enterprising family groups of gardens at different altitudes in addition to their main garden close to the village.

Warfare adds further to the complex interplay of determinants of Yanomamo garden location. The members of a village prefer to have their gardens close to the village, where they are easier to defend, and this may lead them to plant in less desirable land (Smole 1976: 107, 244). But the opposite also occurs (Smole 1976: 239): "[The Docodicoro-teri] had become so dissatisfied with gardening on the low, alluvial terraces near the safety of the *shabono* [village] that they cleared a large new garden high in the mountains about four miles to the south. This brought them much closer to their enemies (the Bashobaca-teri), but they took the risk because they felt there was no place closer to the *shabono* that was as good for growing the *cowata* plantain."

The major food crops of the Yanomamo are plantains, sweet manioc, peach palm fruits, taro-like cocoyams (*Xanthosoma*), and yams (*Dioscorea*). Plantains, a very efficient source of carbohydrates and by far the major food in the diet, must be transplanted by cuttings from the roots. This requires a tremendous effort if a new garden is to be distant from the current one, or if a group must move suddenly and will have to transport the larger roots that produce a crop more quickly. Both these conditions are most likely to occur when a group has been defeated in war.

Plantains produce only once in a period of months, and cannot be stored. They are planted at different times of the year as a man's labor time permits, so the plants ripen at different times, spreading the harvest throughout the year (Chagnon 1983: 71). But exactly when and in what quantities plantains will ripen is difficult to foresee, so that sometimes there are no plantains even where the average production of plantains is ample for the population. Friendly villages even out such local fluctuations by inviting each other to feast when their own supplies of plantains are excessive (see Biocca 1971: 27, 45; Smole 1976: 104, 106, 129, 141-42). Smole (1976: 193-94) suggests that there are no actual surpluses of plantains, since all those not eaten by their growers are eventually consumed in feasts. Plantain gardens continue to produce food for many years with minimal care.

Sweet manioc, cocoyams, and yams help fill the periods when plantains are insufficient. The fruits of the peach palm (*Guilielma* sp.) are seasonal, tending to ripen in January and February (Anduze 1960: 215); with twice the protein of plantains and manioc and ten to 40 times the quantity of fat, they are a highly prized food. According to Chagnon (1983: 70-71), "This palm is an exception to my earlier generalization that it takes a great many palm fruits to get a full belly. Peach palm fruits (*rasha*) have relatively small seed (some have no seed at all) and a very large amount of mealy flesh, about the texture of boiled potatoes, rich in oil, and very tasty." These palms also have an exterior wood so hard that it is nearly impossible to drive a nail into it. The Yanomamo use this wood to fashion their bows, several types of arrowheads, and their fighting clubs (*nabrushi*).

Peach palm fruit is, along with plantain, a "real food," and hence suitable fare for feasting. In Chagnon's words (1983: 71), "Families usually plant one or several of these trees every time a garden is cleared, and the trees produce very large crops of fruits for many years after the gardens have been abandoned. Thus, by remaining in a general area the peach palm crops can be easily and conveniently harvested, and yield enormous quantities of tasty, nutritious fruit." Since peach palms grow better at lower altitudes, and since their fruit is seasonal, they are less commonly served at feasts than plantains and are enjoyed less frequently in villages at higher elevations. But the availability of peach palm fruits in old gardens at a distance from the village permits families to forage for extended periods away from the village when plantains are scarce and wild foods become important in the diet (Smole 1976: 155).

One difference between Yanomamo and Machiguenga gardens must be emphasized. The Machiguenga garden's useful life is only a few years at most. Yanomamo gardens, by contrast, with their stands of plantains and peach palm, are major long-term capital improvements that yield harvestable foods for many years after their initial cultivation. Old gardens are an important resource for the Yanomamo. Because of old gardens a village's territory increases in richness through time and is not to be abandoned lightly.

Scarcity in Yanomamo Ecology. We have referred to various forms of scarcity among the Yanomamo: of large game animals, of top-quality agricultural land, of preferred wild plant foods, of particular foods or raw materials that do not grow well in their gardens, and periodically

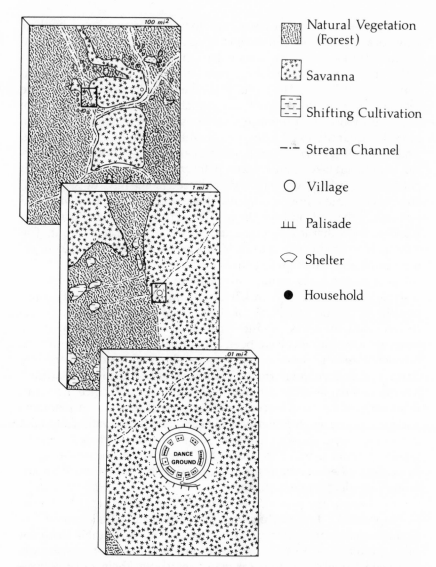

Fig. 4. Settlement pattern of the highland Yanomamo. Family groups cluster together in small villages for defense. Despite a fairly low population density, the environment has been severely degraded and economically barren savannas dominate the landscape.

of their most favored agricultural products, plantains and peach palm fruits. The most dramatic and significant evidence of scarcity in the highlands, however, is the destruction of the forest by overintensive agriculture that has resulted in widespread savannas. This condition is most severe in the areas of densest, longest-term settlement (Smole 1976: 203, 208). In most cases these savannas are the remnants of old gardens. Many have the regular shapes and straight edges of gardens, and some adjoin old gardens that may be savannas in the making. The cooler, drier highland climate may hasten the development of savannas in some highland areas. But as Smole (1976: 208-9, 254) makes clear, areas that were rich gardenlands within the memory of living Yanomamo are now sterile savannas.

Smole (1976: 210) describes three "zones of impact." Near the village the environment has been completely domesticated in a "fragmented zone of intensive use," and savannas often border the villages (Fig. 4). Within "easy reach" of the village, say a one-day trip, is a "zone of intensive foraging" from which wild foods are substantially depleted; this zone regenerates after the village has been relocated. Beyond that is the much less intensively used "zone of hunting and sporadic collecting." The savannas, of course, are a fourth zone, and one that is growing.

Another aspect of scarcity is the local distribution of certain highly desired products. For example, the plants that provide the hallucinogenic drug *ebena* are unevenly distributed, and many villages are unable to procure their own; the same is true for arrow cane, curare, bamboo for quivers, and peach palm. Villages near abundant supplies of these products specialize in preparing them as items for trade (Arvelo-Jimenez 1984; Chagnon 1983: 46-50; Smole 1976: 70-71). The unequal distribution and unreliable productivity of many products make trade an important economic activity among the Yanomamo.

Social Organization

Yanomamo society resembles Machiguenga society in that the family is primary and kinship is the basic means by which social life is integrated and structured. As we shall see, however, the Yanomamo have a further level of social integration not found in family-level societies: village and intervillage alliances.

The Household. The family household is the basic economic unit of the Yanomamo. Although in contrast to the family-level societies, single households always live in large groups, the Yanomamo household is a highly autonomous unit. Within the village each family's space is carefully demarcated and contains its own hearth, sleeping area, and goods. Similarly, although villages appear to have large communal farms, each man's separate garden is clearly demarcated and protected by strict rules against theft. Despite the vastness of the forest and the spaciousness of the village, with its large central clearing, Yanomamo families crowd into tiny spaces, where they hang their hammocks side by side or even stack them one above the other. According to Smole (1976: 67), "It is not at all unusual for a family of five to occupy a space of approximately ten by twelve feet, which means that an individual has about twenty-five square feet of living space." In these close quarters children learn to control their selfish impulses, and in particular to be generous (Biocca 1971: 137-38, 159). Children are indulgently reared, although some "fierce" (*waiteri*) men fly into rages and impulsively beat their wives or children, occasionally injuring them seriously. Parents are reassured by having children, especially sons, to care for them in their old age, and a woman's only defense against an abusive husband is to have her brothers nearby to protect her (see Biocca 1971: 95).

The Teri. The smallest groups observed living alone among the Yanomamo number 30-35 people, about as many as the largest groups normally found among the Machiguenga. No Yanomamo household can live apart from some form of larger group, called a *teri*, which is an extended family or collection of extended families occupying a single village. All teri are named, usually after a feature of the landscape, for "the teri name is geographic" (Smole 1976: 52, 57).

The village, or shabono, is essentially a large circle of lean-tos with roofs of palm leaves sloping up from the ground to a height of 15-20 feet. The center is open to the sky, and the ground is reserved for public events, as when the village hosts a feast for an ally. The individual households are laid out around the circle of the dance ground and under the slanting roof. The slanting roof of the village encloses and fortifies the teri; people can only enter and leave by a narrow gate.

The teri often comprises two intermarrying extended families, each equivalent to a single Machiguenga hamlet (cf. Wilbert 1972: 46). Thus in the Parima highlands the average teri contains 70-75 members. The men of such a teri are either brothers or brothers-in-law, fathers or sons, or uncles or nephews to each other. In larger teri, however, many men are only classificatory kin; they are not biologically close and tend not to interact very much. True brothers, bound by strong family feeling, and true brothers-in-law, who have actually "given" women to each other's group, are very close: they live in adjacent parts of the shabono; plant their gardens side by side, sharing the different microecological zones of the garden area; and leave the shabono together on hunting and foraging ventures (Smole 1976: 67, 94, 158, 188-89; Chagnon 1983: 67, 131).

The larger teri comprise several smaller teri that have come together into a single large shabono for security in wartime. While there, they take on the name of the teri identified with that territory. The solidarity of a Yanomamo teri depends on the density of kinship and marriage ties among the members. Chagnon (1983: 110-45) shows that although the classificatory kinship system pairs many men as "brothers" or "brothers-in-law," a man's closest associates are those to whom he is most closely related genetically and by marriage. Powerful emotions and social sanctions prohibit theft, insults, and violence between close relatives.

A teri increases its solidarity when its members intermarry, a strategy that not only knits distant kin more closely together as affines but also increases the actual degree of genetic relatedness among members. Since close relatives side together in a fight, villages whose members are closely related fight less among themselves and can grow to a larger size, a distinct advantage in times of war. Chagnon shows that the Shamatari, who are greatly feared along the Orinoco (Anduze 1960: 122), have an "average relatedness" equivalent to that between biological first cousins. This is much more relatedness than the neighboring Namoeteri achieve, and allows the Shamatari to live in larger, more stable, and hence more dangerous groups than the Namoeteri.

Although Yanomamo kin groups have been referred to as "lineages" (Chagnon 1983: 127; Smole 1976: 13) and as "clans" (Anduze 1960: 228), these labels suggest more structure than is actually present (Jackson 1975: 320-21; Murphy 1979). Yanomamo kin groups may have a patrilineal bias, but lineal descent from common ancestors is not either a major emphasis or a principle for reckoning property rights in any rigorous way. Nor is there any clear residence rule (Smole 1976: 236). Small teri are stable, cooperative groups by virtue of their close ties of kinship and marriage, but they are not formally "kin groups."

In a very real sense, the teri is a biological group. Mutual support within this group takes many forms, among them assistance in tasks requiring several people, sharing of meat, and helping when a family member is incapacitated. As Chagnon (1983) has argued, this genetic closeness translates into interpersonal support in fights within the village and determines the lines along which the village splits when internal hostilities cannot be resolved. Small teri leave a village temporarily to forage or permanently to live alone or join other groups. As long as they remain members of the same teri, households share the natural resources of the teri territory. Their old and new gardens, however, like their part of the village, remain their own, and no one else may enter unless invited. Households are tied in nets of kinship and marriage to others whom they trust and are loyal to. As a teri grows larger, these bonds become insufficient to hold it together, fights occur, and the teri breaks up into smaller groups.

At some times close allies live together in the same teri; at others they live apart in separate teri. Within a teri their loyalty to one another is based on actual genealogical closeness, ties by marriage, and day-to-day sharing and cooperation, and these same principles apply to trade relations and military alliances between teri. Beyond this, relations between teri rest chiefly on geographical propinquity, the sharing of temporary food surpluses, trade in specialized items, and mutual defense.

Yanomamo occupying adjacent regions try to maintain friendly relations and generally succeed. If war breaks out between two neighboring groups, one will move to a distant location, usually before much fighting takes place. Often neighboring groups are former members of a teri that has broken up; its members now inhabit separate shabono and have different names, but relations between them remain amicable. Such groups visit frequently (Chagnon 1983: 43), invite one another to feasts, share kinship and marriage ties, and in wartime are likely to move together again into a single shabono.

We have seen that supplies of the major staple, plantain, and the nutritionally important peach palm fruit are somewhat unpredictable. Since plantains and peach palm fruits must be eaten when they ripen or they will be wasted, when these foods are abundant a feast is held and the members of friendly teri are invited. Neighbors are the most likely to come, but word is also sent to relatives in distant teri, who may be willing to walk several days to visit and share food. Naturally the guests are expected to reciprocate when they have similar surpluses. This system is so successful that few plantains or peach palm fruits ever rot for lack of eaters (Smole 1976: 40, 187).

Researchers familiar with the Yanomamo have great respect for them as traders, a respect that borders on awed exasperation. They are relentless in demanding things they want and are almost impossible to refuse (Smole 1976: 100; Chagnon 1983: 14-16). They are especially aggressive with strangers, whom they value only for what they can get from them. According to Chagnon (1983: 15), who found himself bullied into making unintended "gifts," "The loss of the possessions bothered me much less than the shock that I was, as far as most of them were concerned, nothing more than a source of desirable items."

Trade is important to the Yanomamo, and since it often involves exchanges between distant and comparatively unrelated villages, aggressive bargaining is common. As we have seen, the ecological basis for trade is regional specialization, but trade is also a significant part of the web of alliances that promote peace in a region. Some division of labor between villages exists even when no ecological differences exist, simply to give villages unique items to trade and thus incorporate them in the trading network (Chagnon 1983: 149).

Typically whenever members of one teri visit another, they expect to trade. Men do most of the visiting and will only visit a teri where they have relatives. After eating and socializing, the guests make the rounds of the village demanding gifts (Biocca 1971: 158, 192). Hosts are expected to be generous; guests do not express thanks, since the gift is expected and "to ask for something is to flatter its owner" (Smole 1976: 237). If hosts are not generous, guests become angry; and their resentment can lead to intergroup hostilities and warfare. To avoid appearing stingy, men may hide their extra machetes, best arrows, or other valuables in the forest when guests are expected (Smole 1976: 102). Hostile "guests" may provoke their hosts by arriving uninvited, eating more than their hosts can afford, and generally demanding unreasonable gifts, as though to test their hosts' readiness to draw the line (Chagnon 1983: 164). Thus trade can contribute modestly to friendship between teri, but may also sow seeds of disappointment and antagonism. Ceremonies and leadership, as we shall see, help to minimize these dangers.

In sum, the Yanomamo economy is centered on the same family-level groups as we examined in the previous section, although food sharing and trade between communities are now more important. As group size increases, the integration of families into the larger group is increasingly fragile. Yet the larger village group does exist. Why? Largely, as we see it, for defense against enemies.

Yanomamo Warfare

The Yanomamo are a paradoxical people. Loving and nurturant family members, they can flare into dangerous outbursts of violence. Frightened of warfare and fully aware of its consequences, they nonetheless allow bitter hostilities to arise and persist over years at the expense of human lives and economic efficiency. They are generous yet envious, honest to a disarming degree yet capable of the ultimate in treacherous deception.

Our students who have seen the Yanomamo films of Asch and Chagnon (Chagnon 1983: 221-22) are invariably fascinated but frequently disturbed and puzzled. How can people be like that? they ask. Some outside observers have even wondered if the Yanomamo are fully human (cf. Chagnon 1983: 205). Indeed they are, as we hope to show in this section, and perhaps particularly human in their vain effort to find more "rational" solutions than interpersonal violence to the predicaments they face.

We must not imagine that the Yanomamo enter into violent conflict lightly. The threat of violence worries them, and they have developed a graded series of responses (discussed below) for heading off its more severe manifestations. Even so Chagnon (1983: 5) reports that in the lowlands at least one-quarter of all adult male deaths result from interpersonal violence. Smole (1976) finds warfare less prominent in the highlands, where some groups have reportedly enjoyed peace for a generation or more. But Helena Valero's account leaves little doubt that the Namoeteri and Shamatari experienced frequent homicides and raids even before they migrated into the lowland contact zone along the Orinoco (Biocca 1971).

Yanomamo of all ages grieve mightily when their dearest relatives are killed (see Biocca 1971: 247, 251, 258-61). Even those not immediately affected by death in the family, however, are affected by a state of war. Labor costs rise markedly as men are killed or wounded, dispatched to build or repair palisades, or posted as watchmen on distant trails to give early warning of an attack. Small teri must move together into a single large village, increasing not only their travel time to their gardens but the possibility that those gardens, and all the labor invested in them, will be lost (Smole 1976: 137).

Any death by violence among the Yanomamo, including deaths by disease that are regarded as caused by sorcery, sends a shudder through the Yanomamo community, testing alliances and highlighting conflicting loyalties. Allies of the contending parties are often fearful of being drawn into the violence; for if their side proves to be the less powerful, they will have to abandon their lands and start over again in a distant region (Smole 1976: 235).

Chagnon (1983: 73-77, 111, 146) documents a general decline in the quality of life during warfare. Although in peacetime the Yanomamo are diffident and careful regarding fecal wastes (Anduze 1960: 228), in wartime they are afraid to leave the village and will defecate into leaves and toss them over the palisade, polluting the immediate environs of the village. Crowded into a village with many comparative strangers, people bicker and squabble endlessly until the threat of violence from within nearly matches the threat from without. When anger threatens to burst into violence, older men and women, as well as the angry man's brothers and wives, attempt to cool him down with such words as these (Biocca 1971: 218): "Oh my son, you must not shoot. You have two male children; one is growing up, the other has only recently appeared. Why do you think of killing? Do you think that killing is a joke? If you kill today, tomorrow your sons will be alone and abandoned. When a man kills, he often has to flee far off, leaving his children behind him, who weep for hunger. Do you not yet know this? Do not remain angry. . . . Do not let yourself be conquered by anger." In the face of such reasonable advice, why do Yanomamo men kill?

The Nature of Yanomamo Warfare. We have seen that the Yanomamo of the highlands are living in locally dense populations, enjoying a relatively comfortable life but aware that the best resources are scarce. Each man is a member of a family that owns valuable resources in old gardens; shares a larger foraging territory with other close family members; and has or hopes to acquire a wife and children, or perhaps two or more wives. He can see that other men experience these same resources as scarce and are doing their best, through intimidation backed up by the threat of outright violence, to acquire or keep them at his expense. Looking ahead, he can see that his access to needed garden land and other territorial resources must be guaranteed. He can only stabilize his position by participating in an alliance with close relatives by birth and marriage, and by showing himself ready to defend his "family estate" by violence if necessary.

This situation puts a premium on men who are strong and fearless. If a man is not temperamentally suited to this role himself, he must seek out and become attached to such a man. The pattern we have seen among family-level groups like the Machiguenga and the !Kung San, where overaggressive men are ostracized or killed by the group, cannot work here. The level of competition has risen to the point where fierce and aggressive men, the *waiteri* men, despite their dangerous natures, are eagerly sought after and invited into the group. Their violence intimidates potential enemies, who are well advised to steer clear. Unfortunately, however, waiteri men are prone to violence and increase the number of violent incidents within and between teri that disturb the peace and make war more likely.

Yanomamo violence has an impulsive quality. Men (and sometimes women) will become enraged and strike out at near kin. Later they will feel regret, but no one seems to hold a grudge if the harm is not great (Biocca 1971: 308). As noted earlier, the Yanomamo have devised a series of escalating mechanisms to control violent impulses. Angry men make lengthy speeches at each other. If these do not serve to dissipate the rage, they enter into chest-pounding duels in which first one, then the other, stands stoically while being struck at full force with a closed fist. If they are too angry still, they may enclose rocks in their fists to intensify their blows.

Beyond this, men fight with clubs (or the flat sides of machetes and axes). These fights are structured events with an audience of kin supporters and leaders who monitor the fight to see that it does not explode into homicide. The combatants must exchange blows in alternating order. If a man falls, a kinsman replaces him. Leaders may intercede and direct some hangers-back to take their turns and share the responsibility for what has become a test of courage between two groups (Chagnon 1983: 164-69).

The Yanomamo say, "We fight in order to become friends again." In this sense the club fight and other duels are "the antithesis of war" (Chagnon 1983: 170), for they occur under carefully controlled conditions and their main purpose is to manage the competitive and hostile feelings between groups before they lead to homicide.

When these mechanisms fail, there is nothing to be done but kill (Chagnon 1983: 174). The successful Yanomamo raid is one in which an enemy is ambushed alone and killed without any of the raiding party being harmed (Chagnon 1983: 185). An especially angry or fierce group may surround a village and wait: since little food is stored in the village, men must eventually emerge and may then be shot. Direct attacks on villages are very dangerous, for the well-armed men inside can see the approaching enemy. The attackers therefore position themselves behind trees at the edge of the clearing and fire arrows into the village. A ten-foot palisade makes direct shots impossible, so they must arch their arrows and a direct hit is a random event. It is during such attacks that women are occasionally wounded and killed.

Yanomamo warfare is remarkably personal: not teri against teri so much as man against man, including the man's family and property (i.e., his "estate"). Men call out insults to each other, claiming their readiness to kill and using the opportunity to utter each other's personal name, a deadly insult. Men are careful to avoid harm to their relatives who live with the enemy. When arrows fall, people examine them and recognize the enemy archer by the unique design of his arrows. If someone is killed, care is taken to identify the killer. The killer must then undergo a ritual of purification, and through gossip everyone, including the deceased's relatives, learns his identity.

Waiteri men who have killed large numbers of men are hated and pursued by their victims' kinsmen. When threatened they may stand in the village clearing, inviting their enemies to shoot. If the threat is a bluff, the enemies will withdraw; if not, the waiteri man may be shot.

The more times a man has killed, the larger the number of vengeful relatives there are to conspire against him (see Biocca 1971: 186ff). It is no wonder, then, that waiteri men tend to die by violence more than other men (Chagnon 1983: 124).

The ultimate in Yanomamo warfare is the "treacherous feast." Powerful hatreds lead one group to feign friendship for another, invite its members to a feast, then fall on them and kill as many as possible. One group was massacred only during the third feast by a group that had made "friends" with them during two previous feasts, lulling them into carelessness. This outcome is uncommon, since such a degree of organization is difficult for most Yanomamo. Teri disunity is usually such that some members have no idea that others are planning to kill their guests. When they find out, they may warn the intended victims, but so confusing is their world that the victims may not believe their warnings (Biocca 1971: 53, 54, 190).

Social Responses to Warfare. Chagnon (1983: 148) vividly portrays the Yanomamo as masters of the politics of "brinkmanship." Each group must establish its reputation for toughness or it will be bullied and exploited, yet groups that are too fierce frighten other groups and have trouble finding allies. In the most dramatic instance, the men of two groups who wish to ally must face each other in duels where they attempt to prove their indomitability by giving and receiving painful blows; yet they must not allow themselves to be provoked to kill or cause serious injury lest they destroy the possibility of an alliance and create new enemies instead.

"Brinkmanship" is an apt term if we do not infer too much purpose or policy behind it. The duel or club fight is in truth the outer limit of the political economy, beyond which the means of social integration are far outweighed by mistrust and hostility. The Yanomamo do not create these fights as deliberate policy; on the contrary, they do everything they can to expand the circle of peace and cooperation outward from their communities, and the fight is the tangible sign of their inability to expand it farther.

The dramatic changes we see in the Yanomamo, as contrasted to family-level societies, are the formation of villages and the expanded role of ceremonies and leaders. These changes are to be understood as responses to the prominence of warfare and the threat of violent death.

The construction of a shabono is a good metaphor for the relationship between family and village. To a visitor the shabono appears to be a communal structure, yet each separate household constructs its own shelter; it is only because shelters are built immediately adjacent to each other, and with the goal in mind of creating a closed circle, that the finished shabono appears to be a communal structure.

Yanomamo villages grow to over 100 members, and regional clusters of villages may even total several hundred members (Smole 1976: 55, 231). Between the smallest teri of 30 members and the largest teri of perhaps 300, any intermediate size may occur. In fact the size of the shabono varies consistently within limits (Chagnon 1968a, 1983). On the one hand the village must have at least 80 to 100 people to allow for adequate defense. A larger village is militarily stronger: more resistant to attack and more successful in raids. But on the other hand, as we have seen, the larger villages are the most subject to uncontrolled social frictions. Village headmen work constantly to smooth out the many hostilities, but a sense of common economic interest in the group is largely absent.

Within and beyond the village take place ceremonies that simultaneously express seething tensions and seek to resolve them. In Yanomamo ceremonies several purposes are served simultaneously: food and other goods are distributed to equalize seasonal and geographical variations in abundance, social relations between old allies are reinforced, and possible new alliances are explored. All these functions depend to a degree on the skill of leaders.

Invitations to a feast are not issued by one teri to another, but by a specific individual in one teri to specific individuals in other teri. These individuals are the headmen of their own family groups, and may or may not have larger followings. Some will accept the invitation; others, for a variety of reasons, may refuse. The groups being integrated by a feast are not large villages but bits and pieces of several villages. Socially the feast is a mosaic composed of only some of the larger number of family groups in a region.

One way to define a Yanomamo teri would be as a group that follows a common leader, or *tushaua*. In the smaller teri the tushaua is simply the head of the dominant family, but in larger teri one man generally stands for the group, is addressed on behalf of all group members, and issues commands to do the group's work. That his commands are often ignored, and that other headmen in his group, also called tushaua, offer other advice or lead their groups in other directions, are signs that his authority is limited by the autonomy of small teri, a remnant of the !Kung attitude that "we are all headmen." But the tushaua is a force to be reckoned with in Yanomamo society, with important functions for and impact on the group. He does not intrude much into the domestic economy, except by influencing where a teri settles and plants its gardens. His major role is managing intergroup relations, keeping the peace where possible and leading the men to war when necessary.

A tushaua attempts to resolve disputes within his teri. He proposes solutions to problems and attempts to reason with the parties involved in disputes. He often invokes general principles such as "You have too many wives already—there are men here who have none." And he intervenes to control dangerous situations: "Let him speak! Let no one point his arrow at him; let everyone keep their arrows in their hands" (Biocca 1971: 37, 110). Leaders are also expected to be more generous than others (Biocca 1971: 216), and for this purpose they plant gardens of above average size (Chagnon 1983: 67). As the official host for intervillage feasts, the tushaua is at the center of the integrative efforts those feasts represent.

On the other hand, a tushaua is expected to be a leader in war. The tushaua orders the palisades built and posts guards along the trails from enemy teri. He calls for men to join him in battle, tells them where to camp and how to avoid detection during a raid, and takes the lead in actual violence. Yanomamo men often seem reluctant to fight or to stand firm in the face of continuous opposition (cf. Biocca 1971: 59). A leader is expected to take the first shot at the enemy and to risk his own safety.

Thus, leaders "are simultaneously peacemakers and valiant warriors. . . . The thin line between friendship and animosity must be traversed by village leaders" (Chagnon 1983: 6-7). This is a delicate balance to strike, and leaders approach the task in various ways. Some are mild-mannered, cool, and competent; others are flamboyant and domineering (Chagnon 1983: 26).

A leader who has killed too often generates such a network of vengeful enemies that he is not likely to live long. According to Helena Valero (Biocca 1971: 193), when the tushaua Rohariwe was invited to what he anticipated could be a treacherous feast, he said: "I think they will kill me. I am going so that no one may believe that I am afraid. I am going so that they may kill me. I am killing many people; even the women and old ones are angry with me. It is better that the Namoeteri kill me."

A certain pessimism or sense of futility, then, is felt by the man who has killed too often (cf. Biocca 1971: 226-47). It is as though he senses that the violence has gotten beyond his control, and in a deeper sense this may be true. Chagnon (1983: 188) documents a case in which the relatively mild-mannered leader of a defeated group, now bullied and despised by the "friendly" teri that gave them shelter, had to become more violent in order to defend his group. He was forced into fierceness against his will by the implacable pressure of fierceness around him.

Proximate Causes of Yanomamo Warfare. Chagnon's (1983) data emphasize the capture of women as the major motivation for war; Smole (1976: 50, 232) sees suspicion of sorcery and the consequent wish for revenge as central; and Helena Valero provides ample case examples of both these motivations (Biocca 1971: 29-41 *passim*, 98, 133, 186-88, 293). Since these are immediate causes, given by the participants themselves, we call them "proximate causes" (cf. Hames 1982: 421-22). As clues to the conditions and events that precipitate war, proximate causes are invaluable guides to understanding the process of growing antagonisms and violent outcomes, as this section will make clear.

As *explanations* of warfare, however, proximate causes are generally unsatisfactory. For one thing, people engaged in warfare often list many different reasons for fighting, leading only to the conclusion that war has many causes, some unrelated to others. We believe, by contrast, that Yanomamo warfare, and warfare in general, can best be understood within the framework of a single theory.

A second shortcoming of "proximate causes" as an explanation of Yanomamo warfare is that whereas the same sources of interpersonal conflict are present in all the family-level societies we reviewed in Chapters 2 and 3, in none of those societies do sexual jealousy and revenge wishes result in endemic raiding. Similarly, in the villages of complex chiefdoms and states that we will examine in Chapters 9-12, these motivations are powerful but do not lead to local warfare, and the warfare that does occur in those societies is qualitatively different from Yanomamo warfare. We postulate, then, that Yanomamo warfare has some deeper cause or causes, a question to which we shall return after examining three proximate causes.

1. Since the Yanomamo are often viewed as "the fierce people" (Chagnon 1983: subtitle), it might appear that warfare is an inevitable consequence of their psychology. Especially dominant in warfare are waiteri men—violent, aggressive men who have come to characterize the Yanomamo to many anthropologists. They are protective toward their own kin and allies, but are exploitative toward others outside their orbit of cooperation and trust. Strong groups bully weak groups, and appropriate their women and other resources. For example, having driven off a group of men from their village, waiteri warriors taunted one of the angry fleeing wives: "So much the worse for you, that you have no arrows and you have a husband who's afraid!" (Biocca 1971: 33, 108-9.)

Groups must appear fierce or lose the respect of others and be bullied (Chagnon 1983: 148-51, 181). A broken and defeated group, the Pishaanseteri, tried to recruit a brave man, Akawe, to bolster their reputation: "You are *waiteri*; you are famous everywhere; you have killed Waika, you have fought against Shiriana. . . . If you kill a Shamatari, we will give you one of our women; you will stay here with us" (Biocca 1971: 316). As this story implies, many, if not most, Yanomamo men are actually frightened of violence. They put up a fierce front, but when the dueling or fighting is about to begin they hang back or find excuses (Chagnon 1983: 183). A truly waiteri man, one not afraid to die and prepared to kill, is necessary to a group to build its reputation for violence.

2. The Yanomamo frequently give revenge as their motive for raiding other groups (Biocca 1971: 40). Yet revenge as an ultimate cause of warfare presupposes the violence it is supposed to explain: one homicide is assumed to lead to another in a perpetual cycle of revenge. But why do family-level societies like the Machiguenga manage isolated homicides without further violence, whereas the Yanomamo cannot? Furthermore, as we shall see, the Yanomamo use ceremonial occasions to remember the dead and renew their passion for revenge. Why do they go to such lengths to keep alive motives for war when the real costs of warfare are so high?

3. Yanomamo men frequently announce their intention to raid other groups and steal their women (Biocca 1971: *passim*). When Chagnon (1983: 86) mentioned to some Yanomamo men Harris's theory that they fight over hunting territories, the men laughed, saying, "Even though we do like meat, we like women a whole lot more!"

Yanomamo raiders try to avoid killing women and girls, and more than once spared Helena Valero's life; "Leave her: it's a girl; we won't kill the females. Let's take the women away with us and make them give us sons" (Biocca 1971: 34). The women they value are women of childbearing age. Old women are not worth fighting over; indeed, an old garden is called an "old woman" because of its barrenness. Being virtually immune from harm in war, old women are very useful for carrying messages between enemies and retrieving the dead during battles.

Many Yanomamo men experience difficulty obtaining wives (Biocca 1971: 41; Chagnon 1983: 142-45). Obtaining a wife often involves negotiations between the man and the girl's parents, and men with high social standing and strong kin networks are more successful. Husbands tend to be much older than their wives, and, with polygyny, many young men are without wives. Raids are in part efforts by young men to obtain wives for themselves and to begin to reproduce. Since captured wives usually escape or may be stolen by their original husbands or other men in subsequent raids, there is an endless cycle of raids and counterraids. Although in some areas there seem to be enough women for men seeking wives (Smole 1976: 50), Yanomamo everywhere do capture women when at war, and competition among fierce men for available women is characteristic.

The difficulty with this as an ultimate explanation of Yanomamo warfare, however, is that whereas everywhere there is some kind of competition among men over fertile women, this competition does not everywhere lead to warfare. Why, then, do the Yanomamo permit or require aggressive men to rout or kill each other in order to obtain reproductive rights over women?

The Ultimate Cause of Yanomamo Warfare. Each of the three proximate causes we have identified—basic ferocity, revenge, and capture of women—is inadequate as an ultimate cause of Yanomamo warfare because it is a universal human characteristic, not peculiar to the Yanomamo. Where the Yanomamo differ from the family-level societies we have examined is in having crossed the threshold from sporadic violence, deliberately isolated and contained, to endemic violence that feeds upon itself in a never-ending cycle of new homicidal acts.

This higher level of endemic violence seen among the Yanomamo is directly related to interpersonal and intergroup competition over scanty resources, of which we find evidence in their greater concern with the definition, defense, and violent capture of territory. Conflicts over access to and distribution of scarce resources keep interpersonal hostilities simmering among the Yanomamo. And it is the comparatively simple political structure—for the Yanomamo remain close to the family level of sociocultural integration—that accounts for the frequency with which these hostilities boil over in impulsive violence, cruelty, and treachery.

A good deal of interpersonal friction arises from the ownership and distribution of resources. Yanomamo are enjoined to be generous with friends and relatives; being ungenerous is taken as a sign of hostility and breeds mistrust. Yet rules guarantee each individual control over his or her production. To enter another's house or garden, even only to take firewood (Chagnon 1983: 68), is considered theft and infuriates the property owner. When demands are placed on them, the Yanomamo are faced with the choice of acceding and giving up things of value, or of taking a stand and risking the disappointment and enmity of others.

Distributions of food within a teri are continual sources of bickering and jealousy. If they are not countered by the positive feelings and reinforcing experiences of close family life, they can give rise to grudges that accumulate into bitter resentment; and in the volatile atmosphere of a Yanomamo village during wartime, persistent resentment can lead to violence (Biocca 1971: 84-86; Smole 1976: 244).

Jealousy and suspicion between teri is even more productive of violence. Members of a teri do steal from another teri's garden and hoard their own trade goods. Women often grumble about the greed of other teri. To quote Helena Valero (Biocca 1971: 206), "Meanwhile the Namoeteri women began to say that the Mahekototeri had many things, many machetes, but that they did not give them away; that when they came, they ate so much and their stomachs were never full; that the more they ate the more they wanted to eat; that they themselves were angry with them." In this case, although the Namoeteri headman wanted an alliance with the Mahekototeri, the women's grumbling prompted a faction of the Namoeteri to warn the Mahekototeri that an attack was imminent, ruining the opportunity for an alliance. In more serious cases, for example when a garden has been seized or destroyed, women will harp at their men and incite them to kill (Biocca 1971: 219).

Because the Yanomamo have been most often compared with more complex groups in Africa and New Guinea, their actual degree of territoriality has been underemphasized. The Yanomamo are significantly more territorial than any of the groups we examined in Chapters 2 and 3. Each teri is associated with a geographical space, generally bounded by prominent features like rivers or watersheds (Smole 1976: 26-27, 231). Because neighboring teri are friendly, members move freely in the extended zones of hunting and foraging distant from the shabono.

As we have seen, when a friendly teri joins others in one village for security, it takes the name of the group in whose territory the village is situated. But it remains the proprietor of its own territory; members continue to plant gardens there, and return there when the large teri fissions (Smole 1976: 234). Why are the Yanomamo attached to territories that are more distinctly defined than the "home ranges" of family-level societies?

The answer is that territories are valuable possessions, full of necessary raw materials, "capital improvements," and resources for future development. This is the main reason why the Yanomamo do not move to villages far from their old ones except when routed by their enemies (Chagnon 1983: 70).

Although getting rid of hostile neighbors is rarely offered as a proximate reason for raiding another teri, warfare does frequently lead to the permanent displacement of a teri from the immediate neighborhood of its enemy (Biocca 1971: 98, 103, 209; Smole 1976: 235, 236). When hostilities cease, however, and the people of the displaced teri have confidence in a lasting peace, they may take the opportunity to move back into superior lands closer to their former enemies (Smole 1976: 93-94).

Yanomamo warfare is not directly aimed at seizing territory as such. In some highland areas warfare is relatively uncommon, and many groups have been stable for generations. But this is because they have formed large villages and territorial alliances and present a formidable obstacle to their enemies.

In areas where warfare is more common, an uprooted group may aggressively displace a weak group because of its own desperate need for new territory. In one instructive example, after the Namoeteri under their leader Fusiwe split into four separate teri, one of them, the Pishaanseteri (Bisaasi-teri), built their shabono provocatively near the Namoeteri garden. The Namoeteri proper were now a small group, and when the Pishaanseteri began stealing their crops and destroying their tobacco plants, some Namoeteri counseled Fusiwe to abandon the garden, but Fusiwe became enraged, saying, "They are asking me to kill them."

The two groups tried to defuse the growing hostility by a club fight. Afterward, Fusiwe stated: "No, I am not angry. You have struck me and my blood is flowing, but I bear no wrath against you." The Pishaanseteri leader's brother, however, replied: "You must go away; you must leave this roca; here we must live. Go and live with the Patanaweteri; we must be the masters of this place." As hostilities escalated, the Pishaanseteri enlarged their ambitions: "We wish to kill the Patanaweteri [including Fusiwe's Namoeteri]; we alone will remain, we, the Pishaanseteri, the most waiteri of all." They did kill Fusiwe and scatter his group, but a conspiracy of many teri now hostile to the Pishaanseteri eventually joined to massacre most of them at a treacherous feast. The survivors then set out on a long search for new territory that finally ended in the Orinoco lowlands (Biocca 1971: 217-50, 302; see also Chagnon 1983: 152-53).

In sum, we hold that Yanomamo warfare is a tragic failure. It is tragic in the classic sense because it is the inevitable outcome of "flaws" or contradictions in the character of the Yanomamo themselves. In a Yanomamo myth, humans were created when one of the ancestors shot Moon in the belly. In Chagnon's words (1983: 95), "His blood fell to earth and changed into Men, but Men who were inherently *waiteri*: fierce. Where the blood was 'thickest,' the men who were created there were very ferocious and they nearly exterminated each other in their wars. Where the droplets fell or where the blood 'thinned out' by mixing with water, they fought less and did not exterminate each other, that is, they seemed to have a more controllable amount of inherent violence." Control of violence is central to the Yanomamo: they know that uncontrolled violence leads to annihilation. Their warfare is not adaptive but, rather, is the failure of adaptation. The Yanomamo are strong-willed family members with estates of real material importance to defend. Their sense of self-interest leads them into alliances that distribute seasonally scarce wild and domesticated produce and extend the region of peace surrounding them. But that same sense of self-interest is offended when allies are not generous (Chagnon 1983: 163), and a feeling that one is being taken advantage of sets in. In order to position themselves for competitive advantage in a scarce environment, men must give the appearance of ferocity and be prepared to back it up with action.

This sets the stage for waiteri men to dominate the scene. Men that in family-level societies would be taught restraint or expelled from the group, among the Yanomamo gain extra wives and a following of men. But, being waiteri, they are truly fearless and expose themselves and those around them to danger: despite efforts to restrain them, they lose control and maim or kill other men, bringing the wrath of their victims' families down on themselves and their close relatives and inflicting on everyone the costly consequences of a state of war. There is seemingly no alternative, since less combative groups are bullied and exploited by stronger groups who covet their women or want to displace them from their lands.

Clearly, then, the ultimate cause of Yanomamo warfare is what Carneiro (1970b) has called geographical circumscription. The Yanomamo of the highlands are surrounded by a lowland into which it has been impossible to flee until very recently. Their highlands are a poor environment of limited possibilities, one in which the control of territories encompassing past, present, and future resources is essential to an adequate quality of life. With nowhere to run the Yanomamo were forced to stand and defend themselves, grouping into villages and alliances, defining their territories and rigorously distinguishing friend from foe.

The Yanomamo have often been compared with more complex groups such as those we describe in later chapters (e.g., Chagnon 1980; Ramos 1972: 127-31). Such a comparison is one-sided, for it emphasizes the relative abundance of wild resources enjoyed by the Yanomamo and the spontaneity and individualism of Yanomamo warfare, making it seem primitive, irrational, and lacking in political structure in comparison with more organized forms of warfare. In comparison with the family-level societies we have examined previously, however, we are impressed not with what the Yanomamo lack but with what they have achieved.

Conclusions

The Yanomamo are in essential ways like a family-level society. The greatest degree of economic interdependence is found in the teri, territorial groups and owners of improved agricultural lands who anticipate the corporate kin groups of later chapters. But because the Yanomamo are "crowded" in their landscape in comparison with true family-level societies, a fundamental and far-reaching transformation has taken place: they can no longer avoid resource competition simply by moving elsewhere, and brave, aggressive men are now treated as valuable allies rather than as dangerous outcasts.

Competition and fierceness are an explosive compound that endangers the teri's well-being. The Yanomamo understand this and do all they can to avoid war. But the inevitable disappointments, sense of injustice, and suspicion that arise from exchanges between non-kin often overcome the limited economic benefits of intervillage trade, leaving teri vulnerable to attack by hostile and remorseless enemies seeking wives or lands for themselves. A teri that does not itself become fearsome in defense of its "estate," embracing and rewarding brave, violent men, has no place to hide and no future.

SIX

The Village and the Clan

IN CHAPTER 5 we examined the causes of economic and political integration beyond the family level. With the Yanomamo the need for defense of the "family estate," both as a region of improved gardenlands and as a set of women in whom men claim reproductive rights, brought village life. We also found that risk-aversive food sharing and intercommunity trade reinforced the patterns of alliance and leadership that emerged from defensive military arrangements.

In this chapter we continue to explore the complex determinants of suprafamily economic integration. Our first case, Eskimo whalers, is of interest because multifamily communities here depend, not on warfare and defense, but on relatively large capital investments in the technology of food production and storage, and on the political management of food sharing and trade between multifamily production units. Our second case, the Maring of New Guinea, illustrates the group-building role of territorial defense uncovered in the Yanomamo case, but under such a greater population density that the importance of property in land comes to the fore. Noteworthy are the Maring clans, kin-based corporate groups that are common in stateless society. Clans hold and proclaim title to land founded on sacred rights and on group might. Finally, we examine a pastoral group, the Northern Turkana. In this case although group formation was undoubtedly influenced initially by raiding and the capture of herds, we find the loose and widely scattered networks of suprafamily units today depending primarily on risk-spreading in a highly unpredictable environment.

With the addition of these three cases to the Yanomamo, we may examine in further detail how each of the four processes of economic integration—defense, risk aversion, capital investment in technology, and trade—creates economic interdependence even in societies only slightly more complex than family-level societies.

Case 6. The Eskimos of the North Slope of Alaska

The North Slope Eskimos offer a remarkably clear example of factors that lead to the formation of a village-level economy. This case is especially revealing because although all North Slope Eskimos belong to the same cultural and linguistic group, only those living on the coast and engaged in cooperative whale hunting (the Tareumiut) have a developed village economy. The inland Eskimos (Nunamiut) are typical family-level foragers; much like the Nganasan (Case 4), they aggregate in groups beyond the camp level only for semiannual caribou drives or, less commonly, to pass the winter in the security of a settled neighborhood.

The Environment and the Economy

The Tareumiut, or "people of the sea" (Spencer 1959), and the Nunamiut, "people of the land" (Gubser 1965), occupy separate niches in the North Slope habitat, a region of some 70,000 square miles, all within the Arctic Circle, descending from the Brooks Range north through foothills and coastal plains to the Arctic Ocean. Despite its desolate appearance as a treeless tundra, the North Slope offers a wide range of animal foods for a small, scattered foraging population. Along the coast are whales, walruses, seals, and polar bears. Inland are the highly valued caribou, along with grizzly bears, mountain sheep, moose, and ptarmigans. Ordinarily there is adequate food to sustain aboriginal population densities of about one person per 20 square miles; but there is a wide seasonal fluctuation in the availability of foodstuffs, and there are unpredictable annual variations in the migration patterns of the most important game mammals, the caribou and the whale.

Both on the coast and inland the spring migrations of whales and caribou are times of anticipated food abundance. As the polar ice pack breaks up, whales migrate close to shore where boatmen can hunt them. Inland the caribou aggregate in herds of hundreds or thousands that flow through passes in the Brooks Range to pasture in the meadows of the North Slope. As summer approaches, the migrations end and game scatters. The snows melt, and although the region is a true desert, receiving only about six inches of rain per year, moisture evaporates slowly. Hence the landscape is a maze of bogs and pools resting on permafrost. During a two-month period of 24-hour sunshine in summer the land blooms and all land animals fatten, but the first snow may fall in late August and by early October the land has frozen over. In the fall there are smaller migrations of caribou, and occasionally whales.

By November winter brings the "hungry time." Winter has its advantages: at temperatures of −10°F to −30°F the snow and ice are well suited for sleds and foot travel; and there is much leisure time and, in settled areas, intense socializing. But winter is a difficult time for hunting, since the game can see, hear, and smell for great distances across the barren snow and are difficult to stalk. Gubser (1965: 260) reports that a man can hear another's footfalls in the snow for distances of well over a mile. Game animals are widely scattered and may not be seen for weeks on end. People are accordingly forced to eat less desirable foods like fish, which is felt to be lacking in oil and hence inferior, or even fox. (Foxes are normally hunted or trapped only for skins, their meat being discarded or fed to dogs.) "With their level of technology and in the environment in which they lived, it was impossible to ensure a sufficient surplus of food to last each family throughout every winter" (Chance 1966: 2).

For the Nunamiut, as for the Nganasan, a household's food supply depends almost entirely on successful hunting, and the diet is dominated by caribou meat and fat. Caribou hides, antler, sinew, and bone provide most other materials needed, including for tents and clothing. Women procure firewood (a major resource scarcity), fetch water (which is melted from blocks of snow in winter), prepare food, and manufacture clothing.

Most of the year the Nunamiut roam in individual families or extended family camps that often split into individual households and go their separate ways for a time before reassembling. In this period caribou also travel in small, widely scattered groups, which the Eskimo conceptualize as nuclear and extended families.

When the spring and fall caribou migrations occur, several camps assemble in predetermined areas for cooperative hunts. As with the Nganasan, successful hunters are called on to organize the collective activities of the caribou drive. But much hunting at this time remains individualistic, and in years when large caribou herds fail to materialize, the Nunamiut simply disperse in pursuit of small herds.

Although summer is a fairly easy time, the take of caribou in the spring and fall is rarely enough to last through the winter, and this poses a dilemma. On the one hand, a family may remain in the vicinity of other families for the winter; since people are required to share food when asked, no one will starve as long as his neighbors are well provisioned. On the other hand, scarce supplies of game and firewood are quickly exhausted in the vicinity of a settled community and inconvenience, hunger, and the constant importunities of neighbors may drive an enterprising family into the lonely tundra, where it need not share whatever food and firewood it obtains. Alternatively, such an isolated family may find nothing to eat for weeks at a time, and may starve.

Warfare, as organized intergroup aggression, does not exist among the Nunamiut, although it was common in other Eskimo areas (Nelson 1899: 327-30; Oswalt 1979: 194-97). As in other family-level societies, fighting and occasional homicide take place, especially over women. Men might try to seize a woman, especially if her kin are perceived as weak, and extramarital affairs and fiercely jealous husbands are common (Spencer 1959: 78). Homicides resulting from such disputes must be revenged and can result in continual feuding. Strangers are also regarded with suspicion and may be pummeled and humiliated if they enter the home range of another group. This pattern of feuding and systematic hostility directed toward outsiders is an apparent break from the emphasis on nonaggression that characterizes a family-level society. Why does it occur? As we shall see, the shift from networking to fighting characterizes local group formation; it is probably an outgrowth of one or another form of intensification that changes the relative costs and benefits of defending territories. Despite this development, networking both for trade purposes and to handle risk continues to be of great importance to the Eskimos.

The Tareumiut economy is very different, although, like the Nunamiut, they do some foraging and hunting of caribou, especially in the summer and the fall. The Tareumiut live in sturdy sod houses in permanent winter villages of 200-300 members located far apart along the Arctic coast. The economy is centered on whales. A successful village can take 15 or more whales in a spring season, producing hundreds of tons of meat and blubber. Unlike the Nunamiut, who dry surplus caribou meat and store small quantities for later consumption, the Tareumiut arduously dig ice cellars out of the permafrost and store large quantities of frozen food for the winter.

In addition to whales the Tareumiut take quantities of walruses and seals, but they must have whales to survive. An adult eats 7-8 pounds of meat per day. Dogs, which are necessary for transport among both Eskimo groups (like the tame reindeer of the Nganasan), must also be fed. Spencer (1959: 141) cites a report that in 1883 a group of 30 people was observed to consume 18,500 pounds of meat in 75 days, an average of about eight pounds per person each day. Despite huge takes of meat and fat in good years, famine is an ever-present threat. When in some years the whales fail to follow their customary routes, the Tareumiut depend more heavily on walruses and ultimately on seals, whose meat is not appreciated but is most reliably in supply (Chance 1966: 9, 36).

TABLE 6
North Slope Eskimo Trade

From coast to inland	From inland to coast
Oil	Caribou hides
Skins (seal, walrus)	Pelts (wolf, fox,
Pottery	wolverine, sheep,
Wooden vessels	musk-ox)
Stone, slate	Horns and antlers
Bags	Caribou legs
Ivory	Pitch
Driftwood	Manufactures of wood
Boat frames	and stone
Muktuk (whale skin	Pemmican and berries
and fat)	

SOURCE: Spencer 1959: 204-5.

The two Eskimo groups are intimately bound by the need for trade. The Tareumiut need more caribou for tents, clothing, and tools, and the Nunamiut covet seal oil for fuel as well as food. In addition many other products are traded (see Table 6): for example, the Nunamiut consider whale blubber an excellent food, while the Tareumiut seek fox, wolf, and wolverine pelts for clothing. Trade at a great distance is important to the Eskimo economy and is often surprisingly well organized. For example, the Tareumiut prepare pre-cut strips of seal hide in standard bundles of 20 as a popular trade item.

Social Organization

Among both Eskimo groups the nuclear family is the basic residential and productive unit. Two or three families might build houses next to each other, and occasionally two houses might share a common entrance tunnel, but food is stored and cooked separately by each. There is a great emphasis on harmony and unity in the primary family group. Spouses are chosen in part on the basis of their compatibility with other family members; indeed, the most common reason given for suicide is that the victim could no longer tolerate living with a "troublemaker." Kin ties remain the strongest basis of social relations beyond the household. Kin are free to visit and ask help from one another, but kin have the strongest relations when they live nearby.

Nunamiut social rules require food sharing within the hamlet group and between exchange partners. Yet ownership of food is carefully noted, and all hunting weapons are identified by personal marks to avoid disputes over who killed what animal in communal hunts. Even spouse exchange is permissible, being viewed as a form of reciprocity in men's property rights over their wives' sexuality. The most respect goes to calm, hardworking, generous men who have no wish "to place themselves above the heads of others."

Local groups among the Nunamiut are known by the name of their usual home territory; for example, the inhabitants of the Utokak River area are called Utokagmiut. Perhaps 200-300 people will identify their home range in such an area. Since strangers may be physically abused if they enter another home range, people establish partnerships throughout the inland region so as to make possible visiting and hunting outside their home territory. Each household member has a unique set of friendship ties that may be activated when needed, ties that are frequently reinforced by the exchange of gifts, trade items, and sexual access to wives. These voluntary dyadic ties are of great importance in integrating households beyond the immediate neighborhood.

The Tareumiut and the Nunamiut enter trade partnerships with each other, and meet in designated places each summer for trade. As many as 500 people might congregate at such temporary marketplaces or "trading emporia" (Spencer 1959: 198). The quantities traded are often large: for example, two men might trade hundreds of caribou skins for dozens of pokes of seal oil. Not everyone participates in this trade directly, but when traders return home they find a ready demand, and distribution soon takes place throughout the community via kin and friendship ties.

The village economy of the Tareumiut is based on cooperative whaling and the distribution of stored food. Although kinsmen prefer to work on the same boat, nonkinsmen must often work together as boat crews, and several boats from one village may cooperate in taking a whale. Whale hunters form voluntary associations under the leadership of an *umealiq* ("boat owner"; pl., *umealit*), who organizes the labor necessary to acquire and maintain a large whaling boat. The umealiq must be a knowledgeable and successful whaler to acquire and hold a following, and must be able to integrate the diverse personalities of specialists (helmsman, harpooner) into a smoothly functioning unit. Followers must trust their umealiq and fellow crew members because a capsized boat in arctic waters rarely has survivors (in fact, few Eskimos know how to swim). The umealiq ensures that the whale is properly butchered and distributed among the hunters.

An umealiq must provide for the security of his followers even in a bad season. All families have ice cellars for storage, but an umealiq has a larger cellar corresponding to his greater responsibilities, and this cellar serves as a sort of social security fund from which his followers may draw supplies. In the early spring, before the next whale hunt, he cleans out his cellar and feasts his followers with the remnants of last year's catch. In addition he is expected to provide clothing and other items to his followers in exchange for their loyalty. Furthermore, an umealiq establishes ties with other umealit in the village, from whom he can call up reserves of food when his own boat has a run of bad luck. Thus the Tareumiut remain together throughout the winter, enjoying a degree of food security unknown among the Nunamiut, whom they criticize for occasionally abandoning aged or infirm relatives during a scarce winter, saying, "They are just like animals, they let everybody die" (Spencer 1959: 95). Finally, the umealiq plays an important role in economic integration beyond the village level. Men who trade frequently and in large quantities tend to be called "umealiq" whether or not they own a boat.

The highest development of intervillage dependence among the Tareumiut is found in the Messenger Feast, an important ceremonial occasion with considerable elaboration. An umealiq who believes he has available—in his own stores and in those of his allied umealit—a large surplus of food and other wealth, invites the umealit of other villages for a potlatch-style feast (see Chapter 7). These feasts include footraces and other forms of ceremonial competition, and are marked by large displays of competitive generosity. At a later date the guests are expected to reciprocate with a feast of their own. We shall reserve our analysis of the potlatch for Chapter 7, here simply noting that the Messenger Feast functions to distribute large surpluses throughout the coast (and inland) and to finance interpersonal and intercommunity competition.

Eskimo social life is imbued with competition and comparison, but among the Nunamiut social pressure to be "honest and patient" keeps the powers of would-be leaders in check. According to Chance (1966: 73), "Nobody ever tells an Eskimo what to do. But some people are smarter than others and can give good advice. They are the leaders." Competitive athletics, acrobatics, dancing, singing, and joking are pastimes for the long winter night, but the emphasis is on each displaying his own strengths and showing admiration for the strengths of others. This is what we have learned to expect from family-level societies.

"Eskimos are not 'self-effacing': they are honest about their own talents and achievements. Their emphasis rather is on the control of aggression" (Chance 1966: 65-66, 78). When hostile feelings threaten to erupt into aggressive action, it is safest to leave for a while until feelings cool. Habitually aggressive men are ostracized from the community, a strong measure given the difficulties of surviving alone in the winter. Nonetheless, when hostility does lead to homicide, the victim's kinsmen unite for revenge, instituting an interfamily feud that can be difficult to stop in a system with no political controls beyond the family level.

The Tareumiut, in sum, illustrate the situation where warfare is less important to multifamily cooperation than risk-aversive food sharing and capital investment in food-producing technology. The absence or lesser scope of such conditions among the closely related Nunamiut is reflected in their status as true family-level communities. The focus of the Tareumiut political economy is clearly on the umealit, leaders who coordinate the manufacture, use, and maintenance of the whaling technology and the distribution of the massive catch it makes possible. As surely as among the Yanomamo, who cannot live except under the defensive umbrella provided by their brothers and brothers-in-law as commanded by their tushaua, Tareumiut families could not survive apart from the cooperating group of relatives and associates brought together in the boat house under the direction of the umealiq.

Case 7. The Tsembaga Maring of New Guinea

The Tsembaga, an archetypical acephalous society (Rappaport 1967: 8, 10), are one of about 30 politically autonomous Maring groups living in the highland fringe of central Papua New Guinea (Clarke 1966, 1971; Rappaport 1967; Buchbinder 1973; Lowman 1980). Some 7,000 Maring speakers live in the montane zones of the steep Jimi and Simbai valleys that border the Bismarck range, tending swidden gardens, raising pigs, and foraging for wild foods. Until the 1950's the Maring remained beyond direct Western contact, and their ethnography offers a rare opportunity to view a tribal society as it functioned in a world of stateless groups.

Even more so than the Yanomamo the Tsembaga live in a crowded landscape with hostile warring neighbors, are organized into clans and local groups, and have elaborate ceremonies. Higher population density has led to intensification and direct competition for land, resulting in persistent warfare among neighbors for lack of regional mechanisms to mediate intergroup disputes. To counter the threat of incursion, battle, and possible death, each family must join both a clan, as a mechanism for asserting rights to land, and a local group for cooperative mutual defense. Ceremonies help to symbolize and unite these larger groups, and also enable the Tsembaga to reach out regionally for allies.

The Environment and the Economy

The Tsembaga live in a rugged, mountainous environment that ranges from about 2,000 feet along the narrow valley floors to about 7,000 feet along the mountain crest. Slopes average about 20 degrees below 5,000 feet but become steeper at higher elevations. Small streams cascade down the slopes to join the main river at the base. The climate is generally tropical and humid. At 4,750 feet Rappaport (1967: 32-33) recorded 154 inches of annual rainfall well distributed through the year and a uniform warm temperature ranging from 60-65°F at night to 75-80°F during the day. At higher elevations temperatures drop and the mountains become enveloped in clouds.

Two primary forest zones have been described by Clarke (1971) for the Maring region. Above 5,000 feet large *pandanus* trees dominate the plant community. Below 5,000 feet remnant forest stands show a more diverse forest community, with trees over 100 feet in height and a scrubby herbaceous undergrowth. Most of the primary forest below 5,000 feet has been cleared for gardens, and this area now is a patchwork of active slash-and-burn agricultural plots and secondary forest growth. Primary forest is mainly restricted to the lowest elevations near the rivers and to the highest elevations.

Population density in the Maring region is considerably higher than that of the Yanomamo. Overall density is about 35 people per square mile (7,000 in 200 square miles), and Figure 5 shows an environment filling up with settled hamlets. Rappaport (1967: 14) records about 200 Tsembaga in their territory of 3.2 square miles, or about 60 per square mile. Looking at the problem diachronically, Lowman (1980: 15) describes a cycle of population growth and decline that interrelates resource pressure, warfare, marriage patterns, and disease. In this cycle the Tsembaga, recently defeated in battle, are probably on the decline, and Rappaport estimates an earlier peak population of between 250 and 300 (80-95 per square mile).

Population density is a key variable in our evolutionary model. Clearly the Tsembaga are more closely spaced than any family-level society, but their relatively low density when compared to Big Man systems in New Guinea's Highland Core (Chapter 7) is equally important. Why is their density not higher? Mainly, it seems, because of environmental and epidemiological factors (see especially Lowman

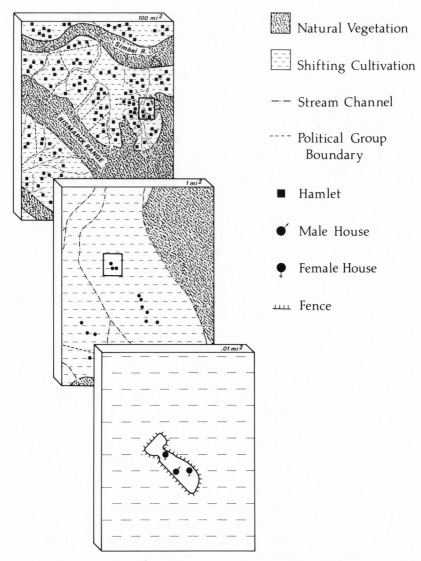

Natural Vegetation

Shifting Cultivation

—— Stream Channel

---- Political Group
Boundary

■ Hamlet

⚂ Male House

♀ Female House

⊥⊥⊥ Fence

Fig. 5. Settlement pattern of the Maring. Except for mountain ridges and valley floors the landscape has been transformed into a mosaic of gardens and secondary growth. Scattered hamlets are protected with fences because each local group's territory abuts enemy lands.

1980). The steep mountain slopes are apparently quite vulnerable to erosion and nutrient depletion, which limits opportunities for intensification; and at lower elevations endemic malaria has restrained human population growth.

The Tsembaga's subsistence economy is predicated on a population small enough to be supported with a diversity of domesticates and some wild foods. Plant foods, consisting of tubers, other vegetables, and fruits, make up nearly 99 percent of the total diet by weight (Rappaport 1967: 73), with the tubers, namely taros, yams, and sweet potatoes, providing the starchy staple. This diet is much more varied than that of higher-altitude New Guinean groups such as the Mae Enga of Chapter 7, and under ordinary circumstances is considered adequate (Rappaport 1967: 74-75). Young children and women also get some protein from rats, frogs, nestlings, and grubs. Meat, although a very minor part of the diet, is obtained by hunting wild pigs and marsupials and by raising pigs and chickens.

To supply this dietary mix the Tsembaga have created a complex environmental mosaic. At all times they seek to have available a mixture of vegetation at every stage of succession from newly cleared fields to forest fallow. They maintain this ecological diversity by using a long-fallow cycle in horticulture, which in turn is made possible by their comparatively low population density and limited production requirements. With a simple technology, forests are cleared for swidden fields, domesticated pigs are herded, and wild foods are foraged.

Shifting cultivation in the secondary forest (elevation 3,000-5,200 feet) is the dominant production strategy. Tsembaga gardens provide a diversity of crops: taros (*Colocasia* and *Xanthosoma*), sweet potatoes, yams, bananas, manioc, sugarcane, various greens, and other vegetables. At higher elevations sweet potato becomes increasingly important in the gardens, providing up to about 70 percent of the yield by calories.

According to Rappaport (1967: Tables 3-5), lower-elevation swidden fields yield about 12.8 million calories per hectare, higher-elevation fields about 11.3 million. Based on estimates of the energy costs of clearing, fencing, weeding, harvesting, and transporting (but not food processing), the ratio of yields to costs are 16.5:1 for the lower gardens and 16:1 for the higher gardens. These ratios are virtually identical, a point not emphasized by Rappaport but exactly what we would predict for a subsistence economy trying to minimize production costs.

The preparation, planting, and harvest of a swidden garden are done by a man and a woman working jointly. Men are mainly respon-

sible for initial clearing, fencing, and some planting. Women then do
the bulk of the planting, weeding, harvesting, and transporting of the
harvest. Generally the garden is the cooperative work project of a nuclear family, although men and women do make gardens with unmarried siblings, unmarried in-laws, and widowed parents (Rappaport
1967: 43).

After a field is initially cleared, the brush is burnt and the field
fenced to protect crops from roaming pigs. Planting takes place immediately after burning, and the standard pattern of intercropping
creates a complex artificial plant community with complementary
species of different height, speed of growth, and depth of roots. Most
important for the Tsembaga is the relatively long period of yield created by such a planting design. Most yields from yam, manioc, sweet
potato, and taro are available throughout the period from 24 to 66
weeks after planting. Some vegetables are available sooner, and other
crops, notably sugarcane and banana, continue to yield for another
year or longer.

After the main harvest period the field reverts gradually to secondary growth as the harvest of the longest-yielding crops continues. At
the same time a couple will prepare a new field, usually adjacent to
the previous field. A progression of old fields forms a swath across the
landscape. At lower elevations the fallow cycle is about 15 years long,
at higher elevations up to 45.

Silviculture is an interesting secondary agricultural strategy practiced by the Tsembaga and other Maring (Rappaport 1967: 55-56;
Clarke 1971; Lowman 1980: 59-62). Two species of trees are commonly
planted as individually owned orchards in the lower elevations of the
Tsembaga territory. Ambiam (Gnetum gnemon) provides an edible young
green leaf, and komba (Pandanus conoideus) provides a fruit rich in oil
as well as protein and niacin (Hipsley and Kirk 1965: 39). A group defeated in warfare will have their trees destroyed, making it harder for
them to live in the territory when they reoccupy it. The group that
defeated the Tsembaga cut down their groves, and this may be why
they make less use of tree crops than other Maring populations (Rappaport 1967: 55).

The Tsembaga raise not only pigs and chickens but captured cas-
sowaries (Rappaport 1967: 56-71; Lowman 1980: 78-97). Pigs are by far
the most important domesticated animal; although they account for
less than 1 percent by weight of the Tsembaga diet, they are a major
source of proteins and fats. They are primarily a ceremonial food,
eaten at major intergroup pig slaughter ceremonies and at health crisis
ceremonies.

Pigs are owned and managed by a male-female pair, commonly a married couple. Men obtain the pigs in trade and from the wild; women raise and tend them as pets. Mature pigs are allowed largely to forage unwatched, but are kept bound to their family by daily rations of garbage and sweet potatoes. Herds remain small, in part because the practice of castrating males means that females can only be impregnated by wild pigs, in part because ritual slaughters keep their numbers down.

Vayda et al. (1961: 71) suggested that pigs in New Guinea act as living storehouses for surplus food produced in good years, making it possible for that food to be eaten in bad years in the form of meat. But Rappaport (1967: 59-68) has shown that pigs are a poor storehouse for energy, since they require about one calorie of energy expenditure by the Tsembaga for each calorie returned in food. Indeed, each pig eats about 0.15 acres' worth of sweet potatoes when the herd is at its maximum density; as Rappaport emphasizes, this is the size of garden required to support one human! Thus the Tsembaga's massive expenditure of labor on pigs is clearly not for calories but to obtain critical supplies of protein and fat. Pigs are also primitive valuables; the exchange of pig meat by the Tsembaga in politically important ceremonies anticipates developments in the political economy that we will describe in Chapter 7 for the higher-density Big Man societies of New Guinea's central highlands.

Hunting and gathering, so important in societies like the Machiguenga and the Yanomamo, are marginal to the main diet of the Tsembaga. Forests provide building materials and dietary variety, but extensive farming diminishes the forest area and thus the supply of wild foods. Wild pigs and marsupials continue to provide protein and fat, but their overall contribution to the diet is very small.

In sum, the Tsembaga economy is marked by the scarcity of key resources. Prime agricultural land is limited and overused. Wild resources, especially meat, are badly depleted, and pigs, produced for fat and protein, are costly to feed. In this situation of overall scarcity competition is intense.

Warfare is an infrequent yet ominous threat to the daily lives of the Tsembaga, whose small territory is surrounded by enemy lands. Actual episodes of warfare are regulated by the ritual cycle and probably engage a given group directly only once every 12 to 15 years (Rappaport 1967: 156). Battles in the open test the two sides' strength; when a numerical imbalance is seen, the more powerful group will charge and kill anyone they can catch. As Rappaport (1967: 110-17) observes, the proximate cause of warfare recognized by the Tsembaga is revenge for past killings. But as with the Yanomamo the ultimate cause is competition for prime agricultural land, which is in short supply and from which a group that cannot gain sufficient military strength will be permanently displaced.

Trade is an important part of the Tsembaga economy. It involves salt, axes, and other items that some groups have access to only by trading, as well as a full range of valuables, such as pigs, feathers, and shells, that are used in social exchanges and ritual displays.

Social Organization

The settlement pattern of the Tsembaga is dynamic, with a multi-year cycle of aggregation and dispersion synchronized with the cycle of conflict and ceremony. During periods of ceremonially recognized truce, settlements are dispersed as individual households and small hamlets through the territory of a local population. Although generally dispersed, residences remain at the middle elevations where agriculture is most productive; low elevations (malarial) and high elevations (above agriculture) are uninhabited. When the truce ends and war may commence, families move to form a concentrated village-like settlement around the traditional ceremonial ground. Rappaport (1967: 173) sees this concentration as part of the preparation for the major *kaiko* ceremony (described below), but it can also be viewed as a defensive preparation for the expected war.

War comes, and the next stage in settlement reflects its outcome. A victorious or nondefeated group gradually disperses again as the pig population builds up and problems with pigs increase. A defeated group abandons its territory and disperses through the lands of other local groups. It may later try to resettle its territory, but in that event its military weakness requires it to concentrate its population in a defensive settlement. When the Tsembaga returned to their territory following defeat, the whole group of about 200 people lived together in a loose village aggregate covering perhaps 12-15 acres. The explicit reason for maintaining residence together, despite the increased distance to their fields and the damage to nearby gardens caused by pigs, was fear of their enemies (Rappaport 1967: 69).

Settlement in the Maring region responds to opposing forces. Populations come together for defense and related ceremonial activities; they then disperse for easier access to more distant fields and to avoid the destruction of agricultural crops by pigs. This dynamic of concentration and dispersion is like that described for simpler societies, but covers a longer time period and has the critical added factor of periodic warfare.

The Maring generally, as typified by the Tsembaga, consist of hierarchically nested groups that are constantly forming by segmentation and coalescing from necessity. The different levels of organization and the economic and political functions of these levels have been discussed by Lowman (1980: 108-28) and Rappaport (1967: 17-28). For our discussion we recognize a somewhat simplified set of four main levels of organization: the domestic household, the patrilineal household cluster, the clan, and the local territorial group.

The domestic household (Lowman 1980: 111-12) corresponds to the hearth unit composed of a married man and woman with their unmarried children and perhaps additional close relations. Members cooperate in economic activities and share food cooked in a single pot. The division of labor, mainly by sex and age, cross-cuts the household and creates a potentially independent subsistence unit. Males and females share work responsibilities in gardening and husbandry, and eat together from their common produce. The woman lives in a separate house with her unmarried daughters, young sons, and pigs. Living in the men's house but still part of one household are the man, his older sons, and his unmarried brothers. These males have been initiated; they have passed ceremonies of instruction and ordeal to make them men. They must live apart from women, although they eat and work with them.

The patrilineal household cluster is a hamlet-sized informal grouping composed of domestic units whose males are linked in explicit and known genealogical relations; its male members are no more distant than first cousins. Typically the males live in a single men's house and interact frequently, but the group is unnamed. The group functions as a unit because of the closeness of its internal kin ties and its members' mutual support in economic, ceremonial, and political undertakings. Households live close together, they often share an earth oven, and a protective fence surrounds the hamlet compound. Not far away are the farm garden plots of the households. Available for sharing among members of this minimal residential grouping are land, plant cuttings, and agricultural produce. In social and ceremonial activities, patrilineal relatives often act together, for example in preparing a bridewealth payment or in sacrificing pigs to specific ancestors. Membership in this group is not sharply defined, and new groups are forming constantly by segmentation.

The clan—which unlike the first two groups is not found in family-level societies—is a formal, named, and ceremonially defined social unit that is very important to the Tsembaga. Membership in a Maring clan is putatively patrilineal, although actual genealogical relationships among members are not always traceable. Some immigration is permitted, especially when the man-land ratio is comparatively low and added members would strengthen a group's position. Immigrants are fully incorporated into the clan within two generations (Lowman 1980: 116); for all intents and purposes, ritual participation with the clan group defines membership. The clan is exogamous.

The clans recorded for the Maring in 1966 had an average size of 75 (Lowman 1980: 120), about the size of a Yanomamo teri. The 200 Tsembaga were divided among five clans that actually formed three groupings (two small clans were united with one larger one). The clan does not normally form a co-residential village, but it functions as a unit in economic, political, and ceremonial activities. Economically it controls a territorial strip that runs vertically from the mountain crest to the river and incorporates all the ecological diversity of the Tsembaga area. Formal boundaries to this territory are known and marked, often by natural features like streams or ridges. Clan members individually hold improved agricultural lands like swidden plots and groves, and the lands of the patrilineal subgroups form noncontiguous clusters scattered at different locations in the territorial strip. Most important, the clan defines ownership rights and restricts access to land. Clan members may exchange land with each other, and extensive land exchanges between two neighboring clans represent a major step toward fusion into a single territorial unit.

The clan is also central to all ceremonial and political events. It organizes and serves as host for ceremonies in the central *kaiko* cycle; indeed participation together in these ceremonies, especially the planting of the *rumbim* following war, operationally defines the group. The clan owns a fighting-magic house and its own set of fighting stones (Rappaport 1967: 125; Lowman 1980: 118). Its ritual leader in war is responsible for the house and its stones, and helps coordinate the ceremonies that coalesce the clan into a single fighting unit. Such "war shamans" (Lowman 1980: 119) hold the highest position of leadership among the Maring, and the history of the clan is largely their history.

Ideally a clan corresponds to a territorial division, although, as we have seen, smaller clans may merge with larger ones. The creation of this suprafamily corporate group with leadership and ceremonial integration is a significant departure from family-level society. A still more significant departure is the ritual integration of most Maring clans into a co-defensive territorial group of clans.

The local group or clan cluster of the Maring is a grouping of two to six clans that numbers from 200 to 792 people and averages 380 (Lowman 1980: 125). The Tsembaga with 200 people are at the very bottom of this range, reflecting their weak political position following a recent defeat in warfare. The clan cluster is not named and has no overarching ritual leaders or war houses, but its constituent clans are tightly interrelated by marriage and exchange. The main ceremonies—the planting of the rumbim that establishes a truce, the planting of the stakes that define a clan's territory, and the slaughter of pigs to repay allies and ancestors for their assistance—are synchronized to prepare the clans to act together in territorial definition and defense. Analytically this local group is a kind of "village," and under some circumstances related to its defense its constituent clans will actually come together into a co-residential grouping. As indicated in Figure 5, these local groups are the significant political entity, beyond which is war.

Beyond the local group no institutional structure exists, although there are frequent interactions. Individuals build networks of interpersonal ties through marriage and exchange outside their local group. These ties act as means of personal and group security: they are used to obtain spouses, trade goods, allies in warfare, and refuge in case of defeat. Since these external contacts are both made and reinforced on ceremonial occasions, a person's participation in intergroup ceremonies is central to his networking strategies.

Among the Tsembaga all external relationships on which the local group depends are based on individual ties mediated by group ceremonial presentation. Although a man depends on his group for access to land, for economic support, and for mutual defense, he must achieve special prominence in his group to have access to a regional network that affords contacts, security, and trade opportunities beyond what the local group may provide. The opportunity to shine comes in ceremonies in which men bedeck themselves with fine feathers and shells and display themselves in group dances. Rappaport (1967: 186) describes in detail the elaborate dress and individual display at the main kaiko ceremony:

Adornment [in the public dancing] is painstaking and men often take hours to complete their dressing. Pigments, formerly earth colors of native manufacture, more recently powders of European origin, are applied to the face in designs that are subject to frequent changes in fashion. Beads and shells are worn as necklaces and garters of small cowries encircle the calves. The best orchid fiber waistbands and dress loin cloths enriched with marsupial fur and embellished with dyed purple stripes are put on. The buttocks are covered with masses of accordion-folded leaves of *rumbim* called "*kamp*" and other ornamentals. A bustle, made of dried leaves obtained in trade, which rustles during dancing, is attached on top of the mass of *kamp* leaves.

Most attention is given to the headdress. A crown of feathers, eagle and parrot being most common, encircle the head. The feathers are attached to a basketry base, which is often hidden by marsupial fur bands, bands made of yellow orchid stems and green beetles, or festoons of small cowrie shells. From the center of the head rises a flexible reed, two or even three feet long, to which is attached a plume made either from feathers or an entire stuffed bird.

A man's success in display reflects his own prestige, which in turn increases (or decreases) the overall desirability of his group as an ally.

The Maring ceremonial cycle is described at length by Rappaport (1967: 133-242; 1971) and Peoples (1982). In briefest outline, hostilities between local populations of Maring are endemic and open warfare is periodic and violent. When it is decided to end open fighting, either because of a major defeat or because of many deaths without a clear outcome, a truce is instigated and ceremonially marked by the ritual planting of the special rumbim plant. Then, following a period of five to 20 years during which the plant grows, warfare is considered impossible. The pig herd is allowed to grow in preparation for the kaiko. When a consensus is reached that it is time to initiate the ceremony, which is designed to repay the assistance of ancestors and allies in past combats, the first step is to plant stakes that mark off the territorial boundaries of the local clan or clan cluster. If a defeated group has not reoccupied its territory and planted its rumbim, the stakes of the victorious clans will be laid out to incorporate the new land; otherwise, the stakes define the same territories as existed before the war. Then the rumbim is uprooted and a major intergroup ceremony is performed at which the group's pig herd is slaughtered and eaten. This ceremony ends the truce; no institutional mechanism is in place to restrict hostilities, and the local groups await the outbreak of war. When this occurs, as it inevitably does, allies recruited through the interpersonal regional networks come together to support the warring groups.

What are we to make of this odd cycle? Three positions are presented by Rappaport, Lowman, and Peoples.

As a cultural ecologist, Rappaport (1967, 1971) sees the kaiko ceremony as a homeostat, or system regulator, which in the absence of group leadership benefits the group by regulating the distribution of human population, the size of the pig herd, the exploitation of wild marsupials, and other variables. Whether or not the participants know it, the ceremonial cycle causes the group to take actions necessary for its survival.

Lowman (1980) disagrees. Rather than a neat pattern of "regulation," she sees periods of rapid population growth and collapse related to a group's success in warfare, marriage, and immigration on the one hand, and eventual severe environmental degradation with overpopulation on the other. Supporting this position, Clarke (1982) argues that any semblance of regulation or equilibrium among the Maring is a result of their simple, individualistic technology and of severe malaria at lower elevations (see Lowman 1980).

Peoples (1982) presents still a third view: namely that ceremonialism is most important in warfare as a means of obtaining and maintaining allies. Peoples addresses the issue of whether the kaiko primarily serves "group advantage" or "individual advantage," and concludes that these two perspectives are not necessarily opposed but can be combined for a more complete understanding of the kaiko.

Although our emphasis is somewhat different from Peoples', we agree that the ceremonial cycle offers both group advantage and individual advantage. Group advantage seems quite clear. The ceremonies are the main way to obtain allies or support outside the group. Given the existence of the ceremonial complex in New Guinea, whose origin has never been clearly explained, participants in the kaiko ceremony have a competitive advantage that allows them to expand at the expense of nonparticipants. This group selection is linked to competitive exclusion in warfare and the "social extinction" of groups lacking the organizational trait (Peoples 1982: 299).

Individual advantage seems equally clear, since in addition to the ongoing networking advantages the ceremony offers, its participants can plausibly equate success in warfare with the number of allies recruited through the ceremonial cycle. Group advantage and individual advantage are thus identical in this instance.

The Tsembaga exhibit both continuity with simpler, family-level societies and important institutional developments beyond the family level. The household and the household cluster remain central for most aspects of production and consumption, but the increasing complexity of life has given rise to two new levels of integration, the clan of perhaps 75 people and the territorial community of several hundred, that unite families with distant kin and nonkin for purposes of corporate ownership and mutual defense. These institutions are maintained by impressive ceremonies but do not have leaders in the modern sense; indeed Tsembaga clans have no recognized position except that of war shaman (Rappaport 1967, 1982).

The importance of the corporate clan and the territorial group, ceremonially integrated, marks the beginnings of what Childe (1936) would have called a Neolithic Society. What caused the development of these institutions?

The most dramatic changes in basic lifeways from, say, the Machiguenga to the Tsembaga are in population density and warfare. In our theory a significant increase in population density leads to a shift in subsistence toward agriculture, restricted access to and competition over limited resources, small group territories, and endemic warfare (cf. Brown and Podolefsky 1976). That is what has happened to the Tsembaga. Their diet is now almost exclusively vegetarian and agricultural, and their environment is almost totally transformed and managed by human groups. Lands are scarce, clearly demarcated, and jealously defended, with the clan restricting access. The territorial group, composed of several clans, must number several hundred for defense purposes, but the territory is small, only a mile or so across, and surrounded by enemies. Access to any resources not available within this small area must be through intergroup trade. The threat of warfare can never be dismissed.

The institutional elaborations of clan and territorial group appear as logical extensions of an exclusionary policy necessitated by the pressure of population on resources. Ceremonies, so important to the Tsembaga, function to define these groups and to interrelate them with other groups for mutual defense. The "domestication" of humans into interdependent social groups and the growth of the political economy are thus closely tied to competition, warfare, and the necessity of group defense for individual survival.

These contrasts with the Machiguenga are of sufficient importance that they should be visible in the pattern of how people spend time. Although time allocation data are lacking for the Tsembaga, a recent study of the Kapanara, a highland group living at a similar population density, allows us to make a rough comparison (Grossman 1984). In Table 7 we see some important differences from the Machiguenga pattern of time use (Table 5). As expected, the time devoted to hunting, fishing, and collecting is much less among the Kapanara, who instead spend four times as many hours in livestock care (pigs, of course) as the Machiguenga do. Also, Kapanara women do much more agricultural work than the men, who are heavily engaged in public ceremonial and recreational activities, in contrast to the Machiguenga. In this case we also see a considerable investment in commercial activi-

TABLE 7
Kapanara (Papua New Guinea) Time Allocation
(Hours per day)[a]

Activity	Men		Women	
Food production				
Hunting	0.2		0.0	
Fishing	0.0		0.0	
Collecting	0.2		0.3	
Agriculture	1.8		3.1	
Livestock	0.2		0.2	
		2.4		3.6
Food preparation	0.8		1.5	
Food consumption	0.4		0.4	
Commercial activities				
Cash cropping	1.0		1.2	
Wage labor	0.6		0.1	
		1.6		1.3
Housework	0.2		0.1	
Manufacture	0.4		0.3	
Social				
Child care	0.1		0.4	
Public ceremony	0.4		0.2	
Public recreation	2.5		1.8	
Education/information	0.5		0.3	
		3.5		2.7
Individual				
Hygiene	0.2		0.3	
"Nothing"	1.2		1.0	
Ill	0.1		0.3	
		1.5		1.6
Other	1.2		0.7	
TOTAL	12.0		12.1	

SOURCE: Grossman 1984.
[a] 12-hour day.

ties (cash cropping and wage labor), new activities that reflect the growing commercialization of the New Guinea highlands in recent decades. Commercial work was undoubtedly less common when Rappaport studied the Tsembaga than it is today, but some of the time that now goes toward commercial projects may then have gone into production of food, including pigs, for ceremonial (as opposed to subsistence) purposes.

Yet the contrast between the Machiguenga and the Tsembaga should not be overdrawn. As Lowman (1984) has emphasized, there is a

broader regional dynamic in Maring society: the highly institutionalized and ceremonialized Tsembaga are not typical of all Maring,
but only of Maring in the higher-density and longer-occupied areas.
Groups that have settled frontier areas where densities are lower and
competition less intense are organized in simpler ways and are more
like family-level societies. They live in hamlets without strong clans
and have less elaborate ceremonies. Like the Yanomamo, the Maring
range along a continuum from family-level to local group organization
depending on local variations in resource availability, population density, and intergroup competition.

Case 8. The Turkana of Kenya

The economy of the Turkana is the individualistic, family-centered
economy now familiar to us, but their society has structural features
that organize and mobilize family and hamlet groups into neighborhoods and regional associations for economic security and defense.
Although their movement patterns and their personal networks of security are very similar to those of the San, their economy is more integrated and complex.

The Environment and the Economy

The Turkana are mobile pastoralists of the eastern Rift Valley of
Kenya (Gulliver 1951, 1955, 1975). The northern part of their region,
on which our description will focus, is hot and dry; rainfall averages
from six to fifteen inches per year, and is highly variable. A "good
year" for pasture comes only once every four or five years; and one
year in every ten a severe drought seriously depletes the Turkana
herds. Rainfall is heaviest from April through August, but may come
suddenly at any time in cloudbursts that fill pools and small watercourses for a few days before the water drains off or evaporates.

The Turkana environment varies from "arid," thorn-brush and
grassland, to "very arid," dwarf shrub rangeland of "low potential"
(Patton 1981: 2). In the northern region mountainous areas and borders of watercourses offer the best grazing lands, and population
tends to concentrate in these areas in the middle and late dry season;
but most Turkana prefer to live in the open plains and will move there
as soon as the rains permit. According to Gulliver (1951: 44), the immediate principle governing Turkana migration is that "browse or
graze which will not last long should be used before that which will
persist, so that maximum use can be made of the total vegetation." As
we shall see, this results in frequent moves by individual homesteads

and a continuous aggregation and dispersion of households as local conditions change.

Gulliver estimated the Turkana population in 1949 at about 80,000, spread over about 24,000 square miles. The average population density is accordingly 3.3 persons per square mile, with the dry plains supporting only about one person per square mile and the moister mountains much higher densities. In 1949 Gulliver visited a temporary mountain "neighborhood" of 20 square miles in which 400 people lived (20 per square mile) along with 2,000 head of cattle, 1,200 camels, and 4,000 sheep and goats. Recent studies (Patton 1981; AMRF 1979) place Turkana population density at about 8 persons per square mile, with a range of from 4 to 26. It is not clear whether these figures are more accurate than Gulliver's estimates, or whether the Turkana population doubled from 1949 to 1979.

For most of the year the Turkana's main foods are milk and meat. The major livestock are cattle, camels, sheep, goats, and donkeys, the last being used primarily for transport but the other four important in the diet. Cattle, and to some extent sheep, require grass for grazing, and hence must be pastured in the best-watered regions, generally the mountains. Camels and goats, by contrast, do well on thorny, brushy "browse," and hence can feed in areas too dry to support cattle; moreover, camels, with their ability to go five or more days without water, can pasture on lands far from water and hence unusable by cattle, which require watering every other day. The pastoralist exploits these differences by dividing his herds in complex, opportunistic ways to make complete use of whatever resources are available at the moment.

In the wet season animals graze freely, milk is abundant, and there is plenty to eat; extra milk can even be preserved by separating and storing the butterfat and by drying the skim milk on skins spread in the sun. Later, in the dry season, animals are lean and milk is scarce. In this season women gather wild plant foods in large quantities, but it remains a time of scarcity and hunger. Water and pasture are the factors that limit population. Rivers are seasonal and permanent springs are few. Water can be obtained by digging in riverbeds in the dry seasons, but in bad years waterholes can be deep: "many women deep" in Turkana terms, since it requires a chain of women to pass buckets from water level up to ground level.

In addition to meat and milk, animals supply most of the homestead's other needs: leather skins for sleeping mats, roofing, drying pans, shields, containers, clothing, and cordage. Women do most manufacturing and food processing, and in rare good years may tend small sorghum or millet gardens near the wet season pastures (in a few areas of the plains). Since young men do most of the herding, older men spend much of their time together in the shade, discussing their herds and the state of pasture.

Pastoralism is the only possible way of life in much of East Africa, thanks to the comparatively high population density and the extreme marginality of the region for rainfall agriculture. The central feature of pastoralism is the concentration of subsistence in movable property, i.e., the family herds. Since the Turkana herds are the envy of neighboring groups, raiding for animals is a constant threat; and many aspects of Turkana social organization are designed to minimize or at least control this threat.

Interestingly, external trade is negligible among the Turkana. Unlike the Kirghiz (Case 11) and the Basseri (Case 14), who have strong exchange bonds to agricultural populations, the Turkana are essentially subsistence pastoralists. As among the San (Case 2) some external and internal exchange certainly takes place, but the security functions of the exchange web appear to outweigh exchange functions.

Social Organization

The basic production unit is the homestead or camp (*awi*), consisting most often of a man, his wives and children, and a small number of other dependents, with a separate sleeping and cooking hut for each wife. The camp is usually enclosed by a fence of thorn brush into which the family herds are led each night and protected from raiding. Each day the herds are taken to pasture by boys and young men, who keep a loose watch while foraging wild foods for themselves or playing together. When pastures are distant boys may sleep out with their herds, and many pass much time alone and away from their homesteads. In some cases, according to Gulliver, a male househead assigns different wives and their sons to different segments of the herd. For much of the year in such cases, and even for years in succession, each wife (along with her children) lives apart from the others and from her husband, who visits her in rotation.

Homesteads are largely self-sufficient and autonomous. For much of the year single homesteads or small hamlets are deliberately scattered to avoid competing with other Turkana for pasture and water. Families may aggregate to make use of the short-lived pastures that spring up in the wet season; and as the dry season progresses they may aggregate again near rivers and in the mountains, where water and pasture are more reliable. The Turkana, however, view themselves as plains dwellers; they describe the mountains as cold, difficult to walk in, and overrun by lions and leopards, and look forward to the time when they can return to the plains. In a good year, when pasture and water are abundant in the plains and a few millet gardens are producing, homesteads that have been apart for months or even years are reunited. The relatively dense aggregations of homesteads (up to 40) that occur in good years, although temporary, are in some ways like villages. There is much feasting and exchange of meat and milk, and major ceremonies are performed.

Although the Turkana lack highly structured kin groups, territories, and a formal political system, they do establish and maintain large networks that amount to a kind of effective community for each homestead. First, hamlet-like groups of close relatives and/or friends live and move together for part of the year. Second, such groups cluster within convenient walking distance of one another, and men in such a cluster meet often to take turns distributing freshly slaughtered meat and to share information on herds and pastures. These two levels of social organization (Gulliver calls them primary and secondary neighborhoods) provide the individual househead with a network of friends through which food and information flow, friends from whom he may beg insistently as a good Turkana should (Gulliver 1951; Patton 1982) and who will cooperate with him in defense against raiding. Although a family is free to move at will, in practice families tend to move with their neighbors and settle near them at new locations.

The Turkana also establish and maintain strong friendship ties at a distance by means of livestock exchanges. True friends are generous with one another, even though they may meet once a year or less. Having friends at a distance helps spread risk: if natural disaster should decimate the herds in one area, each homestead has friends scattered throughout Turkana land to whom they can turn for food and stock to replenish their herds. The sporadic wet season aggregations in the plains are opportunities for homesteads, neighborhoods, and even distant friends to reinforce their networks. During these aggregations marriages and age-set ceremonies consolidate existing ties and create new ones.

In the past, extensive networks were also undoubtedly a response to warfare and defense needs. Raids against other tribes were a normal means of replenishing or increasing one's herds, and through their networks the Turkana could participate in the spoils of raiding parties or seek help against enemy raiders. Even under pacification, at the time of Gulliver's study, men carried spears when traveling, and recently raiding and banditry have once again become common (Dyson-Hudson and McCabe 1985).

The main cement of Turkana social organization, however, is the exchange of livestock. A nuclear family's herds are all owned and managed by the father; and although their daily care falls to women and boys spread over the countryside, there is a strong sense of the essential unity of the family and its herd. Some hamlet groups are the remnants of old extended families whose senior male has died: in such cases the brothers and in-laws continue to live near each other, and because their herds once had a common owner the men continue to feel part of one family. Often, as we have seen, the hamlet-size group also includes friends.

The ties in an individual's network are strengthened by gifts and loans of livestock. The Turkana are closely attached to their stock: they name each animal and know the names not only of their own stock but of their neighbors'. A gift or loan of livestock to a friend is thus a highly personal and symbolic act that will not be forgotten; it lays the foundation for future exchanges. A loan also helps spread risk by placing some animals from the family herd in different microecological zones and subjecting them to different personal styles of herd management.

How extensive is Turkana social structure? On the one hand, there are indicators of "tribal" integration. The Turkana say "we are all brothers" and respect this tribal identity by rarely raiding or using spears against one another (bandits, *igorokos*, are exceptions). They know and acknowledge the "territorial" names of their regions. They also belong to clans, some of them small and localized, others widespread throughout Turkana land. In past times, apparently, whole regions of Turkana mustered thousands of warriors against non-Turkana enemies.

Yet in their daily life the Turkana are not conscious of themselves as a tribe. They have no tribal, territorial, or clan leaders, no corporate groups, and no genealogical reckoning beyond the grandparent level. They are highly individualistic and tend to migrate within circumscribed areas; even close-knit extended families usually separate at times in response to their individual needs. A great many factors, among them the availability of pasture, the mix of livestock, the amount of labor available to the family, the current location of kinsmen, and the threat of raids, influence migration and set up a complex motion of family units in and out of larger "communities" (Gulliver 1975).

We may view Turkana social structure—embodied in rules, weak though they are, concerning mutual respect, territoriality, clanship, age-sets, and bridewealth—as providing a set of opportunities for the individual Turkana homestead. Since a highly unpredictable environment makes the serious depletion of herds a constant possibility, family autonomy, however powerful a cultural ideal, will not work in practice and suprafamily ties are essential. Out of all the possible ties of kinship, marriage, friendship, and neighborhood, the Turkana select some rather than others for emphasis, and solidify them through exchanges of livestock and seasonal feasts. In this way each homestead is essentially free to exploit constantly changing resources, yet retains an extensive social network that can be activated in times of insecurity and danger.

Conclusions

Let us now briefly consider the formation of village-level societies in terms of the fundamental evolutionary processes of intensification, integration, and stratification.

Intensification of the subsistence economy is a prominent feature of the four cases discussed in Chapters 5 and 6. Higher population pressure on food resources causes significant changes in the diet and in the amount of work necessary to meet dietary requirements. In areas with soils suitable for agriculture the dominance of slash-and-burn agriculture in food production is clear. We have documented population pressure on the land among the Yanomamo, but the Maring, with a population density of up to 80 persons per square mile, are the extreme case. Virtually the whole environment of the Maring has been transformed by the agricultural cycle; wild foods are now comparatively minor, probably well below 1 percent of the diet by weight. Protein from meat sources comes largely from domesticated pigs rather than hunted animals, and is obtained only at the cost of considerable labor.

In areas where agriculture is more marginal or impossible, specific environmental conditions offer a variety of alternatives. The Turkana, in the dry East African savanna, combine mixed pastoralism with occasional agriculture. The Eskimos in the extreme Arctic depend on high-yielding but seasonal resources that require storage. Intensification in the subsistence economy can therefore take a number of forms, including increased dependency on agriculture, extensive herd management, and specialized hunting. As we have seen, these forms of intensification and the accompanying integration lead to dramatic changes in the group's social life.

Integration involves the development of suprafamily structures that link families to corporate groups (clans and lineages), organize the corporate groups into residential aggregates corresponding to villages, and interconnect these local groups in extensive interpersonal networks of exchange and personal support. An essential feature of these higher levels of integration is the reliance on ceremonies to define groups and their interrelationships. Another feature, less prominent but always present in some form, is group leadership in the person of the headman responsible for specific ceremonial and economic tasks.

Why do suprafamily organizations with ceremonialism and leadership develop? The answer is implicit at the family level. In the chapters dealing with family-level organization we described a basic contrast in interfamily relationships corresponding to the subsistence strategies being used. For food gatherers resources are basically predictable and their procurement is largely an individual matter; since interfamily relationships are basically competitive, population is generally dispersed, coming together primarily to exploit periodic windfalls of plants or game. For hunters, by contrast, resources are more unpredictable and procurement tasks may require cooperation among several families; camp-level groups are accordingly formed, and families maintain exchange networks to other camps.

This contrast between basic subsistence modes continues to characterize societies at the local group level. In agricultural groups the primary cause of organizational elaboration appears to be defensive needs. As seen in both the Yanomamo and Tsembaga cases, a relatively high population density leads to competition between local populations for the control of such productive resources as prime agricultural land and foraging territories. The formation of the corporate group, the teri or clan, makes it possible to bar the group's land to outsiders and regulate its use by clan members; and the organization of clans into a ceremonially synchronized territorial group makes it possible to defend the territory against neighboring groups.

The more narrowly *economic* causes of group formation seem much less central in agricultural populations. At the local group level of organizational complexity, agricultural technology is simple and does not require cooperative group activities. In the Yanomamo case the seasonal scarcity or oversupply of plantains and peach palm fruits argues for intergroup arrangements; but even in their case the risks are not high, and the economic functions of the teri seem clearly secondary to its defense functions. What appears to happen is that economic activities such as commodity exchanges are handled institutionally in the same way as alliance building, and act to reinforce the more basic relationship.

In hunting and fishing economies and among pastoralists, economic causes are more prominent in promoting group formation and regional networking. Among the Eskimos village organization is directly necessary to the whale hunt; networking within and beyond the village is equally necessary because of the unpredictability of the food supply. Similarly, whatever its origins, regional networking among the Turkana today appears chiefly as a response to the high risk of herd depletion.

Stratification involves the differential control of productive resources, and little evidence for it exists at the local group level. By and large, individuals acquire and exploit their own resources. Except in cases of economic cooperation where a leader controls the necessary technology, as in the Eskimo whale hunt, and in the immediate instance of warfare, where attack and defense (including their economic aspects) are coordinated by prominent men, leadership carries no connotation of economic control.

Yet within the more complex social organization of these societies, as contrasted with family-level societies, are undeniably contained the basic elements of stratification. Individuals compete with each other for prominence and recognized status in the displays and games found in all these societies, notably in the group dancing at the kaiko ceremonies of the Tsembaga and in the Eskimo song contests. As we have seen, this competition has important economic implications, for it contributes to an individual's success in forming networks. And tied as it is to underlying economic and political factors, it foreshadows the development of competitive leadership that we discuss in the next chapter.

SEVEN

The Corporate Group and the Big Man Collectivity

WE NOW EXAMINE the factors that favor the rise of economically powerful "Big Men" among subsistence-oriented producers. The Big Man is a *local* leader, one who makes decisions for the local group and represents it in major intergroup ceremonies. As Big Man systems we will consider together the highly dynamic Big Men of Highland New Guinea and the somewhat more institutionalized "chiefs" of the Kirghiz of Afghanistan and the Indian fishermen of the Northwest Coast of North America. Although the systems are structured differently, they are remarkably similar in terms of social, political, and economic behavior.

In the past the emergence of Big Men has been attributed to surplus food production. Although surplus production is certainly necessary to support Big Men's political activities, we must still ask why food producers forgo leisure in order to generate a surplus in the first place. That is, why are people willing to accept the burden of supporting Big Men and their expensive public displays of wealth and status?

Big Men characteristically manage the economy beyond their own local group. They organize and direct the intergroup ceremonies, accompanied by large-scale, coordinated gift giving, that are critical for a group's prestige and desirability as an ally and exchange partner. They organize external trade and may be important traders. In general, the Big Man acts as a group's spokesman, dealing with other Big Men to organize political and economic relations in the loose association of communities known as the intergroup collectivity.

The Big Man's decisions on behalf of his group inevitably entail a certain loss of family-level autonomy among his followers. True, the Big Man must please his followers or lose their support, but while in power he restricts their options by his domination of the systems of production and distribution.

In the three ethnographic cases that follow, the hunter-gatherer Northwest Coast Indians, the horticultural Central Enga of Highland New Guinea, and the pastoralist Kirghiz of Afghanistan, we examine the different contexts in which strong leaders develop in the kind of village-level societies described in Chapters 5 and 6. The importance of external relations is seen in all cases, but the particular combinations of warfare, trade, and diplomacy differ. In accordance with the pattern identified in earlier chapters, the importance of leadership in other than defensive matters is of secondary concern in the horticultural case (Central Enga), but is of considerable importance among Northwest Coast groups (where the subsistence economy depends on fish and animal resources) and among the trade-oriented Kirghiz.

Case 9. Indian Fishermen of the Northwest Coast

The native societies of the Northwest Coast of North America hold an immense fascination for the Western observer. Their beautiful art, their elaborate technology, the unexpected extent and complexity of their political life, and above all their competitive, entrepreneurial, and seemingly "capitalistic" economy strike many responsive chords. That these parallels to modern society should be found among hunter-gatherers employing a "stone age" technology has led many observers to question whether any evolutionary theory can explain Northwest Coast economic life.

In this section we shall examine the relations between environment, technology, the social organization of production, and the political economy on the Northwest Coast in an effort to explain this seemingly aberrant economic system.

The Environment and the Economy

Most observers agree that the Northwest Coast environment is capable of providing amply for a hunter-gatherer population (Drucker and Heizer 1967). The coast is significantly more productive than the interior, and population densities and village sizes are larger on the coast. Despite local variations in the abundance of certain foodstuffs, the general pattern of food procurement is similar throughout the region, which runs from the Olympic peninsula to southern Alaska. Communities on the coast are oriented toward marine and estuary resources. The seacoast offers eleven saltwater fish, including halibut, cod, herring, and flounder; sea mammals, including sea otters, sea lions, porpoises, and occasionally whales;* waterfowl and shore birds; mussels, clams, and other shellfish; and seaweed and other plants.

Inland a comparable diversity is found. Seasonal runs of salmon and candlefish are major food sources; land mammals provide "a hunter's paradise" (Oberg 1973: 8) with white-tailed deer, mountain goats, bears, moose, bighorn sheep, caribou, and other species hunted for furs as well as meat; geese, ducks, and other fowl abound in some seasons; and a wide variety of berries, roots, and other plants are available.

Owing to the unusual productivity of the natural ecosystem, population densities along the Northwest Coast run from one to two persons per square mile, perhaps the highest density achieved by any ethnographically known hunter-gatherer people. Although at such densities we expect population pressure on wild resources, it is not certain that Northwest Coast peoples have experienced any significant food scarcity (Codere 1950; Drucker and Heizer 1967; Driver 1969).

Considerable evidence, however, suggests that the people expect and fear food scarcity and make serious efforts to avoid it. For one thing, the people themselves tell tales in which communities in the past suffered from hunger (for example, Boas 1910: 139; *People of Ksan* 1980: 13). For another, enormous amounts of food are stored for winter, a time when food is scarce and hunger a real possibility. We know also that the region's supply of wild foods varies greatly from year to year. Just as the Eskimo and the Nganasan can never be sure how many caribou will come their way, so Northwest Coast people cannot be sure of the supply of salmon, which can be hugely abundant one year and quite scarce the next for reasons completely beyond the control of local fishermen (Donald and Mitchell 1975). Finally, some groups like the Kwakiutl (Boas 1966: 17) work hard to intensify the production of gathered resources, for example by clearing areas in which edible plant species such as clover and cinquefoil are collected or by burning over berry patches to raise their yield.

The evidence suggests that truly huge surpluses are likely to occur only seasonally and in the good years. Given the comparatively high dietary requirements of the region's comparatively large population, food scarcity and even famines periodically threaten during the winter months.

Despite the size and complexity of Northwest Coast societies, indi-

---

*Whales are seldom hunted, except by the Nootka; but a beached whale is a great windfall and the occasion for a feast.

viduals in small family groups procure their own food for most of the year. Depending on local circumstances (coast or interior, large river or small, etc.), the annual round is approximately as follows.

In March and April people from separate local groups come together for the great candlefish runs. Candlefish are oily: it is said that one can insert a wick in a dried candlefish, light it, and burn it like a candle. In the early spring millions of candlefish run and intense labor is invested in harvesting them and rendering their oil, which is then stored for home consumption and trade. The oil is a valuable preservative and additive to dried foods. Being storable, it plays such an important part in the political economy that its attraction for people is "like the lure of gold" (*People of Ksan* 1980: 89).

In the late spring and the summer people scatter in family and camp groups, similar to those of the Nunamiut and the Shoshone, to hunt, fish, and collect roots and greens. Coastal groups forage for shellfish and seaweed along the shore and hunt sea mammals in small canoes in coastal waters and among nearby islands. This period is described as one of ease and plenty.

In August and September the tempo of production speeds up as berries come into season and the salmon runs begin. Berries are collected in large quantities, carefully dried on finely crafted racks, and packed in large boxes, sometimes covered with oil, for winter consumption. Salmon runs, like candlefish runs, require a heavy labor investment in obtaining and preserving fish. In both cases the fish in good years are more numerous than the people can handle; as a result, the greater the labor invested, the larger the harvest, with little or no declining marginal productivity.

Once this period is past, people aggregate in winter villages, where they pass the season manufacturing and repairing boats, tools, clothing, and the like. There are some hunting expeditions, but people live mainly on stored foods. This is a period of intense socializing and ceremonial activity. By early spring people are tired of stored foods, many of which have begun to deteriorate and are no longer palatable, and eager to move out of the winter village and resume family-level foraging.

The Indians of the Northwest Coast are masters of woodworking. They build large, sturdy houses, canoes both small and large, and some fish weirs so large that wooden "pile drivers" are used to set the main posts. On a smaller scale they build smokehouses and sheds furnished with drying racks and tables. Of major economic importance also are large boxes made of cedar planks carefully hewn and sewn together. These are watertight and may be used for cooking by filling them with water and adding hot stones from the fire, or for storing oil, berries, seaweed cakes, and other preserves.

In the smokehouses, and in earthen cellars covered with wood and sod, prodigious amounts of food can be stored (*People of Ksan* 1980; Stewart 1977: 145). Being sedentary each winter, households store quantities of berries, oil, and dried and smoked fish and game, along with pelts, bone, and horn used in manufacture. Some households accumulate substantial wealth in capital and consumer goods. This capacity for wealth, and for the growth of *differences* in wealth between individuals, is greater than for any other known hunter-gatherer society. Clearly, then, the importance of storage in the Northwest Coast economy is a precondition to the development of social differentiation (see Suttles 1968).

The considerable accumulated wealth and highly productive localized fisheries were natural targets for raiding; and warfare was in fact present and at times brutal. According to Barnett (1968: 104), "The Kwakiutl say that before the white man came they fought with weapons; now they fight with property. This is a consequence of white interference with their wars, slave taking, and headhunting."

In the late eighteenth century, at the time of the first important contacts with whites, warfare was apparently endemic on the Northwest Coast. Body armor, in the form of heavy leather cloaks or "coats of mail" made from wooden slats bound by sinew, was widely used, and the technology of battle-axes and clubs was well developed (Gunther 1972: *passim*). From all indications "true warfare, aimed at driving out or exterminating another lineage or family in order to acquire its lands and goods, was a well-established practice in the North" (Drucker 1955: 136). Warfare could involve long-distance raids aimed at capturing booty and slaves, but competition over resources lay behind much of it. In Drucker's words (1965: 75), "There is considerable evidence that the coast carried the maximum possible population in prehistoric times, particularly in the northern half of the area. That is, the ample natural food resources were fully exploited within the limits of the native technology. The traditions are replete with accounts of groups driven out of their homes and lands, and of the hardships suffered before they found new homes."

A group that could not maintain its strength against powerful neighbors was lost. Drucker (1965: 81) describes one group as being so "ground to bits" between two powerful neighbors that its members traveled in small groups and ate their food raw for fear that their fires would attract roving war parties. "Both sets of enemies were trying to exterminate these people, to take possession of their rich fishing and hunting grounds." At the same time an elaborate system of exchange bound together local populations with special resource opportunities. In particular, as with the Eskimos (Case 6), extensive trade existed between coast and inland groups.

To summarize briefly, the Northwest Coast Indian populations were confronted aboriginally with an exceptional complex of problems related to their intensive hunting, gathering, and fishing economy. The natural environment offers abundance along with unpredictable fluctuations. In some of its areas resources are abundant; in others they are comparatively scarce. Many resources are also highly localized in their distribution. The economy of the Northwest Coast Indians therefore encompassed a remarkable combination of elaborate fishing and storage technology, occasionally fierce warfare, and considerable trade. We now consider their equally remarkable social organization.

Social Organization

Five levels or units of social organization can be discerned: the family, the house group, the lineage, the village, and the supravillage "intergroup collectivity" (Newman 1957). The focus of the family and the house group is on subsistence; house groups form and fragment through the annual cycle as subsistence prospects dictate. By contrast, the lineage, the village, and the larger collectivity are concerned with the political economy and focus on capital investments, ceremonials, exchange, and warfare.

The family is the elemental economic unit, acting independently during summer foraging. Most tools, clothing, food, and manufactures are individually produced and owned, and do not concern any larger group. But families for much of the year are organized into "house groups," with an estimated average size ranging from 7 (Rosman and Rubel 1971: 130) to 25 (Donald and Mitchell 1975: 333), roughly the size of the family hamlets of Part I. House groups pool resources and often eat from a common cooking box.

The house group is not a lineal descent group, yet kinship, traced bilaterally though often with a patrilineal or matrilineal emphasis, is the major determinant of membership. The oldest male of the house group is generally considered its head or "chief," yet he is not of higher social rank than the other adult males. He and his closest kin constitute a more or less permanent residential core, with less closely related persons moving in and out opportunistically as local resources and labor needs fluctuate. This pattern at the household level reflects a larger division in Northwest Coast societies between elites, who are closely bound to productive resources through politically reinforced ties of ownership, and commoners, who roam more or less freely throughout regional territories, "respecting" different Big Men in succession by residing with them for short periods (Newman 1957: 9-12).

House groups have many communal features. The house itself, permanent, secure, and provisioned with stored foods, understandably attracts its members back each winter. Much of the productive capital of the group, including fish weirs, dams, oil-rendering apparatus, drying racks and sheds, and canoes, is produced jointly and held in trust under the househead's control. Labor in fishing, oil-making, berry picking, hunting, sealing, and trading is contributed equally by the members of the house. Labor contributed to a Big Man, for the construction of his weirs and dams or the maintenance of village streets, is a joint effort of individual house groups.

Beyond the house groups are units variously named *numayma*, lineages, and clans. Such groups acknowledge kin relationships among their members and are set apart by the possession of distinctive tokens, crests, and other markers. When all members of a lineage live in a single village, they are co-holders of rights in specific resources such as streams, berry patches, and offshore islands. But membership is fluid: many people are eligible on kinship grounds to join two or more groups, and will join the one most advantageous at the moment. It is also possible for a non-kinsman to buy into a group.

The named lineage can extend across village boundaries. Such a lineage is not a corporate and territorial group, but it may offer valuable links throughout a wide region in which trade and ceremonial exchanges are important. Lineage ties also provide some security in an area where warfare is endemic and destructive.

Large villages contain more than one lineage and may have as many as 500-800 members. The property-owning core of the village population is more or less stable because of the large investment in houses and productive capital. The house is considered sacred, the permanent residence in which ideally one is born, marries, and dies. Since the winter village is the site of these houses and of the major ceremonies and feasts, the Kwakiutl say "the summer is secular, the winter sacred" (Boas 1966: 172).

Yet as we turn to consider how villages are integrated into a single regional economy, we must remember that the village is only loosely united. Its members are primarily loyal to their own house group, and suspicion of others, especially for theft, is rampant (cf. Boas 1910: 70, 138, 148, 153). As we shall see, many "chiefs" vie for the support of the villagers, and even their loyal followers must be constantly browbeaten to yield up their precious subsistence products to the political process.

The key to the Northwest Coast political economy is the Big Man or "chief." Public life provides many opportunities for expressing differences in rank, and for testing and reordering rank. Ultimately a Big Man's rank is a reflection of his "wealth," that is, of the amount of wealth he can accumulate from the group that acknowledges him as leader. To be sure, his functions are complex and extend into areas only partially related to economic life, such as marriage, life-cycle observances, and kinship structure. Viewing just those functions of the Big Man that are central to the economy, however, we can understand a great deal about how Northwest Coast political economy operates and what it does for people. Several points are apparent.

1. The Big Man represents a group, and for many purposes *is* that group. His wealth is the group's wealth, and his rank expresses the cumulative rank of his following. Thus participants in a ceremony often stress that the Big Man is acting not in his own name but in "our name."

The Big Man is invested with titles and emblems representing the group's territories and wealth objects. In a house group or a local group these titles refer to specific fishing grounds, berry patches, sealing rocks, and the like (the ocean and inland hunting regions are not controlled in this fashion). When a Big Man integrates other local groups with his own, he typically buys their emblems or seizes them by force, so that he becomes, albeit in a restricted sense, owner of the group's resources. Although a Big Man may obtain control of a group by force, perhaps even murdering its original leader, in the long run he must depend on its loyalty, which he must earn by bravery, managerial skill, and generosity.

2. The Big Man organizes a complex economy characterized by large-scale capital investments and an elaborate division of labor. His house contains specialists, such as canoe-makers, harpooners, and carpenters, who are paid from his store of wealth. Although he owns these specialists' products, his followers routinely use them in obtaining, processing, and storing food.

In family-level societies it is difficult to organize the construction of large-scale works like dams, weirs, and defensive structures. A leader is needed to persuade people to do work that does not directly benefit the family, and the Big Man uses his wealth and influence to that end.

3. The salmon fisheries, although very rich, can be overexploited in any stream. The weirs can literally close off a stream. The Big Man, as ceremonial specialist, must decide when to open the season; fish

must be allowed through the weirs for use by upriver groups and to spawn. To some degree the ritual cycle regulated by the group leaders provides a critical management function that overrides the tragedy of the commons (Pinkerton 1985; Morrell 1985).

4. The Northwest Coast Big Man must maintain greater stores than others, for which purpose he invests in storage structures. These and the larger buildings needed to house specialists constitute what Netting (1977: 36) describes as "substantial houses full of weighty possessions." Actually, as with Big Men elsewhere, most of the wealth that comes into his house quickly goes out again to meet his followers' expenses.

5. To support his activities the Big Man requires a share of his followers' production. A successful hunter or fisher must give one-fifth to one-half of his catch to his Big Man (Boas 1921: 1333-40). If he does not, he will receive few favors in the future and may even be roughed up (Boas 1921: 1334).

In turn the Big Man "spends" or redistributes his income, returning part of it to his followers through feasts and other generous acts and using part to pay his specialists for their products. Some of these products are directly useful (e.g., canoes and storage facilities); others enhance the prestige of the Big Man and his group (e.g., totem poles and house decorations). Finally, part of the Big Man's income goes to increase his store of prestige goods such as beaten copper valuables and blankets, which are used in ceremonial exchanges.

6. Where warfare is common the Big Man also maintains a retinue of warrior specialists to supplement such warriors as his regular following may provide. A brave and well-armed Big Man is a source of security to his followers—or a source of worry if they fail to meet his demands.

7. Big Men are the prime movers in large, interregional ceremonies like the potlatch. Most ceremonies occur in the early summer or in November and December, following the major periods of food storage. An infinite number of occasions can justify ceremonies, among them the numerous life-cycle events of a Big Man's family—births, naming ceremonies, etc. What determines the actual occurrence of a ceremony, however, is the amount of wealth a Big Man has accumulated. Only if his wealth is ample will a Big Man host a ceremony, since other Big Men will be quick to ridicule him if his feast is less than sumptuous.

Ceremonial occasions are economically complex. Politically they are occasions for Big Men to compete for prestige by giving away, and even destroying, wealth. Envy and humiliation are integral to the feast. According to Boas (1921: 1341-42), Big Men may exhort their followers thus:

I depend on it that you will stand behind me in everything when I contend with the chiefs of the tribes (villages). . . . I want to give a potlatch to the tribes. I have five hundred blankets in my house. Now you will see whether that is enough to invite the tribes with. You will think that five hundred blankets are not enough, and you will treat me as your chief, and you will give me your property for the potlatch, . . . for it will not be in my name. It will be in your name, and you will become famous among the tribes, when it is said that you have given your property for a potlatch, that I may invite the tribes.

The Big Man and his following seek "to flatten" the name of another group by "burying" it beneath piles of gifts. But a similar feeling of competition exists between a Big Man and those of his own followers who may seek followings to rival his. When a potlatch is proposed, each of the Big Man's followers responds to his proposal, standing and speaking in order of rank. One may speak as follows (Boas 1921: 1343): "I am annoyed by our chief, because he asks us too often for property for his potlatch. I shall try to make him ashamed. Therefore, I shall give him one hundred blankets, that we may bury his name under our property. I wish that you give for the potlatch fifty, or forty, or ten pairs of blankets; and from those who are poor, shall come five pairs of blankets."

All this is conducted openly for all to hear and see. Indeed the hosts offer gifts to the audience of a potlatch as a form of payment for "witnessing" the exchanges between Big Men (Barnett 1968: 93). The need for witnesses is to publicize the economic productivity of the group represented by the Big Man, and, as Newman (1957: 86) indicates, to validate or "legalize" transfers of property control from one headman to another.

Despite the emphasis on blankets and coppers as standards of value, the majority of items given away or destroyed at a potlatch are foods, tools, boxes, other useful goods, and services (Barnett 1968: 76, 85-88). These items represent the surplus available for such uses in this storage-oriented society in years of abundance. In years when food is scarce, by contrast, a Big Man would be humiliated to host a potlatch; and of course none is required until the host is ready. In general, groups with the best resource bases are the largest and wealthiest and have the wealthiest Big Men (Donald and Mitchell 1975: 334-35).

Guests who are given storable items save them for their own future ceremonial needs, or use them to pay off debts or make loans between ceremonies. Food is consumed at the feast or carried home. But the potlatch does not guarantee that food from the rich is transferred to the poor (J. Adams 1973): in bad years the rich meet their own needs first out of limited stores, whereas in abundant years even the poor have ample food. In especially abundant years "grease feasts" are held during which competing leaders pour boxes of fish oil on the fire and burn them in a lavish competitive display of wealth.

8. Big Men can also obtain food for their supporters in bad years in exchange for items of wealth accumulated in good years (Vayda 1961: 621), assuming of course that some other group has food to exchange. Valuables thus permit at least some distribution of food from well-supplied to hungry areas, and stores of wealth objects serve as savings accounts or social security deposits against local famines. Boas (1898: 682, quoted in Barnett 1968: 4) likened such wealth to a life insurance policy, since it could be inherited and would protect young children should they be orphaned. The security provided by storing wealth in this way is central to people's willingness to submit to a Big Man's demands.

The owner of valuables not only has access to another group's stores, but can grant another group the right to participate in a seasonal or unexpected food surplus in the owner's territory. Thus a headman who "owned" a particular beach gave out tokens entitling the bearer to a share of blubber from the next beached whale to appear on his beach (Newman 1957: 82), and thus also outsiders could obtain rights to fish for salmon or candlefish along privately owned streams when the owning group had fish to spare during a good run. By the ownership and allocation of rights of access, labor can be moved opportunistically yet not chaotically from windfall to windfall, from one short-term surplus to another, reducing the loss of food that is common when small forager camps come upon temporary surpluses beyond their ability to consume.

9. In addition to the exchange of food for valuables, trade also occurs at a distance, notably between the coast and the interior. Such trade is not conducted by members of individual households, but is usually organized by Big Men, who through their political activities have established ties to Big Men in other ecological zones.

It is important not to exaggerate the degree of rivalry between Big Men. The language of potlatching is aggressive, and the self-serving speeches are intended to make others ashamed. But Big Men are tough and not easily crushed by mere polemics, and they respect the debts they incur through ceremonial exchanges and attempt to pay them back. Over time they build ties of respect and trust (Barnett 1968: 112; Rosman and Rubel 1971: 170) that can be called on in times of need.

As with the Yanomamo (Chapter 5), these ties serve also to create regions of peace within which aggressive competition between populations can be regulated and turned to constructive purposes. In fact, the evidence suggests that when peace was enforced by whites following contact, rivalries that once would have led to open conflict came to be expressed in particularly acrimonious ceremonial competition. Once again, then, warfare must be attributed to the failure of the political economy to integrate communities that lack strong ties of kinship and exchange. At times even the potlatch ceremonials were turned into Yanamamo-style treacherous feasts by enemies who simulated peacemaking efforts in order to lure wary victims to their destruction (Drucker 1965: 80).

Among the fruits of war were captured "slaves," usually women and children. Most slaves were ransomed by their families, but those who remained with their captors could be ordered about, bought and sold, and even killed at their owner's command. Slavery, however, was not an institution in the sense of being a regular part of the economy (Suttles 1968). Slave-owners, even wealthy Big Men, did the same work as their "commoners" and slaves, and ate the same foods. Slaves never amounted to more than a tiny fraction of the population (Drucker 1965: 51-52).

The Big Man on the Northwest Coast embodies the suprafamilial economic interests of his following. He holds and defends title to their resource base, organizes cooperative labor for projects benefiting the group, generates and maintains large capital investments, stores food and wealth against hard times, maintains economic specialists and exchanges their products for shares of the production of nonspecialist households, exercises or delegates military responsibility, and manages intervillage and interregional ceremonies and exchanges that integrate the economy far beyond the family level.

The interregional group, Newman's (1957) "intergroup collectivity," is in fact an association of Big Men in which no single paramount leader dominates, although some are stronger than others by virtue of their resource base and their political, military, and managerial skills. Through public ceremonies they negotiate the continual exchange of power for prestige and prestige for power, which amounts to the exchange of wealth (blankets, coppers) for economic goods (food, technology, labor) and vice versa.

This elaborate and extensive political economy is made possible by an abundance of wild foods concentrated locally and seasonally. But it is also made necessary by high population densities (with a continuous high demand for food), unpredictable fluctuations in food supplies regionally and seasonally, and warfare and raiding for the control of desirable resources. The political system may be viewed as a mechanism for mobilizing a family-centered populace to increase its security against famine and war by producing foods and manufactures beyond their personal needs. Much of this surplus is directly invested in public works and social security by Big Men. The rest is spent in self-promoting displays that maintain their Big Man status in the face of continual competition.

Case 10. The Central Enga of Highland New Guinea

The Central Enga of New Guinea's highland core are in many ways like the acephalous Tsembaga Maring described in Chapter 6, but there are certain dramatic differences between the two that help us understand the further development of the political economy. The process of intensification is particularly salient in this case and will be emphasized in our discussion. As we have seen with the Tsembaga, population growth leads to intensification, intensification to warfare, and warfare to the formation of clans and local groups. Among the Enga, whose population density is double that of the Tsembaga, intensification has resulted in permanent sweet potato gardening on prime land: there is no other reliable way for such a large population to get enough to eat. Warfare is accordingly oriented toward seizing prime land, its frequency has risen dramatically, and this increased frequency has sped the rise to prominence of Big Men, local leaders who orchestrate the exchange and alliance networks of the regional collectivity on which the local group's survival ultimately depends.

The Environment and the Economy

The Central Enga, including the Mae and Raiapu Enga, live in a mountainous region west of the Hagen range in Papua New Guinea, a region of high population density in contrast to the lower-density "fringe" area occupied by the Tsembaga. We have selected them for discussion because excellent data are available on the economy, ceremonialism, and sociopolitical organization of the Mae Enga (Meggitt 1964, 1965, 1972, 1974, 1977), and on the subsistence economy of the closely related Raiapu Enga (Waddell 1972).

The Central Enga, like other Highland groups, were isolated from direct Western contact until recently (Meggitt 1965: 2). Their earliest recorded contact was in 1933; a patrol base was established in their territory in 1942; and in 1948 the first missionaries and miners arrived. The first ethnographer to study the Enga, Mervyn Meggitt, arrived in 1955, only twenty years after sustained contact was initiated.

The Central Enga live in an area of highland rivers and open intermontane valleys. Their land is high, ranging from about 3,900 feet in the grassland valleys to as high as 7,900 feet. Rainfall averages 108 inches a year, and there are 265 days with some rain. "Summer" (November-April) tends to be a little wetter and warmer than the annual average (50-80°F), "winter" (May-October) a little drier and cooler (40-70°F). Droughts occur in the winter, and this can be a time of food shortage.

Plant communities and microclimates vary markedly from one elevation to another. Below 4,600 feet lie the dense rainforests of the lower valleys, virtually uninhabited because of malaria. The zone from 4,600 to 7,500 feet was originally a mid-mountain and valley forest; now cleared for agriculture, it is a mosaic of gardens and fallow plots. Alluvial fans edge the valley floors and are farmed intensively; three-quarters of the population is concentrated here. Above, from 7,500 to 9,500 feet, is a zone of beech forests that shelter game and are important for pig foraging. Still higher is a subalpine cloud zone of little economic use to the local population.

Much of this environmental diversity occurs within a remarkably compact region, with the steep mountain slopes standing directly above the valley floor. As a result the lands of a Central Enga clan, although typically very small (from one to two square miles), cut vertically across all the zones and incorporate a part of each. The intense human use of the environment, however, has greatly diminished the former diversity of plants and animals, and much of the region now consists of grasslands and permanent garden plots. A clan territory among the neighboring Raiapu Enga had only about 5 percent of its land remaining as forest (Waddell 1972: 14).

According to Meggitt, the population density in the core area of the Central Enga ranges from 85 to 250 persons per square mile, near the maximum for Highland New Guinea groups. Figure 6 shows an environment crowded with settlement and heavily transformed with long use. Much higher than the densities of the simpler horticultural societies described in earlier chapters, this man-land ratio has obvious implications for the subsistence economy.

The subsistence economy of the Central Enga, described most fully for the Raiapu Enga (Waddell 1972), is dominated by an intensive form of agriculture involving mounded sweet potato production, some

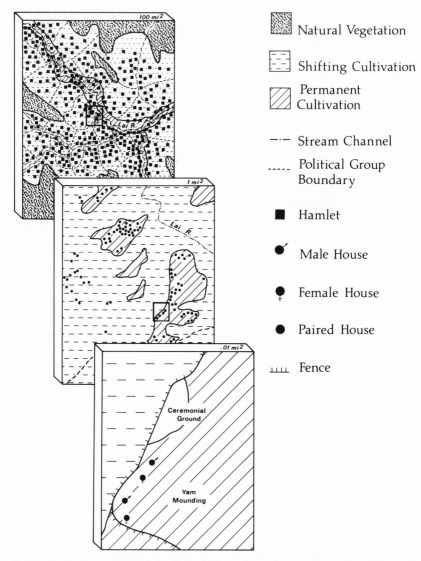

Fig. 6. Settlement pattern of the Central Enga. Apart from uncultivable gullies and ridges, the landscape is filled with gardens. Population is dense and hamlets are everywhere, but they cluster near sweet potato gardens and in defensible locations. Each local group has a ceremonial dance ground.

slash-and-burn farming, and considerable pig husbandry. Because the environment has been degraded by the intensive agricultural use, wild foods are limited and contribute insignificantly to the diet.

The dominant subsistence strategy is year-round sweet potato gardening. In a sample Raiapu community, 62.5 percent of the garden land was in permanent sweet potato production (Waddell 1972: Table 8). The field is made up of mounds about nine feet in diameter; the sweet potatoes are grown in the soft soil of the mound. Following the harvest the mound is pulled apart and earth pushed back around the mound; in the center is placed green manure, consisting of the old sweet potato vines, leaves, and other mulch. When this mulch has begun to decompose, the mound is rebuilt and is ready for replanting. With such artificial fertilization the mounded fields can be kept in constant production, and there is no fallow period (Waddell 1972: 44).

Most of the mounded fields are situated in the lower fans and valleys, where the slope is less than 10 percent (Waddell 1972: Table 9). On the steeper slopes are found shifting swidden gardens that produce a wide variety of crops including yams and bananas (Waddell 1972: Tables 13 and 14). These gardens are similar to the shifting fields of the Tsembaga, which in some ways duplicate natural floristic conditions and use a long fallow cycle (10-14 years) to regain fertility. Swidden production is more important for dietary diversity than for calories, with the gardens constituting only about 20 percent of the total agricultural land (Waddell 1972: Table 8).

Unlike the Tsembaga, the Central Enga do not cultivate trees for food. Some cultivated tree species, however, provide materials used in building and fencing and for other purposes (Waddell 1972: 40), materials obtained from uncultivated trees before the deforestation of the region. And Meggitt (1984) reports the intensive cultivation of the *Casuarina* tree to meet the tremendous need for firewood.

In the highlands pigs are everywhere and usually outnumber humans (Waddell 1972: 61-62). Pigs forage in the hills for food, but like humans depend mainly on cultivated food, especially sweet potatoes (Waddell 1972: 62).

As we have seen, the energetics of pig raising are astonishing and its cost to the farmer is high. Waddell (1972: Table 28) estimates that *49 percent of all agriculture produce goes to the pigs*—more than is eaten by the Enga themselves! Some 438 hours per person per year are devoted to raising food for pigs, which provide less than 2 percent of the total diet by weight: the net gain to the human is incredibly low, only about 40 calories per hour, or about one-twentieth of the calorie production of a *single* sweet potato plant. Of course pig protein and fat, limited from other sources, are essential to the Enga; they *have* to raise pigs. But the high cost of raising them dramatizes the loss that people incur when intensification makes it necessary to replace hunting by husbandry.

As elsewhere, intensification has also produced major changes in the diet itself: with the shift to permanently cropped fields, the diet has shifted almost exclusively to agricultural products. Sweet potatoes make up as much as 90 percent of the food consumed by the Chimbu and other groups. As a result, Highland populations "experience a high incidence of protein-calorie deficiency among infants and a general deficiency in the protein intake" (Waddell 1972: 122), and are at risk from nutritional diseases. The Raiapu Enga (Waddell 1972: 124-25) alleviate this potential health problem by growing a variety of vegetable foods in their shifting gardens (including several introduced species such as peanuts), and by purchasing food, such as canned fish, that provides both protein and fat. At present the diet appears adequate, except perhaps for young children.

Warfare is the most immediate threat to the Enga, and an ever-present one: a war every two or three years in a relatively small region studied by Meggitt (1977). The small local group's territory is surrounded by enemies or potential enemies, and war can break out at any time. Mortality is high, with an average of four deaths per conflict. Attrition is severe, and groups must maintain a high growth rate to remain politically viable.

Although a wide variety of proximate causes for warfare are given (from rape and stealing to land conflict), Meggitt (1977) argues convincingly that the underlying cause is competition over land. Wars are commonly between neighbors, who are in direct competition; a local group will attack and defeat a weaker group and quickly annex its land. The Enga themselves recognize that wars are caused by competition for agricultural land, especially the limited amount of prime land used for the permanent, intensive cultivation of sweet potatoes. Over half of all Enga wars are explicitly acknowledged to be over land.

Trade in food has traditionally not been a major concern for the Enga, although exchange for axes and salt commonly occurs. There is also a lively exchange of primitive valuables used in the social and political maneuverings of individuals in the large intergroup ceremonies.

Social Organization

Settlement Pattern. The Central Enga have no villages (Meggitt 1965: 3; Waddell 1972: 30-39). Homestead farms, traditionally consisting of paired male and female houses, are spread through a clan's territory, although they tend to cluster on the alluvial benches most suited for sweet potato farming (Fig. 6). The houses are often located between the sweet potato fields and the upper slopes, with their swidden gardens and fallow growth grazed by the pigs, a location that minimizes movement costs, a major labor expenditure in gardening (Waddell 1972: 179). Distance to the sweet potato fields is usually less than 7 minutes and to the swidden fields, 24-30 minutes.

Why did the Central Enga not form villages as seen elsewhere in the Highlands, such as among the Chimbu (Brown 1972)? Presumably for cost reasons: to form a village is to increase the distance to agricultural fields and thus the costs of farming. The Tsembaga formed villages anyway for defense reasons, and this certainly makes sense for Highland groups like the Chimbu, among whom warfare is endemic. Why not, then, for the Enga?

Although the answer is not immediately clear, several differences between the groups may be noted. The Enga's hamlets minimize production costs in transportation; the Chimbu's villages maximize protection against sudden attack. If the importance of defense is the same for both groups, the difference in settlement patterns probably corresponds to a difference in production costs. The Enga depend on mounded sweet potato gardens that utilize bench land spread out along river courses; the dispersed nature of their prime land may make village life prohibitively expensive for them. The Chimbu depend on drained field systems concentrated in the flat valley bottom lands; the concentrated nature of their prime land may make village life feasible for them. Also the bench land of the Enga is heavily dissected by erosion so as to create ridges that are naturally defensible. We should emphasize, however, that the organized local group found in both the Enga and Chimbu regions is much more important than the presence or absence of villages. Villages are good indications, especially for archaeologists, of local group formation; but dispersed hamlets may also be organized politically into local groups where environmental conditions make hamlets economically preferable to villages.

Turning to social organization proper, we consider four levels of organization: the household, the clan segment, the clan with its Big Man, and the intergroup collectivity. As in our previous case, organization at the two lower levels responds to subsistence problems and centers on food procurement activities and the division of labor; organization at the higher levels responds to problems in the political economy and centers on defense and economic interdependence.

The Household. The primary social and economic unit is normally the family (Waddell 1972: 20), typically the elemental family with an average of 4.5 members: a woman, her husband, and their children—and of course their pigs.

The Central Enga perform most subsistence activities individually; work groups rarely exceed two or three persons (Waddell 1972: 103). Gardening, especially in the sweet potato fields, is highly individualistic and does not require or encourage large work groups. Women perform the routine subsistence tasks of farming, especially in the sweet potato gardens, cooking, and child care. They are the primary domestic planners, and the gardens are considered their domain (Meggitt 1965: 246). They provide 92 percent of the work in the sweet potato fields and 80 percent of the work in the swidden gardens, excluding the important "male" crop of yams (Waddell 1972: 98). Men's work is more irregular and includes the periodic clearing and farming of swidden gardens, care of the yams, house-building, and numerous public activities (Waddell 1972: Table 25).

Land is owned directly by the household. At marriage a man receives land from his family estate and establishes an independent household. This land, typically including both sweet potato fields and swidden lands, is then managed by the husband and wife working together. Although alienation of the land is restricted and requires the consent of concerned patrilineal kin, the household otherwise retains control of its land.

The amount of land farmed by a household is a direct reflection of its size: the larger the number of consumers in the household, the greater the land area under cultivation (Meggitt 1974: n. 43). In short, the extent of agricultural activity is determined largely by the subsistence needs of the household.

Despite the intensity of the subsistence economy, the traditional technology is simple and personal, relying heavily on the woman's digging stick and net carrying bag and the man's stone axe. Each family has its own productive tools, which it either makes or obtains by trade.

This sketch of the household and its subsistence economy fits closely Sahlins' (1972) model of the Domestic Mode of Production. The household is the primary unit of production and consumption; it has direct control over the main factors of producton—labor, land, and technology—and gears production to meet its own needs.

The basic household organization is, however, somewhat more complex. Despite the economic interdependence of the sexes, males fear females and express a deep antipathy toward them as a threat to maleness and health (Meggitt 1964). Male and female residences are accordingly separate. The woman's house, the basic household center, houses the woman, her children, and her pigs. The man's house among the Mae is ideally the residence of the males from one patrilineage (Meggitt 1965: 20, 22), but among the Raiapu the man's house is individual and paired with his wife's (Waddell 1972: 34). Among the Mae, therefore, household clusters appear to consist of a number of women's houses around a single men's house; in contrast, the Raiapu show a pattern of isolated farmsteads.

Polygynous marriages are common in the Central Enga, especially for Big Men. A man with two or more wives divides his production-consumption role among separate households each composed of a wife, her children, and perhaps other attached persons. Each wife's household is a separate domestic economy. Because women do most production work, a man who seeks to increase his agricultural production so as to finance his political ambitions can do so by marrying many wives. As we shall see, however, access to wives depends on the accumulation of considerable wealth through affinal exchanges, and on access to productive land.

The Clan Segment. Households are organized into patrilineal groupings that are segments of the territorially based clans. Although Meggitt's (1965) structural analysis is perhaps too rigid, we will describe the operation of what he sees as two levels of group formation below the clan, namely the patrilineage and the subclan.

Patrilineages are "people of one blood" named for a founder from whom descent is traced (Meggitt 1965: 16). The founder is said to have been usually the "father's father's father of older living men" (Meggitt 1965: 16-17), and actual kin relationships among lineage members are known. Meggitt refers to the patrilineage as "a quasi-domestic grouping" (1965: 17) whose existence is not readily apparent to an outsider. For the Mae Enga the men's house is typically composed of the members of a patrilineage (Meggitt 1965: 20, 22), but this men's house has no ceremonial importance (cf. Meggitt 1965: 235).

The patrilineage is a group of households with closely related male heads that help each other in specific economic and social situations. The few activities requiring labor outside the household typically involve males from a local group for such things as clearing swidden gardens, building fences, and building houses (see Waddell 1972: 106). A person's patrilineage "brothers" are responsible for helping him when he needs help (Meggitt 1965: 244); should he be disabled, for example, they will prepare his gardens and rebuild his house for him. "Brothers" are also a man's most reliable source of support in arranging marriage exchanges and the like.

Patrilineages range in size from 4 to 68 members, with a mean size of 33 (Meggitt 1965: 5-18). In our terms this is a hamlet-size group of very much the same sort as other hamlet groups we have discussed: essentially an extension of close kin bonds to achieve subsistence and security purposes that are important to the smaller and more vulnerable nuclear family but beyond its reach.

The subclan, by contrast, is a larger unit, organized along political and ceremonial lines, whose members are putatively descendants of a son of the clan founder. A subclan owns a dance ground and a sacred grove of trees, and plays a major role in external exchanges and in political matters. The Mae Enga social system imposes heavy payments, such as bridewealth payments and death payments, on individuals on ceremonial occasions (see Meggitt 1965: 110-27), and these obligatory payments require contributions from a supporting group, the subclan. Similarly, as we shall see, the primary support of a man's climb to Big Man status comes from his subclan.

Though the individual is the focus of bridewealth payments and other social exchanges, such exchanges, with their accompanying public display, also reflect back on the subclan group as a whole. As with the Tsembaga, a man needs a broad regional network of interpersonal ties to provide him with wives, nonlocal trade goods, security in cases of local disaster, and political support in competitive exchanges. A man's subclan plays an essential role in helping him establish his regional network. In return, the successes of any subclan member in regional networking increase the status of the subclan and its members' desirability as partners for persons in other groups. Because individual status translates directly into group status, the support given by subclan members is clearly part of a more general strategy for building up their own personal networks.

Competition among subclans occurs, as we shall see, and is for dominance in the political affairs of the clan. The subclan is also the point of cleavage for the formation of new clans by segmentation. Subclans among the Mae Enga range in size from 45 to 145 members, with a mean of 90 (Meggitt 1965: Table 7), about the size of the Tsembaga clan grouping.

The Clan and Its Big Man.   The clan is politically the most important group among the Central Enga. Defined by its carefully demarcated territory (Fig. 6), the clan is a defensive group, protecting the claims of its members against outsiders. The clan is also politically autonomous, being the largest group to act as a group in both warfare and ceremony. It is led by a Big Man, who speaks for it in external affairs and works internally to mobilize it for ceremonial and political action.

The clan is first and foremost a corporate entity, restricting access to land. It is putatively patrilineal, with rights to land in clan territory reflecting a reckoning of male descent lines thought to derive from a common founding ancestor. As we have seen, where good land is in short supply the rules for allocating such land place a premium on lineal descent. Individuals who are not patrilineal kin can become attached to a clan and gain access to land, but only where the clan has ample land and needs more settlers for security reasons. The clan is supposed to be, and largely is, exogamous, with wives coming from other localized clan groups as part of a regional system of exchange and alliance. Meggitt (1965: 9) estimates average clan size for the Mae Enga as 350 persons (range 100-1,000), roughly the size of the Tsembaga territorial group.

As a group the Enga clan owns a main dance ground and plot with an ancestral-cult house (Meggitt 1965: 227). The dance ground, thought to have been cleared by the founding ancestors, is the focus for ceremonial exchanges with other territorial clans involving death and homicide payments and the dramatic competitive exchanges of the Te ceremonial cycle.

Apart from ownership of the ceremonial ground and joint defensive action, the clan asserts itself as a discrete group in certain ceremonies, in the clan meeting, and in the action of its leader, the dominant clan Big Man. In a number of ceremonies, notably the *sadaru*, group identification is made explicit. The sadaru is the ritual exclusion of bachelors, in which males are instructed in defense against female pollution (Meggitt 1964; Waddell 1972: 87). This occasion involves "four nights of seclusion and instruction in a special house erected in a remote part of the territory. . . . At the 'emergence festival,' when the bachelors return fully adorned and chanting from their mountain retreat, members of the host clan distribute food to the large number of visitors present" (Waddell 1972: 87).

The bachelors are a cohort of male patrilineal kin, united as a group in this seclusion, who are to become the next generation of household heads and political actors. They are presented publicly at the main dance ground by their clan to visitors from neighboring clans who are to be their affines, trade partners, allies, and of course potential enemies. The fineness of this display of the clan's future prospects is important to the process of clan members' economic and political maneuvering in the region.

Although the right of the household to independence is valued, as expressed in the statement "Each man makes his own decisions" (Sackschewsky 1970: 52), there are times when the group must act together, as in warfare and Te exchanges, and in these matters the clan meeting is crucial. All active males of the group affected, a clan or subclan segment, meet to discuss the issue and come to a consensus. Non-kin have very limited rights in such a meeting; women and children are excluded. The consensus reached by the meeting applies to all those who participated in it.

Leadership, a key ingredient in group action, is seen clearly in the clan meeting and related ceremonial and political events. The Big Man, though his status is highest, need not be the one to call the meeting, nor is his word considered binding to the group. Rather the Big Man is a man of renown—known for success in political and economic affairs and listened to because of his demonstrated ability to influence individual action, his control over wealth and exchange, and his public speaking ability.

The Big Man is both an individual entrepreneur and a group spokesman. In the first role he uses the resources available to him through the manipulation of his extensive interpersonal network based on marriage, alliance, and exchange. By aggressive and calculated action he comes to control a high percentage of the exchange and production of valuables, notably pigs, which are important in all social exchanges of the group. In the second role, as spokesman for the group, he exhorts its component units to work together for the group's survival and the general good of all its members.

The selection of the clan Big Man demonstrates this dual nature (Meggitt 1967). As we have seen, each clan is composed of a number of subclans. From the males of a subclan, on the basis of personal qualities of leadership and calculation and with support from patrilineage brothers, one emerges as Big Man in matters requiring subclan action, such as the collection of marriage exchange payments and the initiation of ceremonies. Subclan leaders then compete with each other for leadership of the clan and the status of primary Big Man. In part a man's ability to achieve and hold this status depends on the size of his immediate support group, i.e., his close relatives; but he must also broaden his support to receive assistance from other subclans and ultimately from other clan members. He does this by such means as offering to help raise marriage payments for a member of another subclan and thereby placing that person and his patrilineal relatives in debt to him. Another would-be clan leader may make the same offer, or a more generous one. This is the way the two compete for supporters.

There appears to be a push-pull dynamic in the activity of group leaders. Subclans and clans must have an effective leader to serve their interests in interclan relationships involving marriage, exchange, and defensive alliance. A group thus pushes a potential candidate forward. In turn the attractiveness of real control over wealth, power, and women (for it is the Big Man who is polygynous) motivates the leader to act in such a way as to maximize his personal power and reproductive success.

The clan as a unit exists chiefly to cope with the external relationships of warfare, defense, alliance, and exchange. To understand the clan and its leader is thus to understand their place in the regional system of competition and cooperation. We will therefore first briefly sketch the nature of regional interaction before returning to the place of the clan, the ceremonies of integration and interrelationship, and the emergence of the Big Man.

The Intergroup Collectivity. Frequent and vicious warfare is the most salient feature of interclan relationships. All people outside the clan are potential enemies, and all land outside the clan's small territory is potentially hostile. Perhaps half a mile beyond a person's home lies an alien world fraught with the risks of sudden death. According to Meggitt (1977: 44), "In the past all movement outside one's own clan territory was hazardous, and in general men made such excursions only in armed groups and for compelling reasons, in particular to attend distributions of wealth, to negotiate exchange transactions, to trade, and to assist friends and relatives in battle. Casual social visiting by lone men was not common, not only because it exposed the wayfarer to the dangers of ambush and murder en route, but also because it violated Mae notions of personal privacy and group security."

The threat or promise of warfare is central to all clan decisions. A large clan, with power in numbers and a shortage of land, looks for an excuse to attack a weaker neighboring clan and seize its land. A small clan, weak in numbers and vulnerable to attack, must encourage settlement by nonpatrilineal relatives to swell its defense force. Since the losers in war lose everything, a household's control of essential resources depends on its clan's political power and success.

A group's ability to defend its territory or seize new territory depends primarily on how large a fighting force it can field. This depends both on its own size and on how many allies it can recruit for a confrontation. Clan size is determined in part by demographic factors; the individual fertility of members can have a dramatic effect, with some clans growing rapidly while others are declining. As we have seen, one strategy available to a small or declining clan with land to spare is to accept nonpatrilineal kin as members (cf. Meggitt 1965). By contrast, strict patrilineal kin rules, whereby a man receives land only from his father's clan, are maintained if the group is large and its density high. This correlation between the percentage of patrilineal kin and population density supports the more general proposition that lineality relates to subsistence intensification.

Successful clans tend to become larger, in part because a clan's success in regional exchange and warfare increases its members' ability to obtain wives and thus the clan's reproductive potential. Success quite literally breeds success and failure failure, producing relatively rapid upswings and downswings respectively in clans' fortunes.

It was once widely assumed that warfare acts as a negative feedback mechanism to regulate demographic growth. Thus as population grows resource shortages occur and warfare over resources increases, producing rising mortality rates that keep the population down. This appears *not* to happen among the Central Enga (see Meggitt 1977: 112), where people try to have as many children as possible so as to provide males as warriors and females for the regional exchanges that are critical to alliances. In effect, an increase in warfare has intensified the pressure to expand a group's population in competition with neighboring groups.

Another important factor in arranging alliances is a clan's reputation as a reliable and beneficial confederate. Success in obtaining allies is tied to success in a set of related ceremonies of exchange involving marriage, death compensation, life-cycle displays, and the regional Te exchange cycle. In each ceremonial setting the reputation of the individual and the group is displayed publicly by group size, personal adornment, and the exchange of primitive valuables. This is seen most clearly in the Te.

The Te is a cycle of competitive exchanges that link up many Central Enga clans (Meggitt 1972, 1974). Its main participants are a number of clans linked as an exchange line but with a number of alternative paths (Meggitt 1974: Diagrams 2 and 3). Other clans peripheral to this main line are joined to it through personal exchange relationships with clan members of the main line clans.

Starting at one end of the chain, initiatory gifts of small pigs, marsupials, pork, salt, axes, and other valuables are given as individual exchanges from one partner to the next down the chain of clans. After this pattern of giving has continued for a time, individuals from the clan at the initiating end begin to demand repayment in pigs. As this signal passes through the system, individuals start to amass pigs to be given away alive at a series of massive ceremonial occasions accompanied by display and oratory. This series of ceremonies, which involve major interclan presentations, begins at the opposite end of the chain and proceeds in a wavelike action to the beginning, taking six to nine months to complete. The clans that begin the main gifting ceremonies then start to demand repayment, and those at the opposite end start slaughtering perhaps half the pigs that they had amassed and giving the pork in an elaborate interclan ceremony to the next clan in line. All gifts from members of one clan to the next clan in the chain are thus displayed and given together to maximize the visual effect of scale and to identify the coordinated group action. These clan-level ceremonial presentations are coordinated by the clan Big Men.

The Big Men also orchestrate the interclan negotiations to end hostilities between local groups and to make homicide payments. When it becomes clear that fighting has gone on too long with mounting casualties and no clear outcome, the Big Men must call for a large meeting of the opposing groups to exchange quantities of pork, to settle claims for homicides, and thus to reestablish peace (Meggitt 1977: 20). Most important in this peace process is the payment of homicide compensation. Each death in battle must be paid for by the enemy who killed the man and by the ally who encouraged him to fight. To end a war, responsibilities for each casualty must be assigned and the compensation payment made. For an individual, obviously, payment received for the death of a kinsman should be large and the payment given for a killing should be small. With such opposed interests between warring factions, ending a fight is not easy. But the Big Man can arrange satisfactory payments by arguing that they must be generous to reflect well on the group's prestige. At these meetings the rhetoric of the Big Men is militant as they vilify their opponents, but their action is clearly to appease and mediate and thus to reestablish the status quo of the regional collectivity.

The Big Man's role in these ceremonies is to coordinate the clan's presentation of gifts and payments, to make the gifts and payments, and the occasion itself, as impressive as possible, and thus to put on a show that reflects well, first on the Big Man as leader and organizer, second on his clan as a powerful group, and third on individual clan members who seek to maintain and expand their networks of in-law exchanges and trade partnerships.

The survival of the group depends directly on its profile in these competitive exchanges. Who will become an ally to an unsuccessful clan, a partner to an ineffective Big Man, or an in-law to a nonparticipant? Prestige gained in the Te translates directly into successfully obtaining allies, trade partners, and wives. Success in the Te thus translates into success in other social and political realms and ultimately affects the survival of the group and its participating households.

The Enga clan sits in a hostile social environment among armed neighbors who are anxious to seize its land. Its productive and reproductive success depends on its defensive posture as a group and its recruitment of allies. These in turn depend on its success in the organized intergroup relations orchestrated by the clan Big Man and presented in the dramatic intergroup ceremony.

In the group ceremony and the economic and political maneuverings of the Big Men we discover a well-developed political economy. Goods are mobilized from the constituent households to support a set of actions that are basic both to the rise to power of an individual Big Man and to the long-term political survival of the local group.

In the life history of 'Elota, a Solomon Islands Big Man, Keesing (1983) observes that interclan regional relationships in the Solomons did not result in constant wars; as long as a clan was considered powerful, it could live at peace *most* of the time. Although an offense against the clan must be met with anger and a show of aggression, it was usually thought prudent to accept compensation in wealth rather than take violent action. A measure of balance in political power, carefully maintained and portrayed in ceremony by the region's Big Men, was the basis for peace. Thus warfare in the Big Man systems of New Guinea and Melanesia can once again be seen as a failure of the political process by which intergroup relationships are negotiated and maintained.

The reason for strong leaders among the Enga seems straightforward: the local group could not function without them. It seems odd at first, especially considering the very high population density and the intensely competitive attitude of Big Men, that a single regional Big Man has not emerged from the competition and transformed the society into a chiefdom. The fact is, however, that a chief cannot govern effectively without economic control and the conditions for economic control are absent in the Highlands: storage is unnecessary, technology is simple, and technology and trade are broadly based rather than concentrated. In contrast to the chiefdoms to be examined in Chapters 9 and 10, an Enga chiefdom would have no way to exert control over the basic factors of production.

Case 11. The Kirghiz of Northeastern Afghanistan

Within a single generation the Kirghiz of northeastern Afghanistan were transformed from a predominantly family-level herding society into a society with strong local leaders. The conditions that brought about this swift and dramatic evolution are clearly identifiable and shed considerable light on the other Big Man systems described in this chapter.

The Environment and the Economy

The Kirghiz (Shahrani 1979) are nomadic pastoralists of the Pamir zone of Afghanistan, near the borders of China and the Soviet Union. They inhabit broad, flat intermontane valleys at altitudes of over 12,000 feet, above the limits of agriculture. Average annual precipitation is below 6 inches, and there are fewer than 30 frost-free days each year. Vegetation is sparse, and the environment is made especially inhospitable by the persistent harsh winds.

Historically the Pamir lay on the trade route known as the Silk Road, connecting China with the Middle East. It is not devoid of resources. When Marco Polo passed through on his way to China, he was impressed by the abundance of wild sheep, and today there are also mountain goats, Tibetan wolves, brown bears, marmots, hares, and turkeys. Mountain streams feed marshes and lakes where pasture is seasonally abundant. Peat bogs provide fuel for cooking and heating.

Formerly the Pamir was a favorite summer pasture of the Kirghiz. In July and August the days are hot, and pasture grows lushly in alpine meadows on the valley floor. But over the long winter the pasture dries out; the meadows are covered by snow, and the winds are bitterly cold. In past winters the Kirghiz retreated with their herds to lower pastures in China and Russia. By 1950, however, the borders had been closed, with some Kirghiz isolated on the Afghan side. This political change forced them to intensify their use of the Pamir in order to live there year round.

The Kirghiz of the Pamir now number approximately 1,800 persons living at a population density of roughly one per square mile. This group pastures a mixed herd of some 40,000 sheep and goats (in which sheep outnumber goats by three to one), 4,000 yaks, and small numbers of camels and horses. As among the Turkana (Chapter 6), herds of different composition have different requirements and can make use of contrasting microenvironments. Sheep and goats are always herded together because they complement each other's feeding habits; since sheep graze and goats browse, they do not compete directly for food. In the winter sheep have the advantage of being able to paw away snow to reach the frozen undergrowth, and goats stay close to them to find food. In the summer, when sheep by themselves would tend to graze too long in one spot and so destroy pasture, goats move on quickly, the sheep follow them, and overgrazing is minimized. The Kirghiz recognize this complementarity and deliberately maintain mixed herds.

Yaks are native to the area, well adapted to the cold and the high altitude, and able to exploit pasture that the other species cannot. The Kirghiz herd only about 4,000 yaks, but because of their large size and rich milk they make a major contribution to the diet.

Like the Turkana the Kirghiz are careful to use short-lived pastures first, leaving the more permanent pastures as security for the scarcest times. In the winter, whenever a bit of pasture is exposed by the wind, herders move quickly to exploit it before it is covered by new snow. The south side of the valley is in shade much of the time; the north side is sunny. Since the prevailing winds blow from north to south, during the long winter the south side is in shadow and piled high with drifted snow. At this time the Kirghiz are scattered in small family groups throughout the north side of the valley. The best pastures are found there and on the valley floor near water, but both are used as sparingly as possible in the winter. Only in the spring, when sheep and goats give birth, do the Kirghiz move their herds into the richest pastures in order to strengthen their animals for birthing.

In the summer families move to the south side to make intensive use of the short time in which those pastures are available. This is a time of abundance. There is little competition over pasture, and settlements are larger. Then as fall approaches they move onto the valley floor for a month or so, slowly making their way north to the winter encampments.

Since the closing of the borders, the Kirghiz have begun to intensify their use of pastures. They allow the richest, best-watered pastures to grow all summer, and then harvest and store fodder for winter. Led by their khan, they have begun also to irrigate pasture land and to fertilize with dung.

The Kirghiz produce much of their own food. Meat and dairy products are important in the diet, particularly in the four or five warmest months. Milk is processed into yoghurt and "sour milk," which, when salted, can be frozen and stored for winter. Cheese is made and dried for storage, and butter can be clarified and stored for several years in bags made from goat or sheep stomachs. Meat is eaten frequently, especially on multicamp ceremonial occasions. Wild foods are of little importance except among the poorest families, and vegetables are rarely eaten. Trade in foodstuffs, however, is essential to the household economy, as we shall see.

Although one might expect the animal herds to be a strong temptation to raiders, raiding does not appear to exist. Why? Apparently for two reasons. On the one hand, the khan is powerful enough locally to resolve disputes among the Kirghiz themselves. On the other hand, the existence of powerful states capable of regulating borders and punishing outlaws prevents external peoples from attacking Kirghiz herds.

At the time the borders were closed, the demand for animal products in the agricultural areas of Afghanistan had grown markedly. Afghanistan's population growth seems to have required considerable expansion in agriculture at the expense of open lands that once pastured wild or domesticated animals. Each year the Kirghiz export about 5,000 sheep and goats, 200 yaks, 7,000 kilograms of clarified butter, and many hides, ropes, felt blankets, and the like, acquiring in return agricultural produce, tea (which they consume in prodigious quantities), metal and wood products (including yurt frames), opium, and many other outside goods. Their dietary staples are now mainly wheat and other grains obtained through trade.

To summarize, the intensification of pastoralism has resulted in a number of significant shifts in the Kirghiz economy. These include incurring the considerable risks of stock raising year-round in a marginal environment, using new methods of intensification such as irrigation and fertilization, and exchanging animal products for agricultural produce and other goods available from a settled agricultural population. As with the Nganasan (Case 4), the development of this exchange on a systematic basis has made the Kirghiz into specialist producers in a broader market economy.

Social Organization

The basic social unit of production is the household. An average Kirghiz household consists of 5.5 persons, 120 sheep and goats, 12 yaks, a horse, and one or more dogs. A single herdsman can manage a herd of several hundred animals by himself, and a herd of over 100 sheep and goats is sufficient to meet basic subsistence needs. The household usually inhabits a yurt constructed from wood and straw and wrapped in large felt covers. Recently, however, families have begun to build winter homes of stone and earth, constructed on land claimed as property by the household or kin group.

Eighty percent of Kirghiz households are nuclear, some with unmarried or elderly members attached; the rest are either extended families or polygynous. The household is an integrated unit sharing a single hearth, and is largely independent. A senior male acts as its spokesman, but all adult members, male and female, have a voice in economic decisions. Typically two or more households form a group known as an *aiel* (camp) or a *gorow* (corral, indicating a common shelter for their herds). These camps grow larger in summer and smaller in winter. They usually consist of patrilineally related households, but in any case are fairly stable units that claim territory and share responsibilities for hospitality. The camp has a leader, a wealthy and respected man who mediates disputes within the group and represents it in intergroup ceremonies and conflicts.

Beyond the camp, patrilineages and neighborhoods form loose cooperative and ceremonial units. Patrilineal relations are important in marriage, particularly among the wealthy, who see endogamy as a way of keeping wealth within the larger kin group. Since the closing of the borders, kin relations have taken on added significance as groups of kinsmen lay claim to territories and regulate their use, becoming as a result corporate kin groups.

The border closings have greatly increased the stratification and political centralization of the Kirghiz. Formerly they moved freely through the Pamir largely as independent family camps, although leaders existed for specific functions in ceremonial exchange and dispute resolution. But after 1950 camps and kin clusters laid claim to strips of land transecting the valley to ensure their access to all the microenvironments they need for year-round subsistence. And with the construction of permanent houses, corrals, and irrigation works, the ownership of carefully defined plots of land has become common.

One result has been the differential accumulation of wealth. Formerly such wealth differences as existed were primarily a matter of age: young couples with small herds would join the camps of wealthy relatives for whom they could work while building their own herds. Now, however, a few households with very large herds are surrounded by dependent households with only a few animals each; two-thirds of all households now own few or no animals, and a mere 5 percent own 80 percent of all sheep and goats. A few men with exceptional skills in managing both animals and people have gained control of the herds. The dependent households obtain animals and care for them under the direction and patronage of the wealthy men. If their own animals fail to survive a harsh winter, a not infrequent occurrence, the wealthy man provides them with food and new animals.

This control of herds by an elite is a response to the intensification of production in the Pamir. When households could leave the zone during winter, they did not experience such great risks of losing animals. The wealthy man now functions as a risk-avoider who scatters animals throughout the Pamir. When disaster strikes in one place, he moves in resources from another location, becoming a main source of security for his dependents. He also identifies poor herders and corrects their mistakes or else withdraws his support.

A further element in the centralization of power by leaders has been their role in external trade, from which they derive significant wealth. Wealthy leaders of camps and lineages engage in this trade for much of the winter, when they travel to agricultural areas for barter. The khan, the acknowledged spokesman for the whole Kirghiz group in the Pamir, actively develops trade relations with outside markets. He operates with his people's wholehearted support because the non-Kirghiz itinerant traders are accused of cheating the Kirghiz and exploiting them by encouraging the use of opium.

In sum, since the closing of the borders, an intensified use of the Pamir has resulted in a number of important interlinked changes:

1. The Kirghiz are more careful in their management of existing pasture, carefully preserving for winter use the rich pastures formerly enjoyed in summer.

2. They have begun to invest in capital improvements, such as secure shelters for people and animals, irrigation works, and buildings for storing fodder.

3. These improvements have led to more disputes over access to particular campsites and pastures, which are negotiated between group leaders and if necessary adjudicated by the khan.

4. Herd control has become centralized in the hands of a few Big Men, who act as risk-spreaders and skilled managers and are responsible for the welfare of their dependent households.

5. Trade, now central to the household economy, makes a good living possible but tends also to be handled by the wealthy.

Thus the intensification of production has been accompanied by increased economic integration and stratification. Economic integration among the Kirghiz is accomplished mainly by wealthy men, who, by centralizing herd control, pool the region's resources in a system of social security. Since the restricted range of movement means that the average household must supplement its animal food supply with inexpensive grains, integration with the outside world through trade is also important. And here, too, wealthy men, who possess horses and camels for transport, play a central role. The increased stratification of the political economy appears as a natural consequence of economic intensification and integration.

Conclusions

In Chapters 5 and 6 we examined village-level groups in which group leaderships as such played no significant part. For the groups in this chapter, leaderlessness is no longer an option; strong leadership is required to integrate the village-sized community into a regional economy, especially in the "intergroup collectivities" of the Enga and the Northwest Coast fishermen. (The Kirghiz differ because they are surrounded, not by numerous communities of roughly equal size and power, but by extremely powerful national economies that force the Kirghiz khan to be more a broker between his people and the superordinate political economy than a Big Man in the classic sense.) In order to understand the causes for this growth beyond the comparatively acephalous village-level economy, let us examine the three dimensions of intensification, integration, and stratification.

The intensification of production is a powerful agent of change among the Northwest Coast fishermen, the Enga, and the Kirghiz, although its specific form varies from case to case. On the Northwest Coast intensification has made it possible to harvest the often stupendous seasonal or unpredictable supplies of candlefish and salmon, to store these foods against periods when food is scarce, and to distribute local surpluses to areas that are temporarily experiencing shortages. This achievement, which depends upon such capital investments as traps, weirs, drying racks, storehouses, and watertight boxes, reduces the amount of wild food that escapes capture and thus allows the total population density to increase to remarkable levels for a hunter-gatherer people.

With the Mae Enga intensification has meant a shift toward total domestication of the environment. The forests have been cut and wherever possible turned into permanent fields. Production has come to focus on a single highly efficient crop, sweet potatoes. The most dramatic rise in production costs is seen in pig husbandry, where half the sweet potatoes go to compensate for the loss of wild game that elsewhere lives off the land.

The Kirghiz, a pastoral people used to following an extensive migration route through seasonally rich pastures, were suddenly forced to occupy a single comparatively poor area along that route. They have responded by fertilizing, irrigating, and harvesting and storing fodder in order to survive throughout the year, where formerly all they needed was a few good months of summer pasture.

The different forms of intensification create somewhat different organizational needs and hence variants on the Big Man system. In the forager economies of the Northwest Coast, leaders are needed chiefly to manage the high risks involved in the pursuit of migratory species, to supply equipment needed for the periodic large-scale procurement and processing of wild food, and to negotiate alliances and peacekeeping arrangements. The group leader also directs the major intergroup ceremonies, which are essential to a group's prestige and its members' ability to form regional exchange networks, and is obligated to support his followers when they individually experience economic difficulties.

Among the pastoral Kirghiz, where technology is relatively simple, leaders are needed chiefly to spread risks and to conduct the external trade on which the subsistence economy depends.

The horticultural Enga need leaders primarily for political maneuvering and the regulation of war. The group leader of the Central Enga is a politician par excellence, orchestrating the group performance at intercommunity ceremonies in such a way as to keep old allies and obtain new ones. In the New Guinean world of constant intraregional warfare, the leader, as negotiator of intergroup alliances and of intergroup peace, is essential to the group's access to agricultural land. The formation of corporate groups discussed in Chapters 5 and 6 is the first step to restricting access to productive resources. The next step is to be able to defend this right of access against constant threats of aggression, and this can be done most effectively by an acknowledged leader.

Among the Northwest Coast fishermen and the Enga, as among the Tsembaga and Yanomamo, warfare represents the outer edge of the political economy. In the current cases the population is larger and more interdependent and the political economy more complex. Yet in all these cases, simple and complex alike, warfare is not so much the outcome of deliberate policy as the breakdown of policy, proceeding from leaders' ultimate inability to restrain the competitive, acquisitive impulses of strong, family-centered individuals in the interest of the greater good of the community.

Depending as it does on differential control over strategic resources, stratification is clearly evident, albeit in incipient form, in Big Man societies. In all cases the Big Man controls resources, such as smoked fish, pigs, or herds of sheep and goats, that help him to spread the risks of food production far beyond the family level. In other ways the Big Man's economic control varies among the three cases: control of technology in the hunter-gatherer economy, control of long-distance exchange in the pastoralist economy, and control of intergroup exchange ceremonies in the horticultural society. But in each case leadership involves economic management and manipulation for individual as well as group advantage. As we shall see in Chapters 9 and 10, the further evolution of the political economy into true chiefdoms depends on more elaborate forms of economic control.

EIGHT

The Economy of the Local Group

IN CHAPTERS 5-7 we examined societies characterized by economic integration beyond the family level. Although the family remains of primary significance in everyday living, its economic behavior can no longer be explained without reference to considerations that extend far beyond individual families, and even beyond the territories of the local group.

Unlike family-level societies, the societies described in Chapters 5-7 have relatively high population densities. To be sure, these high densities characteristically reflect a greater "environmental potential" than that enjoyed by comparable family-level groups, e.g., the Northwest Coast Indians as against the Shoshone, the Mae Enga as against the Machiguenga. But it is more significant that the economies of these societies are of a different order, with much greater emphasis on the social management of risk, technology, warfare, and (to a lesser extent) trade.

A local group is a politically independent and coordinated unit with a population of roughly 100-500. Its features characteristically include local corporate segments (clans and lineages), property rights over productive capital, territoriality, intergroup warfare, ceremonially based suprafamily exchange networks, primitive valuables, and group leadership. Each feature may serve several functions, and features may differ in importance from society to society; taken together, however, these closely interrelated features represent a social formation radically different from the family-level society.

The main elements of societies organized as local groups are the following:

1. Population density is usually low to moderate but sometimes quite high, as among the Mae Enga (with 85-250 persons per square

mile). It is characteristically above one person per square mile, and thus higher than family-level densities.

2. The environmental conditions associated with local groups are highly variable, from Arctic coasts to tropical rainforests. Some local groups, either naturally or through cultural improvements, enjoy more productive and diverse environments than others; but overall productivity appears to be greater for local groups than for family-level societies. Resources tend to be seasonally concentrated or capable of higher yields if suitable capital improvements are brought to bear.

3. The technology consists primarily of individual tools such as the digging stick and the harpoon. Some key technologies, especially in intensified hunter-gatherer and pastoral economies, are owned by individuals but used by a much wider group; examples include fishing weirs, whaling canoes, animal corrals, and shipping carts. Food is most commonly domesticated, but high-yielding seasonal wild foods are also important to some groups. In either case food is stored for considerable time in cellars or bins, or unharvested in the gardens, or in the form of livestock.

4. The settlement pattern is comparatively sedentary. Hunter-gatherers continue to aggregate and disperse in an annual cycle, but they found winter villages and live in them for many months while eating stored foods. In agricultural societies, villages or hamlet clusters are located near especially productive lands and occupied for many years consecutively; pastoral groups are more mobile because of the needs of their herds. For defense reasons villages and hamlets may be surrounded by palisades or other protection. Within the village or hamlet cluster is usually a dance ground used by the group for ceremonies.

5. Social organization has several levels, and characteristically the following three: the family level, involved in primary production and security; the local group, including corporate units (clans or lineages) organized as defensive and economic entities; and the intergroup collectivity, consisting of multiple local groups bound by ceremonial and economic exchanges.

6. Territoriality is variable, but territory is carefully demarcated. Among hunter-gatherers the most highly productive seasonal locations, such as the fish runs, are the spots most prized and carefully defended. Boundaries are carefully marked in agricultural zones, especially when population densities are high. Pastoralists are typically subdivided into many smaller units moving independently and opportunistically through the area to make best use of pasture.

7. Warfare is endemic and linked to territorial defense. In lower-density societies, such as the Yanomamo, warfare consists mainly of raiding designed to present an outward image of fierceness that keeps potential enemies at a distance. For higher-density groups like the Tsembaga or the Enga, boundaries are sacred and defended in organized battles against any advance. The region beyond the local group's territory is hostile, filled with enemy warriors and sorcerers.

8. Ceremonialism is pervasive. It functions both to define local groups and their corporate segments, and to create and maintain regional intergroup relations whereby groups can obtain allies in war, marriage partners, and exchange goods.

9. Leadership varies from headmen to Big Men. Certain men distinguish themselves by their fierceness or diplomatic skills and become recognized leaders for their local groups. These leaders are essential to regulating the regional relations between groups and also important in economic management.

We have identified several analytically discrete primary economic responses that reflect distinct processes of evolutionary change, namely, risk avoidance, technology, warfare, and trade. These responses vary in importance from case to case, and their different combinations generate multilinear paths of evolutionary development.

We next consider the development of the local group from the three perspectives taken throughout the book.

The Ecological Perspective

In the societies considered in Chapters 5-7 a degree of household autonomy is sacrificed in return for the greater security offered by a larger group against both starvation and enemy attack. If we contrast the Tareumiut to the Nunamiut or the Nganasan, the Turkana to the !Kung or the Shoshone, or the Tsembaga and the Mae Enga to the Machiguenga, we find that greater population densities are being maintained under roughly analogous resource conditions. As we now see, this is only possible by solving basic problems involving risk management, technology, warfare, and trade.

Warfare is a response to competition over productive resources that occurs at the fringe of the political group, where integration is weakest. It is the most prevalent of our four factors at the local group level, and is especially important among farmers. In three cases (Yanomamo, Tsembaga, and Central Enga) warfare is the dominant cause of local group formation. This is particularly clear in the Yanomamo case, which has several features marking it as transitional between the family level and the local group, notably a low population density, relatively ample resources, a relatively fragile village structure that continually fragments from internal disputes, a small polity size (150–250), and the presence of several headmen in a single village. Were it not for raiding and the consequent necessity of intergroup ceremonies to obtain allies, why form a local group?

For the Tsembaga in the Highlands of New Guinea, where good land is scarce and its coveters many, the small local group (about 200 people) and its tiny territory (three square miles) are the only guarantees of safety and subsistence. The local group is composed of several corporate clans ceremonially integrated in defensive territorial units; other local groups are politically independent but may be allied through individual interpersonal ties of marriage and exchange and through intergroup ceremonials. The Central Enga also have competition, warfare, corporate clans, local defensive groups, and intergroup networks and ceremonies; in this case, however, a much higher population density and increased competition over productive agricultural lands have led to increased reliance on the group and its regional allies.

From the Yanomamo treacherous feast to the Central Enga Te ceremony, all the warlike groups use public ceremonials as opportunities to display group strength, test old alliances, and build new ones. At the very least a family must have friends to turn to for help when attacked, and to flee to if their group is defeated. These are the same friends and relatives with whom a family engages in ceremonial gift exchange, marriage alliances, and reciprocal hospitality.

To what extent does warfare as such encourage the development of sustained leadership? To be sure, fierce warriors and successful war leaders gain prestige from their efforts, but they tend to fall back into ordinary lives once a war is over. In the ethnographic literature on leadership, other skills, particularly in oratory and diplomacy, are generally seen as more central to attaining political power than prowess in war.

The Yanomamo, Tsembaga, and Central Enga cases document a continuum in the evolution of tribal societies. Increasing warfare, itself an outcome of competition, can by itself cause local group formation and the development of leadership and ceremonialism. But warfare by itself does not present the conditions of control necessary for the further evolutionary development that we will see in Chapters 9 and 10.

Among hunter-gatherers and pastoralists, factors other than warfare (but similarly linked to intensification) cause group formation. Risk management is the most important. Risk management is important in local group formation when unpredictable local and seasonal variations cause sharp fluctuations in the food supply. The most dramatic examples arise when a population depends on wild or domesticated animal resources. In these cases, risk management involves using political channels to call in labor from neighboring or distant local groups to harvest, consume, and where possible store foods in temporary abundance, allowing as little as possible to go to waste; and when food is scarce, distributing the accumulated stores through the same channels. This permits a larger population density than in family-level societies, where surpluses occur that the community is not able to use.

We see risk management most prominently in the Turkana, Tareumiut, Northwest Coast, and Kirghiz cases. A Turkana family spreads the risks of losing animals as a result of locally poor pasture, disease, or raids by scattering some of its animals across the landscape in the herds of its many kinsmen and friends, some of whom may be faring much better. Similarly, when pastures in the plains are temporarily lush, Turkana families come in from areas of more permanent pasture to harvest the local windfall before it is lost to drought.

The Tareumiut manage risk by harvesting and storing whale meat and blubber, which are abundant for only a short spring season. Thanks to cooperative boat crews and the coordinated efforts of crew leaders, huge amounts of food are produced, stored, and shared over the scarce winter months, allowing the Tareumiut to boast, "We don't let people starve." And when local resources fail, food may be available from the extensive ties that Tareumiut households and boat-owners have with other whaling villages on the coast and even with nomadic caribou hunters in the interior.

Risk management has been elaborated into a fine art among Northwest Coast groups. Through the political allocation of rights of usufruct, given physical expression in tokens and emblems, Big Men control the exploitation of temporary abundance, attempting to ensure that as much as possible is consumed or stored. Recall the Big Man who issued tokens allowing access to the next whale stranded on his group's beach: all coastal groups welcome this sort of windfall, but the Northwest Coast groups organize it to a higher degree than elsewhere. With their vast capacity for storage in great cellars, smokehouses, and watertight boxes, and their frequent ceremonial allocation of useful items in exchange for valuables, they spread the risks of a sudden failure of the food supply across many local groups, and even across "tribal" and linguistic boundaries.

Among the Kirghiz the necessity of using high-altitude summer pastures for year-round subsistence has strengthened the role of the khan, who is increasingly becoming "owner" of the group's animals in the sense of managing the risks of herd depletion in the interests of the group. In addition to allocating animals carefully among his followers according to their managerial skills, he also distributes animals to families who fall victim to the sudden losses that are inevitable in such a harsh environment.

The process common to each of these cases, then, is an integration of the economy beyond the family level, spreading the risks of local failure to a larger security network. In the next section we will examine the structural elaboration of the political economy that makes this integration possible.

Capital investment in technology, although less important than defense and risk management at this level, may strengthen the economic basis of the local group in particular situations. Among the Tareumiut and on the Northwest Coast, large whaling boats, fishing weirs, storage cellars, and drying racks are beyond the means of independent families. Leaders are needed to exhort people to work, to oversee the placement and maintenance of the equipment, and to direct its use. These leaders are also the natural guardians of the food produced, part of which should now be regarded as the socialized production of the group, not simply the aggregate subsistence production of individual households. A trend to a large-scale technology is especially evident where seasonal variation and storage are important. Not much is invested in productive capital by agricultural groups like the Yanomamo, the Tsembaga, and the Mae Enga.

Trade is less important as an agent of evolution at this level, except perhaps when a local group engages in trade with a more complex society, as the Kirghiz do. Trade between groups of roughly equal complexity is perhaps best understood as a subordinate aspect of risk management or alliance formation. Exchange is embedded in broader economic, social, and political relations of which commodity distribution is usually of secondary importance (Dalton 1977).

The Structural Perspective

In discussing the family-level economy we took special note of the comparative absence of public ceremonies. When anything like public ceremonial (as opposed to ritual, which may be quite elaborate but private and individualized) occurs among such groups as the !Kung, the Shoshone, or the Machiguenga, it lacks elaboration and occurs with little formal leadership or procedure.

Not so in the local group. Ceremonies remain few in the household and the hamlet, but at the group level ceremonialism is pervasive. Ceremonies are essential to sustaining the intensification of production, the integration of multifamily economic communities, and the control of the political economy by leaders, and hence are fundamental to the continued existence of the families in the community. Of particular importance, intensification and competition resulting from higher population densities create potentially disastrous resource depletion, as described for the Turkana, and the tragedy of war. The ceremonial cycles, actively manipulated by the new leaders, offer some possibility for group and regional management especially, as seen in the intergroup collectivity of Big Men.

The ceremony is the focal point of the political economy, where the behind-the-scenes work of accumulation and power brokerage comes to fruition. Here we see on display several structural features that are absent or little developed at the family level: the creation and reinforcement of formal ties beyond the family group; the public proclamation and validation of the group's superior political status; and the advertisement and exchange of property ownership by means of primitive valuables or money.

Because of the importance of public ceremonies in conferring prestige, especially in Big Man societies, some have termed the economy supporting these ceremonies a "prestige economy" (Herskovits 1952: 464-65). Yet since in these societies, as elsewhere, political and economic power reside not in wealth objects as such but in the control of access to resources, labor, and economic goods, the "prestige economy" and the "political economy" are in fact one and the same. An individual and his support group gain prestige in a competitive intergroup ceremony to the degree that the leader can demonstrate his ability to mobilize his supporters to provide him with goods, labor, and military support. His group's ceremonial performance proclaims both its economic and military might and its leader's ability to commit its resources in intergroup enterprises. In effect, prestige is latent power, or the promise of power.

A central feature of ceremonial practice in all local groups is the public affirmation of the group as a body. Family-level societies have little need for such affirmation, since the family's interdependence is evident on a daily basis and is rooted in primary kinship ties. It is different when the interdependence is between fifty or a hundred families who may not know each other intimately or like each other very well, and who may not be disposed to submit gracefully to the sacrifices group life entails. The ceremony not only provides the opportunity to heal factional wounds through dances, contests, and feasts, but focuses the attention of all on the material exchanges that are its real, integrating core.

Structurally the formation of clans and lineages distinguishes the organization of the local group from the simpler family-level society. Clans and lineages are kin-based corporate units; they own land and other resources together and restrict outsiders' access to their estate. In a world without overarching regional legal institutions to guarantee land use rights, the clan proclaims the legitimacy of its claim, often basing it on a strong religious or ancestral bond. As we have argued in the Tsembaga and Central Enga cases, the clan develops in response to increasing competition over scarce resources; in essence this competition makes clear the utility of a fair-sized landholding group based on biologically rooted principles of kinship and reciprocity. Sometimes, as with the Central Enga, the clan is the main defensive unit and politically autonomous; sometimes, as with the Tsembaga, clans or clan segments form local group associations for mutual defense.

The leader of the local group represents his group at the intergroup ceremony, which other members of the group may also attend. He compels their material support by reminding them that he acts "in their name." His strength is their strength, and can be translated into whatever the group needs to defend its interests in a competitive environment: food, technology, and access to resources in times of scarcity, prestige and valuables in time of abundance, and a reputation for strength and valor in times of war. Hence, unlike the !Kung, who admire the self-effacing meekness of a cooperative and generous compatriot and who joke that "we are all headmen," the Big Men who integrate local groups into intergroup collectivities proclaim their eminence to all who will listen.

Where the intergroup collectivity is less tightly integrated, as among the Turkana and to an extent the Yanomamo, the headman has less real power than a Big Man and must tread more warily to keep his followers' support. In fact each hamlet-size group usually has its own headman, who acts in a semiautonomous manner to motivate group action. This kind of leadership is more appropriate to the comparatively intimate and egalitarian village-level group than to the large-scale intergroup collectivity.

The public proclamation of ownership, and of the transfer of ownership, also appears to be more intense in a regional collectivity. On the Northwest Coast and in New Guinea a major aspect of ceremonial behavior is the public display of crests, emblems, tokens, valuables, and property markers. What we see here is an early form of legalization of rights to nonlocal resources that are needed to reduce people's vulnerability to sudden local shortfalls. People in family-level societies respond to such shortfalls by shifting residence; in the more densely settled local group this option rarely exists.

The Economic Perspective

The economic orientation depends on some definition of individual self-interest as the motivation of behavior. Family self-interest is an easy extension of individual self-interest; but since no persuasive analogy between "group interest" and "individual interest" can be sustained beyond the family level, group-level economic processes are best explained in terms of the self-interest of individuals and their immediate families.

As we have seen, greater population density is one mark of the local group as distinguished from the family-level society. Other things being equal, the local group is more efficient; it knows how to exploit the land to support greater numbers of people. Yet this increase in "efficiency" does not necessarily preserve or improve the quality of life of the individual, as "economizing behavior" does in theory. On the contrary, the greater competitiveness, regulation, and violence that characterize the local group and the intergroup collectivity make the individual's life decidedly more tense. What we observe is the outcome of two contradictory forms of self-interest in operation: one, the biological imperative to reproduce and leave one's imprint on the gene pool, increasing population but also increasing the competition for resources; the other, the desire to preserve one's easy access to essential resources by reducing the level of competition from other groups in the environment, which means either excluding them, killing them, or drawing them into a mutually beneficial relationship.

Much of the complexity of the local group, as distinct from the family-level society, is attributable to efforts by individuals (and families) to defend their access to resources in the face of greater population pressure. Investing more labor in capital improvements is easily accounted for as a rational effort to get more food from a restricted territory. The importance of ceremonies, many of them altogether foreign to our experience, is too often overlooked. In the local groups that we have covered, we have come to perceive (and respect) ceremonies as an efficient mechanism for arranging the sharing of risks, allocating rather than wasting surpluses, and sensibly assigning the authority to manage the group's resources.

As in family-level societies, individual families still negotiate their participation in the system. They select both their friends and their Big Men according to their expectations of getting fair measure in return for what they give up. Often they are asked to give up more than they think is fair, and disputes break out that threaten the unity of the group; against such times each family has established ties to other communities in which life may be better. In societies in which people are formally associated with lineages, clans, age sets, ranks, and privileges, it is impressive what real freedom they have to shift residence, social affiliations, and political loyalties as may be necessary in the pursuit of their self-interest.

Yet despite the persisting strength of individual self-interest the local group and the intergroup collectivity represent a wholly new level of economic integration, an order of magnitude greater than the family level. The individual family in these systems cannot simply withdraw from the competitive matrix; there is nowhere for it to go. When the fish are not running or the herds are decimated or the enemy is on the rampage, the family that is on its own must surely perish.

Part III

The Regional Polity

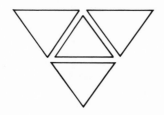

NINE

The Simple Chiefdom

THE EVOLUTION OF CHIEFDOMS is marked by distinctive changes in
the scale of society, in the organization of leadership and stratification, and in the political economy. The scale of society is the most
dramatic change. Chiefdoms are regional systems integrating several local groups within a single polity (Carneiro 1981). For the first
time the polity, defined as a group organized under a single ruling individual or council, extends beyond the village or local group. Often
the community associated with the ruling chief is unusually large in
comparison with nonstratified societies; however, the more dramatic
change is in the size of the population that is united politically.

Archaeologically, chiefdoms succeed the simpler community organization of early Neolithic society. With chiefdoms we see the beginnings of truly large-scale constructions, such as the mound groups of
the Olmec (Bernal 1969; Earle 1976) and of the Mississippian (B. Smith
1978), the ziggurats of the Ubaid (Wright 1984), and the henges and
cursus of the Wessex chiefdoms (Renfrew 1973). These impressive
early monuments testify unambiguously both to the central organization of a large labor force and to the function of a site as a regional
ceremonial and political center. To judge from the ethnographic examples discussed in this chapter and the next, chiefdoms range in
size from the low thousands to the tens of thousands, making them
an order of magnitude larger than simpler polities (cf. Feinman and
Neitzel 1984).

A polity of this size requires hierarchical integration, and we accordingly find a ruling aristocracy, with highly generalized leadership
roles in social, political, and religious affairs, whose members occupy
a hierarchy of offices at both the regional and local community levels.
Looking ahead, this new development, unlike anything we have seen

in simpler, decentralized societies, foreshadows the differentiated and internally specialized regional bureaucracies of the state (Chapter 11).

Within the chiefdom the regional organization is based on an elite class of chiefs, often considered descendants of the gods, who are socially separated and ritually marked. The organization is explicitly conceived as a kin-based community-like organization expanded into a regional governing body. The chiefs are related to each other through descent and marriage, and the idioms of kinship and personal bonds remain central in the political operation of the chiefdom. The tie between the developing economic system and developing social stratification is plain to all, and the chiefs come to dominate the economy as well as the social and political realm.

The evolution of chiefdoms and, later, states depends on the development of systems of finance used to mobilize resources to pay for the operation of new chiefly and state institutions. For purposes of our present analysis, we may dichotomize systems of finance into different forms: staple finance and wealth finance. *Staple finance* (Earle and D'Altroy 1982; D'Altroy and Earle 1985) is a form of what earlier writers (Polanyi 1957) have called "redistribution," a system in which staple foods and craft goods are collected from individual households as a kind of tax or rent. The subsistence products are then distributed directly to those working for the chiefdom (or state) to be used for their support. Such a system, as seen clearly in the Hawaiian and later Inka cases, act in the absence of extensive market exchange.

Wealth finance (D'Altroy and Earle 1985; Brumfiel and Earle 1986) involves the controlled production and distribution of primitive valuables, such as the *kula* valuables of the Trobriand Islanders. These valuables are critically important in establishing a person's social position, as in bridewealth payments, and in gaining personal prestige and associated political office. Valuables are marks of social status that define an individual's political and economic rights in a society. By channeling the distribution of the valuables, ranking chiefs used them almost as a political currency. The development of true currency systems is apparently tied to the later developments of states and associated market systems (cf. Brumfiel 1980).

It is not immediately clear, though, why hierarchically organized, large-scale societies should evolve (see Earle 1978). Why do individuals and local communities give up their autonomy and submit to the demands of a regional ruling elite? Plainly the ruling elites benefit; they have an improved living standard, greater reproductive success, and power to direct human affairs. But what is in it for the commoner?

To answer this question we must look at the two sides of the chiefly contract: service and control.

As we have seen, the services provided by Big Men to local groups include the management of large-scale subsistence activities, the conduct of long-distance trade, the storage of food and wealth objects, and the maintenance of alliances through debt-credit relations. At the regional polity level chiefs provide the analogous services, whose nature varies with the form taken by intensification in various environments. Less variable is the nature of control that rests on *ownership* of critical productive resources, technologies, and religious power.

In the simplest terms a chiefdom is a stratified society based on unequal access to the means of production. This point, stressed by Fried (1967), is essential for understanding the differences between chiefdoms and simpler societies. A chief's control translates into an ability to manipulate the economy in such a way as to derive an investible surplus from it. He is granted the power to control or monopolize economic management under certain specific conditions deriving from the same factors that we have identified as requiring individual families to group together: risk management, technology, warfare, and trade. As population increases, there comes a time when the local group or intergroup collectivity can no longer be relied on to handle these life-and-death matters. Let us now examine the specific conditions that lead to the acquiring of economic control by a regional chief.

We begin with risk management. As we have seen, when population density is sufficiently high to put a population at risk during a shortfall, risk management requires the production of a surplus and, usually, storage. Under some conditions the surplus and its storage can be handled by individual families; when it is centralized under a leader's control, however, risk is averaged across more subsistence producers and the necessary per capita surplus (which we may think of as a kind of insurance premium) is less. The stored surplus also provides the chief with the means to invest in other political and economic ways, on his own behalf or his polity's. Another environmental source of risk arises from the intensification of shifting cultivation in fragile environments; in such an environment the agricultural cycle must be managed by the chief lest it be mismanaged by his people. By reallocating land to subsistence producers annually, as in the Trobriand Islands, a chief is in a strong position to control his group's economy.

Another opportunity for control derives from the ownership of the technology needed to intensify production and to offset rising costs. Irrigation, in particular, can dramatically increase agricultural output and significantly reduce overall costs to the household. The high initial (or fixed) costs of an irrigation project are beyond the means of commoners, and the geographically concentrated pattern of land use makes control comparatively easy. In the Hawaiian case (Chapter 10) irrigation was developed by the chiefs, and rights to cultivate on irrigated subsistence plots were then exchanged for labor on chiefly plots. Irrigation, as we will argue, has a very special relationship to the evolution of social stratification because it permits intensified production at comparatively low average cost.

Other forms of intensified agricultural technology include terracing and field drainage. Certain forms of fishing technology, such as large canoes and fishing weirs, present similar but more limited opportunities for capital improvement and control. With intensification the percentage of total land to which the most cost-effective technique may be applied becomes small enough to be easily defended by a ruling elite with a military capability. Flannery (1969: Table 5) has attempted to diagram the relationships between population density, subsistence intensification, and the percentage of total land surface to which the "most productive" technique can be applied. In 7,000 years of Iran's prehistory, the increase in population density from nearly zero to about 17 persons per square mile correlates with a shift from foraging to irrigated farming and with a decrease of "most productive" land from perhaps 35 percent to 2 percent of the total land surface.

Yet another opportunity for control derives from warfare, a major consideration in our discussion of the evolution of local corporate groups and Big Men. In high-density, nonstratified societies independent local groups are constantly fighting to displace each other from the best lands, and leaders become essential for defense. In chiefdoms warfare continues but involves either a chief's attempt to extend his chiefdom regionally or a would-be chief's attempt to take power in his own region. The goal is less to control land or resources than to control populations and thus their surplus production.

As we have seen, warfare alone does not account for the evolution of stratified societies, even in situations of unusually high population density. The control of warfare as a basis for the regional development of chiefdoms depends either on technological superiority, such as the bronze weaponry of European chiefdoms or the iron weapons of the African states (cf. Goody 1971), or on circumscription, as in the case of the Polynesian chiefdoms, that effectively limits the options of a conquered population (Carneiro 1970b).

The fourth opportunity for control derives from the growing need for large-scale trade, trade that cannot be conducted by individual commoners because it requires politically negotiated agreements, or an elaborate trading technology such as long-distance canoes, or both. As population density grows, external trade may be necessary especially between pastoralists and settled agriculturalists, to obtain inexpensive food products. The development of specialized pastoralists has been called the "secondary products revolution" (Sherratt 1981), as herders provide costly protein and plow animals in return for agricultural products, especially starchy staples. As we have seen with the Kirghiz, this symbiosis appears to develop on the peripheries of state societies with agricultural core populations; the continuing (because fundamental) need for exchange on both sides creates an opportunity for economic control by the elite of the pastoral society, who negotiate legal and economic contracts between the two societies. External trade may also be used for risk management, as discussed below for the Trobriand Islanders; to acquire valuable raw materials, as in the case of the Maya civilization (Rathje 1971); and to build alliance networks. In each case the need for trade provides an opportunity for control.

As we turn now to examine how different chiefdoms are related to these four lines of development, we emphasize that chiefdoms come in many sizes and shapes (see Feinman and Neitzel 1984 for a discussion of this point). The size of a chiefly polity (i.e. a chiefdom) can range greatly in size, from below 1,000 as in the Trobriand Islands chiefdoms to more than 50,000 as in the Hawaiian chiefdoms. At the lower end these chiefdoms are very similar to Big Man systems; at the upper end they approximate states. It is best to conceive of chiefdoms as a continuum in the evolution of social stratification and governing institutions. Generally, as size increases, the economic and social distinctions between rulers and followers widen, and the positions of leadership become increasingly hierarchical in their arrangement. At the simpler end of this spectrum, chiefs provide only a limited number of services and these services vary considerably from case to case; at the upper end chiefs provide a full range of services. More important, the evolution of chiefdoms depends on expanding the scale of institutional integration, and this is possible only with increasingly tight economic and political control. The nature of this control will be particularly clear in the diversity of Polynesian chiefdoms discussed in Chapter 10.

Case 12. The Trobriand Islanders

The Trobriand Islands are a small group of flat coral islands that lie about 120 miles north of New Guinea's eastern tip. In contrast to the large islands of Melanesia, such as the New Hebrides or the Solomons, which provide near-continental environments, the Trobriands' small size, unvaried resources, and physical isolation seem to constrict their human population. As we shall see, however, trade by means of traditional sailing canoes effectively connects the economies of the island world off New Guinea and provides for both local survival and political finance.

Trobriand ethnography has a special place in anthropology because of the seminal field research of Malinowski (1922, 1935), which began soon after pacification. Later studies by Austen (1945), Powell (1960, 1969), and Weiner (1976, 1983) and important reanalyses of Uberoi (1962) and Burton (1975) make the Trobriands a critical case requiring careful consideration in any analysis of chiefdoms.

The present importance of the Trobriand case is for understanding the transition from a Big Man system to a chiefdom. Many of the characteristics of Big Man systems are present in the Trobriands, but hereditary ranking, institutionalized leadership, and some regional centralization suggest the chiefdoms of Polynesia. Why hereditary chiefs and not just Big Men?

The Environment and the Economy

The Trobriand group consists of one dominant island (Kiriwina, 70 square miles) and several other islands (totaling about 12 square miles). Kiriwina has little topographical relief, with about 60 percent of its land surface being low arable soils and the rest swamps and occasional jagged coral outcrops. Many resources, such as clay and stone, are not available on the islands. There are no streams, and fresh water comes from a subsurface lens. Vegetation consists of gardens, secondary brush, small stands of coconut and betel palms near villages, and small remnants of native vegetation. Malinowski notes that "little is left to nature and its spontaneous growth" (1935: 4). Except for swamp areas, the landscape is the product of human use.

The climate is warm and humid. Rainfall is seasonal, with most precipitation falling during the monsoons. Droughts, although not common, are severe and feared. When the monsoons fail, agricultural production fails and famine seizes the islands.

The population density of the Trobriands is quite high for a horticultural population. Powell (1960: 119) calculates it at about 100 persons per square mile, a figure he says has not changed significantly since the turn of the century. Figure 7 shows the island environment crowded with small villages. In general the population is concentrated near the cultivable land (about 70 percent of the total), where densities exceed 130 persons per square mile. From Malinowski's (1935) description one sees a crowded landscape transformed by human work.

The subsistence economy combines intensive agriculture and fishing. Foraging is restricted to small quantities of shellfish and crabs found along the shores and marshes. The bulk of the diet consists of root crops, especially yams and taro. The main crop is yams, planted in September and October and available for harvest starting in May and June. Fields are typically prepared new every other year, harvested twice, and then allowed to revert to a bush fallow for three to five years before being used again; their fertility is apparently maintained largely by adding ash to the soil. Following the initial cutting and drying, the vegetation is carefully burned; and then a planting pit is excavated, with care to remove all roots and stone, and filled with the loosened soil and a seed tuber. During the growth period of about eight months the yam crop is tended and weeded. Subsistence yams are then harvested as needed, and exchange yams are harvested and stored in special structures for about six months.

Because of the seasonality and fairly short storage life of yams, a lean period exists during which stored yams from an earlier planting and a mixture of other crops are important. Special taro gardens are common. The staggered plantings of yam and taro through the year provide a long harvest period and some security against unpredictable conditions that might destroy a single crop.

Security in the subsistence economy is paramount to the Trobrianders. The islands, without topographical relief to catch rains and without streams for irrigation, are at risk from periodic droughts. Stories of droughts and famine are commonplace, and food is prominently displayed on all ceremonial occasions—at death, at marriage, and at community dances. To have food gives people a sense of well-being, security, and pride. To lack food "is not only something to be dreaded, but something to be ashamed of" (Malinowski 1935: 82).

There are three main ways of dealing with the threat of food shortage. One is to spread food production through the year by staggering plantings. A second, and perhaps the most important, is systematic overproduction. Encouraged by chiefs and garden magicians and reinforced by the strong ethic of food accumulation as a measure of personal status, the household head routinely strives to produce more than enough food to meet his family's needs. Because good and bad years cannot be foreseen, this extra effort not only enables the household to get by in a bad year but results in large surpluses in average and good years. The implications of these surpluses for political finance will be discussed later.

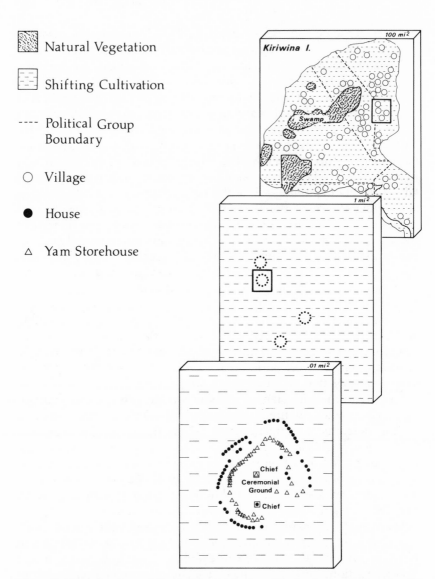

Natural Vegetation

Shifting Cultivation

- - - - Political Group
Boundary

○ Village

● House

△ Yam Storehouse

Kiriwina I.

100 mi²

Swamp

1 mi²

.01 mi²

Chief
Ceremonial
Ground △ △ △
△ △ △
Chief

Fig. 7. Settlement pattern of the Trobriand Islanders. The landscape has been totally transformed by intensive, shifting cultivation. Small villages cluster together and are often linked to a chief's village, where special ceremonies take place in the central dance ground. Trobriand villages generally contain 13–28 houses; the one depicted here, being a chief's village, is considerably larger.

The third way, though its effectiveness has been disputed (Powell 1969), is to distribute food between villages as part of the routine of structured gift exchanges and ceremonial distribution. Although such arrangements may be too limited to prevent shortages in the event of widespread crop failure, they probably provide some averaging effect and in the long run cut down the required surplus for the whole system.

The overall intensity of horticultural production required by the population density, by security needs, and (as we shall see) by status rivalry is seen clearly in the planning and regulation of the gardening cycle. Cultivation is done by a village, involving ten or more families. The hamlet head or chief first meets with his garden magician to decide on the location of the large garden section and to allocate plots within the section to individual households. The section is then cleared, burned, fenced, and planted by the village's males, usually in an organized and concerted effort. The main steps of garden preparation, plant tending, and harvesting are overseen by the garden magician, who carefully evaluates the participants' work and encourages them to exert themselves. In a prominent position in the garden section are "standard plots" (*legwota*) cultivated by distinguished community members, which serve both as the focus for garden ritual and as exemplars for all plots in the section.

Whereas in simpler societies subsistence production is largely a household concern, in the Trobriands significant gardening decisions are taken away from the household and centralized in the ritual specialist. How this arrangement relates to chiefly management and control will be discussed below.

Warfare, although apparently less intense than among the Enga, certainly exists in the Trobriands. Local groups fight each other at least occasionally, especially during famines but also, as we shall see, for explicit political purposes.

The intensification of cultivation has also led to trade in subsistence products. Soils on the islands are a thin layer over the coral, and a soil's fertility depends on its thickness and development. Some areas, especially northern Kiriwina, are considerably more fertile than others. Productivity on the best soils is nearly double that for average soils and four times that of poor soils (Austen 1945: 18). Communities in highly productive areas concentrate on agricultural production; people in communities with more marginal lands are more likely to specialize in fishing or craft activities, trading their products for staples.

Fishing is the most important specialized activity. Except in the best agricultural areas villages tend to be located on the coast, where both inshore fishing in the shallow lagoons and offshore fishing are practiced. Some fish are traded for yams and other produce. The development of this local exchange, sometimes referred to as "fish-and-chips reciprocity," is decentralized and unorganized by chiefs; coastal and interior people trade on an individual basis, using traditional exchange rates. Of course chiefs are indirectly important to trade, since they maintain the intercommunity peace that makes trade possible. Other specialties are also important, notably the manufacture of polished stone axes, baskets, various carved wood objects, and lime for betel chewing.

Trade with other island populations is critically important to the Trobriand economy. Certain goods such as clay pots and stone for axes are not available on the coral islands of the Trobriands. These imported products are economically significant; stone axes, in particular, are essential to efficient horticultural production. Traded food products, such as sago from Dobu, also add variety to the diet and provide a critical food source in years of severe need. The surplus yams available in good years can be used to obtain valuables in the *kula* exchange (described below), and in bad years the valuables can be exchanged, directly or indirectly, for badly needed food.

To summarize, the Trobriand Islands population faces four critical economic problems resulting from intensive production on small coral islands: a high risk of food shortages; intergroup warfare; considerable local variability in subsistence production, necessitating internal exchange; and a pressing need for external trade to obtain food and manufactured products not available locally.

Social Organization

Settlement Pattern. The village of a ranking chief presents the salient unit for analysis, with its characteristic arrangement of household and storage structures (Fig. 7). The village arranges private and public spaces in a way that mirrors the division and integration of the subsistence and political economies. Private space, circling the village, contains the residences and small storage structures of the member households. The household, with its separate residence, enclosed storage building, and work space, is the focus for the domestic economy. The yams from a family's garden plot are harvested as needed and not stored (Weiner 1976); yams received in obligatory exchanges, however, are placed in an enclosed storage structure.

In the center is the public, ceremonial space, where the dance ground, the chief's display storage structures, and the chief's residence are located. On the dance ground are performed ceremonies that define the group's corporate character and display its economic well-being to outsiders. The large central storage structures are constructed with open spaces between side-wall logs to permit viewers to see the concentrated wealth of the chief and his support base. These stores are used to finance such chiefly activities as the hosting of village ceremonies and the construction of trading canoes. The chief's own house, similar to other houses but larger in size, is at the edge of the central dance ground, where it appears to dominate the group activities of the village.

Simpler settlements, without ranking chiefs, have no central public area except for a simple dance ground. Settlements often form clusters, with a chiefly village dominating. On a more regional scale a few chiefs have come to control larger areas, and their villages are the most elaborated. The settlement system is thus hierarchically organized, with larger public dance grounds, display storage structures, and elite residences found at the political centers.

The Family and the Dala. The household is the basic economic unit of subsistence production and consumption. Average household size is only 3.2 (Powell 1960: 119), primarily organized as a nuclear family with a husband, his wife, and those of their unmarried children who have not moved into village bachelor houses. Each household owns its separate house and storage structures and has a separate garden plot in which it grows food for itself and surplus yams for exchange.

The main division of labor is by sex (Malinowski 1929: 24-27). Males do the heavy agricultural activities including clearing, fencing, and planting; they are also the main traders and the specialists in canoe construction and wood carving. Females do the gardening, especially the weeding, collect shellfish, prepare food, care for children, and produce goods that include mats, banana leaf bundles, and skirts made from these bundles (Weiner 1976). By and large, food procurement activities are male-dominated, food preparation female-dominated, and craft work shared but differentiated into male and female crafts. Where communal labor is not involved, the household organizes the basic division of labor in production (Malinowski 1935: 355).

Beyond the household the most important unit is the small village, a small residential population of about 65 persons who typically represent a *dala* (Weiner 1976). The dala is a corporate group that owns a territory used for farming; membership is matrilineal, but residence is virilocal and somewhat complicated. Among those who must reside with the dala's head, or "manager" (Weiner 1976), is his oldest sister's oldest son, who is next in line for hamlet leader. The village also includes some non-dala members who receive land from the hamlet head, notably his own sons. The small village is therefore a composite group of matrilineal kinsmen and supporters and their families.

The village is important both economically and politically. Economically, as we have seen, it organizes and manages food-planting activities. Joined to this managed cycle of horticulture is a land tenure system based on group ownership, with rights of allocation vested in the leader (see particularly Weiner 1976). The dala owns the land, but the dala leader, by controlling its annual allocation, controls effective access to it. A household can obtain land only from the leader, who has considerable freedom in allocating land to non-dala members. This link between the control of land tenure and the developing political economy foreshadows the economic basis of the more institutionalized Polynesian chiefdoms. The small village, rather than a simple kin group, has become a flexible political support group.

In addition the village is organized ritually by its leader. As we have seen, he may designate another villager as his garden magician, but he (the leader) is the "owner" of magic, which is especially important in gardening, and the initiator of ceremonies at the dance ground.

The Local Group. Two to six small villages form a local group or village cluster of some 300 people. This group is highly endogamous; prior to pacification, warfare was prohibited between the constituent hamlets. Intermarriage between hamlets binds the village cluster into a social unit interconnected by many affinal exchanges, notably the annual yam exchanges. Each farmer farms several yam plots, some for his family's support and at least one for exchange.

When a man's daughter or sister marries, a large payment of yams must be given annually to his son-in-law or brother-in-law. Malinowski (1935) analyzed this payment as compensation to the woman for her rights in the subclan territory, which she gives up when she joins her husband's household; Weiner (1976), as we shall see, has a different hypothesis. Whatever the explanation, the pattern of endogamy and affinal exchanges results in high economic interdependence within the cluster. Although these exchanges are not over a large enough region to buffer the group against major economic disruption, they are helpful in the event of local crop failures or the temporary incapacity of a household's work force.

The most important role of the local group cluster is political. The several dala or small villages composing the cluster are ranked socially with respect to each other, and the leader of the highest-ranked dala serves as cluster leader. This is apparently a true office, with explicit leadership responsibilities for coordinating group activities in ceremonial and group defense; although it may not always be filled, the expectation is always to fill it in due course with a deserving candidate. The cluster leader seems to be financed largely by marrying women from the various dala and thus obligating his male in-laws to provide a huge quantity of yams, which are then stored in display yam-houses and used to support ceremonial occasions. By manipulating marriage and exchange ties, a chief can bring a support group into what Malinowski (1935) calls a tributary relationship.

As with the Tsembaga or the Central Enga, one justification for this local territorial group may have been defense. Prior to British pacification warfare was endemic on the Trobriand Islands. The cluster, although itself not a corporate group, was organized as a defensive unit; warfare was prohibited within the cluster and mutual defense required. In the developing political economy warfare between politically powerful chiefs served to establish and maintain a cluster's privileged position. For example, in 1885 the ruling chief of Omarakana made war on a neighboring chief who had refused to give him a wife and by implication accept a tributary relationship; the Omarakana chief won and laid waste the defeated chief's villages (Powell 1960). As we shall argue for the Hawaiian case in Chapter 10, warfare in chiefdoms becomes transformed from simple competition over land to competition over rule and its implied control of land *and* labor.

Regional Relations and the Chiefdom. The importance of political competition between chiefs for regional control of groups helps to distinguish the Trobriand case from the Big Man systems described in Chapter 7. A high-ranking chief can extend his economic support base and area of political control by marrying women from other village clusters, receiving in turn a flow of yams amounting almost to a tribute payment. According to Malinowski (1935), both the permitted number of wives and the size of yam payments depend on the rank of a woman's husband. If the husband is a high-ranking chief, the yam payment is considerably higher and all male members of the woman's dala are required to provide yams. By marrying many females from many dala over a broad region, a high-ranking chief becomes the center of an extensive system of mobilization. The potential extent of this system is illustrated by the powerful chief of Omarakana in the 1930's, who Malinowski (1935) claims had some 80 wives!

Weiner (1976) notes that the flow of yams to chiefs is balanced by important reciprocal flows of goods, and notably by the large distributions of "female wealth," skirts and banana leaf bundles, at the mortuary ceremonies of the dala of a yam-giving male. In essence, a male receiving yams in his wife's name is obligated to purchase female wealth for her to distribute at these ceremonies. Her distribution of wealth at the mortuary ceremony is then a measure of his renown and reliability in a broader system of ceremonial exchange and display.

Sahlins (1963) found two main points of contrast between the ideal types of the Melanesian Big Man system and the Polynesian chiefdom: the size of the polity and the nature of leadership. The Big Man polity is characteristically small (a few hundred); larger units tend simply to segment into independent factions. Leadership is based on a man's personal demonstration of ability in the course of representing his support group in competitive displays (as in the Enga and Northwest Coast examples, Chapter 7). A chiefdom is characteristically larger, and is achieved by organizing local communities in a regional hierarchy based on the inherited rank of their respective leaders. Positions of leadership constitute *offices* with explicit attached rights and obligations. Chiefs thus "come to power" that is vested in an office, rather than building up power, as Big Men do, by amassing a personal following. Social status in chiefdoms is inherited, based on an individual's genealogical position in a social hierarchy, and access to power through office is accordingly confined to specific elite personages.

Leadership among the Trobrianders presents a transitional form between the New Guinean Big Man and the Polynesian chief (Powell 1960). Both locally and regionally social status is based on the established rank of a person's dala, which is itself dichotomized into elite and commoner subgroups. Only a man born in a high-ranked dala can accede to power. There is a hierarchy of officers. The leader of a village's highest-ranked dala (if there is more than one dala) is the village leader; the village leader from the highest-ranked dala of a cluster is the cluster leader. A cluster leader from one of the highest-ranked dala in the region can then use the privileges of his rank to acquire multiple wives and extend his power base regionally to form a support group of up to several thousand members. This pattern of inherited status, established political offices, and regional integration identifies Trobriand society as a chiefdom, but elements of a Big-Man-like system continue to operate, particularly at the fringe of the chief's power.

External Relationships and the Kula. Beyond the cluster level a chief's status is based not only on the rank of his dala, but also on his successful participation in highly politicized ceremonial events, notably competitive yam harvests and kula voyages. As we have seen, a high-ranking chief, married into many local subclans, collects exchange yams from a broad support region; each presentation of yams is accompanied by a ceremonial welcome of the presenter and the display of the yams in massive piles before they are stored away in the chief's yam-houses. Yams are both a direct measure of the productive might of a chief's support group and the main capital with which to finance his future political moves. By displaying them in this fashion, in contrast to the private, enclosed storage houses of the commoners, the chief asserts his power.

Kula voyages are made for purposes of ceremonial exchange between the Trobrianders and other island peoples. In the Trobriands they are organized by a high-ranking chief and involve the obligatory participation of all canoes belonging to chiefs in his kula district. Following a preliminary amassing of valuables and other goods, the canoes set sail, stopping first at a small island where the initiating chief ceremonially distributes food to the participants. On the next day the canoes proceed to the island where the exchange is to take place.

The kula is a well-described traditional system of exchange (Malinowski 1922; Belshaw 1955; Leach and Leach 1983). The islands involved, covering a relatively large extent of ocean off eastern New Guinea (roughly 210 miles from north to south and 270 from east to west), exchange many valuables and utilitarian goods. The most important valuables in Malinowski's time were shell necklaces (*soulava* or *bagi*) and conus shell armband pairs (*mwali*). The two valuables circulated in exchange for each other and in opposite directions: the soulava clockwise and the mwali counterclockwise (Malinowski 1922: Map V). The utilitarian objects included pottery and carved bowls, raw materials such as stone for axes, and agricultural products.

The valuables of the kula underlie political power in the Trobriands. Severely limited in their exchangeability, they form a separate *sphere of exchange* (cf. Bohannan 1955) normally limited to special social occasions. Since most valuables are controlled by chiefs, most other people can obtain valuables initially only from a chief. As argued elsewhere (Earle 1982), the restrictions on the flow of wealth appear to be tied to its role in competitive display and political maneuvering. In particular, the ease with which valuables are exchanged for subsistence goods in nonstratified societies (cf. the Tsembaga case, Chapter 6) is *not* found in the Trobriands.

In a typical large kula voyage a flotilla of canoes from the Trobriands arrives at an island such as Dobu, where they line up according to social rank and are ceremonially greeted by the Dobuans. The Trobrianders then disperse among the Dobuan hamlets to meet their trade partners. In some cases a partner has earlier received a gift from a Trobriander and must now reciprocate with an equivalent valuable; in other cases the Trobriander approaches a partner and solicits a desired valuable with gifts of food or craft products. The Dobuan *may* then give over the valuable with the expectation of a return gift on his next voyage to the Trobriands.

During the actual handing over of valuables a strict decorum obtains; but a show is made of disparaging the quality of a gift received and exaggerating the quality of the gift presented, with a view to increasing the stature of an individual or a group by imputing greater worth to its valuables than to its trading partners'. At the same time as these exchanges of valuables, goods from the different islands are exchanged by barter. Thus kula voyaging creates what is in essence a market, in which people from different regions exchange locally specialized food and craft products with all comers, negotiating exchange equivalence by bargaining.

When their business has been done, the Trobrianders set sail for home, often stopping at several islands on their return. Before landing in the Trobriands the flotilla again stops at the small island, and a special comparative display of the valuables takes place. As described by Malinowski (1922: 375): "From each canoe, a mat or two are spread on the sand beach, and the men put their necklaces on the mat. Thus a long row of valuables lies on the beach, and the members of the expedition walk up and down, admire, and count them. The chiefs would, of course, have always the greatest haul, more especially the one who has been the toli' uvalaku on that expedition." This display is a direct measure of individual success in the kula, and after the voyage word of personal accomplishments and disappointments quickly spreads through the communities.

Competition and display are integral to the political maneuverings of individuals, especially chiefs. By encouraging production and manipulating exchange a chief publicly demonstrates his political prowess and the economic capability of his support group. Success in both production and exchange depends on initiative and manipulation by all participants. In the kula, for example, although valuables travel in prescribed directions, much care and discernment goes into selecting the specific recipient of a valuable from among all those who desire it. In giving his valuables and solicitory gifts a chief calculates the prospective return both in future valuables as such and in increased status for himself and his group.

Although status is ascribed to a leader according to his dala affiliation, his renown can be either augmented or tarnished by his successes and failures in the public display ceremonies; indeed these successes and failures may alter the ranking of dala itself (Uberoi 1962).

There is thus a continuing adjustment of political and social position in Trobriand society as a result of competition among chiefs.

Conclusions

Why did the incipient stratification and institutionalization of political hierarchies evolve in the Trobriands and not in seemingly similar societies? For two reasons, deriving respectively from the political economy and the subsistence economy. First, in the political economy, the social differentiation inherent in the institutionalized leadership on the Trobriands is underwritten by differential access to the means of production and distribution.

External trade, as we have seen, is essential to both the political economy and the subsistence economy, and chiefs are able to monopolize this trade by their ownership of seaworthy sailing canoes (Burton 1975). These trading canoes are technically complex, consisting of a large hollowed-out trunk, a free board, frame and outrigger, a mast, and a pandanus leaf sail; they are 30-35 feet long and capable of carrying a dozen men and heavy loads of goods. Making a trading canoe requires the careful attention of a specialist and considerable manual and ritual labor, and only ranking chiefs, with access to yams and valuables, can afford the expense. Thus the control over production and exchange, made possible largely by control over capital, has led to social stratification and a self-perpetuating elite.

Yet as Malinowski (1935) was quick to recognize, chiefs are equally indispensable in the daily lives of the Trobrianders. Small islands characteristically are ecologically unstable and poor in resources. As a risk management strategy, Trobriand chiefs act as "tribal bankers," investing the surplus made available in a normal year or a good year in capital equipment such as canoes, in foreign trade for nonlocal materials and craft goods, in the political ceremonies that determine individual and group status, and in wealth valuables. In a bad year, when there is no surplus, the chief's management of production guarantees a sufficiency for subsistence needs. Chiefs also, by establishing and maintaining foreign trade relationships through the kula exchange system, provide access to markets essential for the smooth operation of the local economy: markets where in good years surplus food can be traded for a wide range of products and in bad years valuables can be exchanged for food.

The power and elite status of the Trobriand chief depends on the centralization and control of the economy. As we have seen, this control stems in part from the requirements of long-distance exchange and in part from the requirements of risk management. Once control was in their hands, chiefs extended it to include monopolies over the production of certain key resources generally desired by the population, among them coconut (important especially for oils that are scarce in the diet), betel nut (chewed as a stimulant), pigs (a major source of protein and fat), and ground stone axes (important for garden clearing). Coconut and betel palms and pigs were apparently owned exclusively by chiefs (Malinowski 1935; Austen 1945), and axes, made from imported stone, were ground by specialists working for chiefs (Malinowski 1935).

In simplest terms, the subsistence economy of the fragile and isolated Trobriand Islands could not be successfully intensified without leadership to manage the production cycle and external exchange. Such conditions in and of themselves do not produce chiefs; rather the intensification process in certain situations offers possibilities for control. In the Trobriands these possibilities include the land tenure system, the storable surplus, and the capital technology of trade. It is by controlling such elements of the subsistence economy that a chiefdom comes into being and perpetuates itself.

TEN

The Complex Chiefdom

THE TROBRIANDS REPRESENT comparatively simple chiefdoms constructed on the structure and ideology of the kin group (*dala*) and its affinal relationships. In this chapter we examine the more complex chiefdoms of Hawaii and the special case of the Basseri of Iran. Since Polynesia encompasses the full range of chiefdoms from simple to complex, from polities of several hundred people to one of 100,000, it is worth discussing the Polynesians generally before examining in detail the unusual, complex chiefdoms of Hawaii, which represent the fullest extent of Polynesia's evolutionary development.

The scattered islands of Polynesia extend from Tonga and Samoa to Easter Island and from New Zealand to the Hawaiian chain. Across this immense Pacific area are found clusters of islands, some of them 500 miles distant from their nearest major island group. These islands vary in size from the large land mass of New Zealand (102,000 square miles) to the tiny coral islets of the Tuamotus just south of the Equator, and in climate from temperate to tropical. The larger island groups, such as the Hawaiian Islands and the Society Islands, are dominated by chains of volcanic mountain peaks; they vary in size from massive young islands such as Hawaii (4,038 square miles) to small, eroded remnants and coral atolls.

Prior to European contact these isolated lands were colonized and inhabited by the Polynesians, and this uniform origin made for considerable similarities from island to island in language, material culture, subsistence practices, and other traits. For our purpose the most important element of Polynesian cultures is their sociopolitical organization as chiefdoms. Differences between these oceanic polities will help us to understand the processes involved in the development of chiefdoms to the very threshold of statehood.

Fundamental to the organization of Polynesian chiefdoms was the principle of social inequality based on inherited status. Each chiefdom was composed of a conical clan with embedded lineages (Fig. 8). The senior line, shown in thick black, was ideally represented by the first son of the first son of the first son, etc. Junior lines were founded by second or third sons, whose descendants formed the lineages of the clan (cf. Kirchhoff 1955). The rank of a man and his descent group was based on his birth order within the household: the second son's descent line was inferior in rank to the first son's, and so on.

In theory, at least, each person and each lineage in the system had a unique rank based on distance from the senior line; the closer to the senior line, the higher the rank (as shown by the numbers in Fig. 8). In practice rank tended to be restricted to a chiefly group composed of the senior line; junior lines rarely troubled to calculate rank distinctions. Although chiefly titles were usually inherited patrilineally, a line was not exogamous and membership was characteristically cognatic.

Many a junior line, dissatisfied with its inferior rank in the clan, set itself up as an independent local chiefdom. Such chiefdoms competed fiercely for the control of land and commoner populations, and rank became determined by genealogical position and by political domination resulting from conquest warfare (Goldman 1970).

At each level in the social hierarchy of groups, leadership was exercised by the chiefly line of highest rank; a chief from a local group's senior line organized and directed group activities. When local groups were organized as regional entities, the highest-ranked local chief coordinated regional ceremonial cycles and military operations.

The Polynesian chief was both a sacred person, closely linked to ancestral gods and instrumental in ceremonials, and a secular leader, responsible for organizing military action, directing economic activities requiring capital investment, and adjudicating internal disputes. Since any activity requiring the group to act together was in the chief's domain, his secular and religious aspects were closely conjoined and mutually reinforcing.

In the Polynesian chiefdom explicit *offices* of leadership, positions marked by special status, existed whether or not they were filled at any moment. Each group in a hierarchy of nested groups (local community, district, island chiefdom) had such an office, with the offices of subgroups ranked according to status relationships. In Figure 8 numbers 1, 5, 9, and 13 are lineage heads; 5 is subservient to 1 as leader of a larger segment, and 13 subservient to 9; and all are subservient to 1, the paramount chief of the conical clan.

Broadly speaking, each office conveyed to its holder both the right

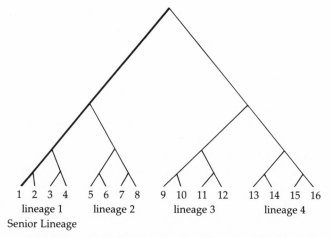

Fig. 8. Conical clan structure of a Polynesian chiefdom

to mobilize labor and goods as needed to support himself and his relatives and to fulfill his obligations, and the obligation to maintain order and productivity within the group. Officeholders were explicitly entrusted with performing ceremonies considered necessary to meet ritual obligations to the gods, maintaining a strong military posture, keeping internal order, and creating and maintaining productive facilities such as irrigation systems, terraced fields, and fishponds.

Polynesian chiefdoms were financed by "redistribution," an elementary form of taxation (Earle 1977). Goods were amassed by the chiefs from commoner production by right of office. Some of these goods were simply kept by the chief for his personal use, but most were used to pay people who worked under the chief's orders at various tasks designed both to meet his obligations as chief and to maintain his position of domination as chief and his imposing lifestyle.

In Polynesia the continuum from simple to complex chiefdoms was well represented (Sahlins 1958; Goldman 1970). Indeed Polynesian ethnography offers an excellent opportunity to consider how an overarching sociopolitical organization can be maintained despite inherent tendencies to fragmentation. As we have seen, political units often split along lineage lines, with the senior line of a lower-ranked lineage, e.g. number 9 in Figure 8, becoming a separate chiefdom in competition with the original senior line, as represented by number 1. What, then, were the conditions that discouraged fragmentation? For it must be these conditions that allowed for the growth in the size of the political system and the development of new institutions to solve the problem created by this new scale.

Well-documented examples of simple chiefdoms in Polynesia are the Marquesans (Handy 1923), the Tongarevans (P. Buck 1932), and the northern New Zealand Maori (Best 1924, 1925; Firth 1929). All these polities were small, typically around 1,000 people, and in a constant state of war with neighboring polities, apparently for the control of land. Population density varied from one to 300 persons per square mile, island size from 15 to 102,000 square miles, and the subsistence economy from shifting agriculture to an intensive mixed economy. Where island size was large (as in the case of the New Zealand Maori), overall population density was low; where island size was small, population density was high. The simple chiefdoms were thus apparently prevented from becoming more complex either by low population density, itself an outcome of marginal agricultural conditions as in the Maori case; by terrain so rugged as to restrict communication, as in the Marquesan case; or by a very small total population, as in the Tongarevan case. In all cases intense competition between groups made leadership necessary, as in the New Guinea Big Man systems. But in all cases inadequate access to basic resources made it impossible for leaders to control the economy, and thus impossible to expand their control to regional populations.

Among Polynesia's typical chiefdoms are Mangaia (P. Buck 1934), Uvea (Burrows 1937), Futuna (Burrows 1936), and Mangareva (P. Buck 1938), each with a maximum population size in the low thousands; and Tahiti (Ellis 1853; Oliver 1974), with populations between 5,000 and 10,000 (cf. Goldman 1970; Sahlins 1958). Population density varied from 48 to 375 persons per square mile, land mass from 8 square miles to 56,000, and the subsistence economy from intensified shifting cultivation to a mixed economy. Here again the larger land masses have the lower population densities. The major factor permitting political elaboration on the smaller islands appears to have been the management or ownership of capital improvements, especially irrigation systems; but elaboration beyond the regional level was impossible for such small populations. On the larger islands, such as Tahiti, the comparatively small-scale technology offered only limited opportunities for the political control of productive capital.

Polynesia's most complex chiefdoms, those of Hawaii and Tonga, were large-scale societies consisting respectively of the large populations of volcanic islands and many separate populations on closely grouped coral islands. The Hawaiian Islands have a large land area (6,446 square miles), but in past times had a comparatively low population density of only about 39 persons per square mile. The Tongan group is much smaller (256 square miles), but had a much higher overall density group (117 persons per square mile). In Hawaii, be-cause individual islands are large and supported sizable populations prehistorically, a large chiefly polity could arise by expanding to in-corporate a whole island. The unusually fertile alluvial soils found along the coast offered an ideal opportunity to develop traditonal irri-gation technology and through it to control the subsistence economy. Indeed the low overall density of the Hawaiian Islands obscures the high density of human population on these coastal soils, which were a concentrated, highly productive resource held by the chiefs and farmed by their dependent commoners. Kirch (1984: Table 10) esti-mates prehistoric population density at 310 persons per square mile of arable land. In contrast, the Tongan islands depended on intensified shifting cultivation and silviculture (Kirch 1980). As in the Trobriands, political control rested on management of the cultivation cycle and on long-distance exchange with neighboring island groups, especially Fiji and Samoa, that were conquered by the expanding chiefdom (Kirch 1984).

Case 13. The Hawaii Islanders

Our main Polynesian case is the Hawaiian Islands. At contact (1778) the seven large islands had a population of 200,000 to 300,000 divided among four large competing chiefdoms. These chiefdoms, which are considered by researchers (Sahlins 1958, 1972; Goldman 1970; Hom-mon 1976; Earle 1978) to represent the highest stage of sociopolitical development in Polynesia, are examined here in an effort to ascertain the factors responsible for the evolution of chiefdoms to the threshold of state society.

The complexity of Hawaiian chiefdoms is seen in their social strati-fication and evolved regional institutions. The society was rigidly di-vided into two classes, commoners and chiefs. The commoners were the rural cultivators, fishermen, and craft producers. Their genealogies were shallow, seldom reckoned past the grandparents' generation, and ranking and interhousehold organization were informal and ad hoc. A number of related households might live close to each other and cooperate in joint economic ventures, but there is no evidence of a lineage or corporate structure for commoners.

By contrast, the chiefs were organized into various ruling descent groups associated with the major islands of Kauai, Oahu, Maui, and Hawaii. In theory, a person's rank, and by extension his rights to office and support by related chiefs, was determined by his distance from the senior line; this distance was not easily computed, however, because a person received status through both his mother and father, and ruling chiefs accordingly retained genealogical specialists to evaluate individual claims to rank and position. Marriage was important as a way of accruing status for offspring. An elite chief could marry only an elite woman; and a high-ranking chief was polygynous, both within his ruling line to solidify the political position of his children, and outside his line to build alliances to other lines. Competition for ruling positions was fierce, and marriage and descent were highly politicized and carefully planned.

Population estimates for the prehistoric Hawaiian chiefdoms range from about 30,000 (Kauai and Niihau) to about 100,000 (Hawaii). Typically a major island and its closely associated smaller islands were ruled as a single complex chiefdom. Although attempts were made to extend a chiefdom by the conquest of other major islands, these attempts usually failed because of the difficulties of controlling such a large and widely separated population.

The Environment and the Economy

The Hawaiian Islands consist of seven major islands located in the north central Pacific, just within the tropics at 20° north latitude. Isolated from all other major land masses by more than 2,000 miles, this chain of islands is composed of the peaks of a volcanic mountain range. The geological age of the islands, and thus the extent of their erosion, varies greatly. The large island of Hawaii still has active volcanoes, and its broad, sloping surfaces have few valleys and permanent streams; by contrast, Kauai is heavily eroded with deep canyons that carry water from the central mountains to the sea. Most soils are volcanic in origin, but the most productive soil is the alluvium found in the eroded valleys and along the coastal plains at the valleys' mouths. The amount of alluvial soil varies greatly, depending on the development of the stream systems (see Kirch 1977; Earle 1980b).

Rainfall is another important factor for environmental variation. At sea level the expected annual rainfall in this area of the Pacific is about 60 inches. The distribution of rain is uneven, however, owing to the islands' sizes. On the windward side of the islands rainfall typically varies from about 60 inches at the coast to 300 inches or more in the central mountains; on the leeward side rainfall is much reduced by a rain shadow and is often below 20 inches. Variation in rainfall and soils is a critical determinant of differences in local subsistence strategies.

As we have seen, population density for the seven islands at contact has been estimated at 39 persons per square mile. But much of the islands' landscape is steep and broken terrain, and much is too high or too dry for the tropical root crops grown by the Hawaiians. At the low elevations where most of the population was concentrated, usually within a mile of the coast and near streams, densities ran to over 300 persons per square mile.

As a result of these locally high population densities the subsistence economy depended primarily on intensive agriculture. Irrigation, terracing, and drainage systems were devised that permitted permanent, year-round cultivation. The dominant crop, taro (*Colocasia esculenta*), was grown in irrigated pondfields wherever available water and soil conditions permitted (Kirch 1977; Earle 1980b). Pondfields, similar to the rice paddies of southeast Asia, consisted of short ditches from the many island streams that led water to terraced fields, each of which was a small pond with taro planted in the mud or on small mounds. The technology was comparatively small-scale and easily managed, and yields were high. A major problem, however, was the concentration of these irrigation systems on the valley bottoms near the sea. A community's agricultural production was periodically wiped out by floods and tidal waves, which indeed continue to plague modern taro systems in the islands today (Earle 1978).

Where irrigation was impractical, short-fallow shifting cultivation was used for taro and in drier locations for sweet potatoes. Other crops included yams, sugarcane, arrowroot, and a number of tree crops, especially breadfruit and banana. Domesticated animals, among them pigs, dogs, and chickens, were also of some importance for protein. Pigs were particularly prized by the chiefs, who appear to have monopolized their husbandry.

Fish were the main source of protein. A variety of fishing techniques were employed, notably the *hukilau*, in which a large fishing party with nets encircled a school of fish in shallow water and dragged their catch to shore. Offshore fishing for pelagic fish, requiring special fishing equipment and large canoes, was also important. Along the shores and in the alluvial flats directly in back of the shore, fishponds were built in which small fry were raised especially for use by the chiefs (Kikuchi 1976). The technology of the ponds was simple, ranging from an enlarged taro pondfield to extensive areas enclosed by sea walls of rock and earth.

Other wild foods, although secondary, provided variety and additional protein. Shellfish and crabs were collected, seabirds were netted at offshore rookeries, feral chickens and pigs were hunted in the mountains, and many wild plant foods were collected. The island

interiors, with little permanent habitation, were an important foraging zone. All in all, the diet of the Hawaiians, although heavily dependent on starchy staple crops, was good on the major counts of calories, protein, and variety.

The varied subsistence economy was of course a response to the environmental diversity of the high volcanic islands of Polynesia. There
were three major zones in close proximity that permitted three very
different types of exploitation: the alluvial bottomlands and gently
sloping uplands for intensive agriculture; the shallow offshore bays
and reefs for productive fishing; and the interior "wild" forests for
hunting and foraging. While Service (1962) has argued that chiefdoms
frequently developed under such conditions to manage exchange
among specialized communities, in fact, as we shall see, trade was
remarkably limited and took place largely within community boundaries (Earle 1977).

Warfare would seem another likely outcome of the high population
densities and the uneven distribution of productive resources from
one locale to the next. As it happens, however, resource competition
between local communities was eliminated by the chiefdom, which
guaranteed ownership rights in land through a system of chiefly land
allocation; and warfare between chiefdoms appears to have been primarily a strategy of political competition.

As we will argue, most problems of production were handled at the
family level or the local community level, although chiefs were undeniably important for managing risk and adjudicating intercommunity
conflicts. More importantly, however, the particular form taken by intensification in Hawaii afforded opportunities for economic control
that formed the basis for the development of social stratification.

Social Organization

Wittfogel (1957: 241) has argued that the development of irrigation
systems in Hawaii required the development of a managerial system
of chiefs and land managers that then formed the basis of Hawaiian
political organization. Alternatively, Service (1962), discussing Polynesia in general, has argued that the environmental diversity of the
islands required a centrally managed exchange system whose managers rose to power as chiefs. As we have argued elsewhere (Earle 1977,
1978), these theories are inadequate because neither irrigation nor exchange posed problems requiring a control system extending beyond
the local community.

The independent nuclear family, organized along standard lines of
division of labor by age and sex, could easily provide most of the labor

needed in irrigation agriculture (Earle 1978). Because of year-round production in the taro fields there were few peak labor periods; and the only labor needs extending beyond the nuclear family would have involved the construction and maintenance of pondfields and their reconstruction following flooding. Irrigation systems were small and limited to a single local community; a system typically served only four or five farmers, and rarely more than twelve. Construction was apparently done by gradual extension, and reconstruction—as today—was handled by small work teams. The historical records suggest that closely related families, brothers and brothers-in-law, lived together along a short ditch and cooperated with each other in its maintenance. Although land managers certainly existed in Hawaii, they were not necessary for the irrigation systems.

As for exchange, the three major resource zones (inshore fishing, lowland farming, and upland foraging) were characteristically very close to each other; in many areas it is less than eight miles from the coast to the mountain peaks. A household located in the lower valley near the sea, as most households were, thus had access to all critical resources and was essentially self-sufficient. Where resource zones were more separated, as in some locales on the big island of Hawaii, households had to specialize to some degree in locally available resources; but because the land of a local community ran as a strip from the central mountains to the sea, exchange took place largely between households of the same community linked by close kinship bonds (Handy and Pukui 1958). Exchange between households of different communities was not extensive and, where desirable, took place at small informal gatherings that functioned as simple markets (Ellis 1963 [1827]: 229-30). The system that the chief did manage, as we describe below, did not function as a commodity exchange system among specialized subsistence producers.

Chiefs and their land managers did, of course, have legitimate managerial functions, notably organizing the community's rebuilding efforts after periodic flooding and tidal wave damage, and controlling intraregional warfare by allocating use rights. We will return to the managerial roles of the chiefs after a discussion of the organization and finance of their domains.

Economic Control and Finance

The distinctive institution that developed in Hawaii was the generalized governing hierarchy of chiefs. Chiefs received positions at three levels according to their rank in the ruling line. At the apex of the sociopolitical hierarchy was the paramount chief. The owner of all lands, he allocated land to his people in return for a portion of the food, craft goods, and raw materials they produced and for their support in war. Competition for the paramountcy was intense, and at the death of a ruling chief the island characteristically split into warring regions supporting rival claimants. The victor in these wars of succession then became the next ruler.

Second in the hierarchy were the district chiefs, high-ranking men of the ruling line with a strong private allegiance to the paramount. They were responsible for disseminating his orders and decisions to the community chiefs, and for mobilizing the goods and labor provided by the communities at his order. The district chief may have taken a share of the goods he mobilized for the paramount, but most of his income came directly from the individual communities allocated to him by the paramount.

Third were the many community chiefs (*ali'i 'ai ahupua'a*: "Chiefs who eat the community"). A chief was usually a close relative of the paramount who had been his supporter; in return for his support he was awarded a community, which provided his income. Each chief in the paramount's support group thus received what amounts to a land grant, but control over this land was dependent on the paramount and was typically reallocated by the paramount's successor. The chief in turn appointed a land manager, who oversaw production in the community and performed a wide range of social, political, and religious duties. Land managers, like chiefs, were not people from the community; rather they were often chiefs' junior kinsmen. Land managers thus held the day-to-day responsibilities for community direction that are held by the local chief in simpler chiefdoms. They were specialists of a sort, but unlike the specialists we shall encounter in Chapter 11, they were not yet members of separate bureaucratic institutions. Their duty was directly to their chief, who was also usually a kinsman.

As in the Trobriand chiefdoms competition for high office was severe, among other things because income depended on position in the hierarchy. The warfare that followed the death of a paramount chief was not only for succession to the paramountcy, but for access to the full range of political offices and income estates. To obtain or retain an estate it was necessary to support a successful contender, and the land records accordingly show the wholesale replacement of chiefs with each succession.

In addition, there was a strong expansionist drive to acquire land by conquest from other island chiefdoms to be granted as estates to the paramount's political supporters and to provide added revenue for the paramount himself. Every paramount chief maintained a small cadre of highly trained military specialists for use in operations against neighboring chiefdoms.

Religious institutions helped to consolidate the chief's control. Throughout the chiefdom were shrines used to house the gods and to accommodate ceremonies conducted by priests drawn from the ruling elite. At large shrines dedicated to Ku, the god of war, ceremonies initiated and supervised by the paramount himself were apparently used to build a consensus for military action among his supporting chiefs. More common were the small community altars used during the annual Makahiki ceremonies. For these ceremonies Lono, the god of land and fertility, traveled around the island accompanied by the paramount chief. At community shrines the chief, acting for the god, performed rites designed to maintain the fertility of the community's land; in return he received food, craft goods, and raw materials. The ritual obligations and significance of the paramount were made explicit by these rites, which embedded the financing of the governing chief in a ceremony to guarantee economic productivity.

The political economy, based on redistribution, was the financial base for the island chiefdoms. As we have seen, redistribution is simply an elementary form of taxation whereby goods mobilized from subsistence producers are used to compensate governmental and warrior personnel, religious functionaries, craft specialists, and other "nonproducers." In Hawaii the redistribution system was comparatively simple. For a specific operation, such as a large ceremony or military campaign, the paramount in consultation with his immediate advisers established the required amount of goods and personnel and assigned quotas for each district. The district chief then allocated his quota among his communities, and the required goods and people were supplied by the households under the direction of the community's land manager.

By such means the Polynesian conical clan that originally organized small total populations by internal ranking was promoted and transformed into a generalized ruling institution. Kinship and direct personal bonds between individuals remained as the internal logic of the system, although in Hawaii the scale had so increased that no one could have known all the people involved. The paramount could perhaps have known all elite personages, who probably numbered no more than a thousand in any chiefdom; but the commoners would have been largely faceless suppliers of the paramount's needs for goods and labor.

As the size of the chiefdom swelled from the low thousands to the tens of thousands, a new level of regional integration was required: one that would tie these faceless but indispensable commoners more securely into the system. A basis was found in the notion of restricted land tenure. Since all lands were owned by the paramount chief, the allocation of community lands to his supporters and the further allocation of small subsistence plots to commoners formed the basis for requiring payments in labor and goods. The chief's control of the basic productive resource of agricultural land was most clear when he introduced such capital improvements as irrigation and terracing. Subsistence plots in irrigated or terraced lands, with their high productivity, were allocated to commoners in exchange for a commitment to work on adjacent land owned by the chief. Each irrigation system was thus a mosaic of chiefly lands providing for political finance and commoner allotments providing for subsistence. An ideology of reciprocity between chief and commoner was thus established; commoners labored for the chief as a kind of "rent" for their subsistence plots.

This ideology of reciprocity can be extended more generally to a duality in the political economy. Subsistence producers, for their part, generated the wealth used by the chief to compensate nonproducing personnel, to invest in capital improvements, to make political payments that extended and consolidated his control, and to finance conquest warfare designed to provide an enlarged income base. The chief's obligations to his subsistence producers were essentially reciprocal: to keep peace within the chiefdom and thus guarantee local communities access to limited resources; to make capital improvements designed to increase yields; to take any steps necessary to keep local families functioning as viable economic units and to discourage their movement to neighboring chiefdoms; and to mediate as necessary between the local community and high-level religious and military institutions.

Hawaiian Prehistory: An Evolutionary Sequence

The evolutionary dynamics of the Hawaiian chiefdoms are best understood by examining their development over time. The history of this development, which can be reconstructed as a result of recent archaeological work, shows clearly the growth and elaboration of a chiefdom through three stages (Cordy 1974, 1981; Hommon 1976; Kirch 1982, 1984).

First was the stage of initial colonization and settlement, roughly A.D. 500-1200. The islands were apparently colonized before A.D. 500 by a small group (a boatload or two, perhaps 50 people) thought to have come from the Marquesas or the Society Islands (Kirch 1974).

They appear to have been purposeful colonists, probably refugees, who carried with them everything needed to reproduce their society. During the first eight centuries settlements were on the coast and the population expanded to perhaps 50,000, gradually filling up this most desirable habitat. At first the diet probably consisted largely of marine species, with game and wild plant foods less important. The early extinction of several endemic flightless birds may indicate resource depletion (cf. Kirch 1983). Also in this period domesticated foods increased in importance (Kirch and Kelley 1975). The archaeological evidence indicates small settlements of fishermen using some horticulture; and minimal evidence for social differentiation suggests that this society was probably organized at the community level, probably as a simple chiefdom.

During the second stage, A.D. 1200-1500, population grew rapidly (into the hundreds of thousands) and settlement expanded rapidly into the interior. Perhaps the best evidence for this process comes from the marginal island of Kaho'olawe (Hommon 1981). Prior to 1400 the population of this small island was concentrated on the coast, but during this period settlements spread to the interior, where in due course perhaps half of the island's people practiced shifting cultivation. By about 1500 the fragile environment of the interior was stripped of vegetation, and its inhabitants retreated to the coast or decamped for other islands. According to Kirch (Kirch and Kelley 1975: 55-64, 180-83), massive land slumps and associated land snails and carbon found in the Halawa valley on Molokai and dating to about 1200 testify to enlarged horticultural development and severe erosion in the uplands in this period. At roughly the same time many endemic bird and land snail species became extinct, probably because of massive changes to their habitat. At comparatively low population densities of perhaps 15-30 persons per square mile, the shifting cultivation of these fragile environments resulted, at least locally, in environmental disaster (Kirch 1984).

During this second stage the archaeological evidence indicates the beginnings of a regional chiefdom. Social stratification is indicated by a differentiation in house sizes and forms, with the larger houses presumably those of the elite (cf. Cordy 1981). During this period religious shrines (*heiau*) appeared and grew in size; since these shrines were linked historically with ceremonies of chiefly legitimization and required a corporate effort to erect, they are seemingly good evidence of chiefly organization. Chiefs were probably important in this period both as managers of intensified horticulture (as in the Trobriands) and as military leaders. Their control was facilitated by their circumscribed island environment, which severely limited the options of a dependent commoner.

During the third stage, 1500-1778, the population may have initially continued to grow but later appears to have stabilized (Kirch 1982: Fig. 2.11; 1984). The major change was in agriculture (Kirch 1985). Shifting cultivation continued and was even expanded in areas of moderate slope, where erosion could be controlled by terracing (Rosendahl 1972). More important was a rapid construction of irrigation complexes throughout the islands; at contact in 1778 irrigation complexes blanketed the coastal plain and lower valley areas. Most irrigation complexes seem to have been constructed in the last two or three centuries before contact, probably in response to the combination of long-term population growth and massive induced erosion (cf. Athens 1984). For whatever reason, the irrigation system with its ease of control formed the economic basis for the complex chiefdoms described at contact.

The prehistory of the Hawaiian Islands documents the evolution of complex chiefdoms over about 14 centuries. This development was based initially on the process of population growth and economic crisis requiring centralized solutions, and then on the continued control of the economy and its use in finance, made possible by the natural circumscription of the island environment and by the elite-owned capital improvements to the subsistence economy, leaving a disgruntled commoner population with few options.

Case 14. The Basseri of Iran

The Basseri (Barth 1964) are organized as a regional chiefdom, with a number of local segments under a single paramount chief. The Basseri chiefdom is based on the management and control of subsistence trade between a settled state and a pastoralist population using marginal lands at the outer limits of state control.

The Environment and the Economy

The Basseri number some 16,000 people living in 3,000 tents along the arid steppes and mountains of Fars province in southern Iran. A well-defined political group under the authority of a paramount chief (khan), they move within a delimited corridor 20-50 miles wide and extending some 300 miles from high mountains near Shiraz in the north to low deserts near Lar in the south. Basseri population density is about two persons per square mile. Both the higher elevations, up to 13,000 feet in the mountains near Kuh-i-Bul, and the lower ones, 2,000-3,000 feet in the deserts near Lar, are unsuitable for agriculture. The sedentary farming populations, which outnumber the pastoralists by about three to one in the whole of Fars province, cluster in the middle altitudes, around 5,000 feet. Small groups of Basseri follow carefully planned migration routes around and through the farming regions to utilize the more extreme environmental zones.

Climate determines the broad outlines of Basseri migration. Although annual rainfall throughout the region averages only 10 inches and agriculture is possible only with irrigation, precipitation is heavier in the higher elevations, where it is stored through the winter and spring in snow packs. In the winter, and in the spring as long as good pasture remains, Basseri camps are found in the low-elevation deserts to the south. As summer approaches and vegetation withers, the camps move northward, following the receding pastures toward ever higher mountains, where melting snows sustain late-season grazing. By autumn the mountain pastures have dried up or been grazed out, and herders must move into the agricultural zone, where they pasture their animals in recently harvested fields before returning south.

The economy centers on the production of meat and milk from mixed herds of sheep and goats. No cattle are raised; donkeys, horses, and camels are kept in small numbers for transport. Sheep and goats reproduce well in this environment, but early frosts and contagious diseases can kill as many as half of them in a bad year.

Milk is not consumed fresh but is immediately processed into sour milk or junket, then either eaten in that form or further processed into cheese. Sour milk may be pressed and sun-dried in the spring, when milk production peaks, and stored for use during the following winter. The Basseri eat meat frequently, always fresh and never preserved by drying or salting. Animal hides and wool become tents, clothing, storage containers, ropes, and other products. Women spend a significant part of their time spinning and weaving.

For all the importance of meat and milk, the Basseri diet is actually dominated by agricultural products obtained through trade with farmers. Wheat is basic; an unleavened bread made from wheat flour is eaten with every meal and is the single most important foodstuff. Sugar, tea, dates, fruits and vegetables, utensils, and many other items are also obtained by trade in exchange for clarified butter, wool, and lambskins.

Some Basseri own plots of agricultural land on which wheat and other cereal crops are raised. A few Basseri actually farm these plots, but most disdain agricultural work and contract with sharecroppers from nearby agricultural villages. The Basseri view these plots as investments in security and economic well-being: a way of banking profits generated by successful herding, and a means of upward mobility for the elite.

Social Organization

Even more so than among subsistence pastoralists such as the Turkana (Case 8), the basic economic unit among the Basseri is the family, which alternates between the settlement patterns of tent (nuclear family household) and camp. The tent typically houses a nuclear family and an occasional added member; it is a self-contained production unit, particularly at that point in its developmental cycle when adolescent sons are available for herding. All the required productive property, including tent, rugs, utensils, sheep and goats, transport animals, and containers, is owned by the individual family, and little is shared with other households.

A tent household needs about 100 goats and sheep for a satisfactory subsistence. These are not shared in reciprocal arrangements with relations or friends in other ecological zones, but are concentrated in a single herd directly supervised by the household head and his sons. Only men with unusually large herds will contract out a part of their herd to poor herdsmen, who then pay a share of meat, milk products, and newborn kids and lambs to their well-to-do patrons. Like peasant farmers (see Chapter 12) the Basseri use the market as a source of security instead of relying on extended social networks. In a good year they can sell surplus animals and buy land, which stores wealth securely and generates income that can be used to replenish herds after a bad year. Since they depend on the market for the agricultural foods that are the mainstay of their diet, what they need most in bad times is money and other secure property.

Soon after a couple is married, the groom's father releases his son's share of the anticipated inheritance from his herd. The newlyweds move into their own tent and strive to become economically autonomous. Since it is inefficient to use a vigorous adult male exclusively for herding a small flock, households commonly cluster in a hamlet-sized group of two to five tents whose occupants travel together and share herding duties. In forming such groups, ties of friendship established over years of mutual aid are as important as kinship. Friendship also provides the basis for trading partnerships with agriculturalists. But the autonomy of the household is primary, and tent groups will disperse and reaggregate as conditions warrant.

In the winter, when tent groups are scattered throughout the sparse low-elevation pastures in the south, large clusters of tents are rare. But at other times, where pastures are richer and more localized, tent groups form camps of 10 to 40 tents. Camps travel together and, although they are liable to fission, are held together by cross-cutting ties of descent and marriage reinforced by endogamy.

Each camp has a recognized leader, but just as the camp has little formal structure, so its leader has little economic or political power. His primary role is to help smooth relations between households, to resolve disagreements over where to locate, and to control the constant pressure from independent-minded household heads to break up the camp. Despite this pressure, camps are fairly stable and enduring units. Not only do ties of kinship and friendship help protect households against economic setbacks, but many families are reluctant to separate from the group out of fear of strangers, who may be thieves. Barth (1964: 47) describes the Basseri's view of their camp as "a small nucleus of human warmth surrounded by evil." Actual violence between camp groups is rare, and warfare is unknown. Indeed, despite the mutual suspicion between camps and the norm of endogamy, one-third of all marriages are between members of different camps and mobility between camps is not uncommon.

To this point we have a picture of the Basseri as autonomous tent households or tent groups in self-centered pursuit of good pasture and largely unfettered by structural constraints on access to particular resources. Friendship is important in day-to-day economic cooperation, and leadership is based less on the control of wealth than on personal distinction and service as a negotiator and compromiser. This description could apply as readily to the pastoralists we have discussed in earlier chapters.

The political economy in this case, however, centers on the *oulad*, a larger group than we have encountered among our other pastoralists. Oulads are territorial units, averaging from 40 to 100 tents (that is, about the size of the local groups examined earlier), in which membership is rigorously determined by patrilineal descent, reckoned simply as direct descent from a distant ancestor and not involving a segmentary system of lineages and sublineages. Within an oulad relations remain informal and economic life centers on tents and camps.

The oulad is easier to understand if we examine the economic role of the Basseri's paramount chief. He is simultaneously a Basseri chief and an elite member of the larger agrarian society. As a member of the elite he is much wealthier than other Basseri, owning thousands of animals, agricultural lands, and even whole villages. He and members of his family own houses in the city of Shiraz and move comfortably in elite urban circles.

One of the chief's functions is to allocate pasture rights to his subjects; the oulad is the corporate unit that receives these rights, in the form of an *il-rah* or "tribal road." The il-rah specifies a definite route for the oulad through the Basseri region's various ecological zones, and the precise pasture locations available to the oulad at each stage of the yearly cycle. Hence it is possible for more than one oulad to pasture its herds in the same location without causing conflict as long as each does so at a different time according to its il-rah. The chief generally allots pasture to oulads according to their traditional il-rahs. But when demographic change causes one oulad to have an excess of pasture relative to the needs of another, the chief calls the heads of the two oulads together and works out new il-rahs to which the members of each oulad must adhere. Since all pastures in southern Iran are owned by somebody, individuals have no access to resources apart from the lands guaranteed to their oulad by the chief.

In the political economy of the Basseri the oulad's territory is somewhat analogous to the "village lands" of the peasant community (Chapter 12). As in a peasant village, the households of an oulad are largely independent, self-contained domestic economies with less risk-sharing and kin-structuring between them than we find among the lineages and clans of the village-level and Big Man societies examined in Chapters 5-7. This is so largely because the superordinate state has taken over two functions that would otherwise be performed by kin groups: the defense of territory, now entrusted to a legal system that protects property rights; and risk-spreading, now entrusted to a market whose existence is protected by the state.

For these reasons the oulad is not the center of negotiation, network building, and conflict resolution that the local group is in less complex societies. It does not even have a leader, but only a spokesman who communicates messages from the chief in his absence. When the chief is present, individual heads of household address their concerns directly to him rather than to an intermediate official.

We view the Basseri chief as having two main functions in the political economy. First, he manages the use of pasture land in order to avert the "tragedy of the commons" (Hardin 1968), the degradation that occurs when family herders compete opportunistically for scarce pasture. He can impose restraints on pasture use that individual herders would not impose on themselves, since without group controls another herder would simply take the pasture for his own herd. He is empowered to impose his will by means of fines and beatings. He alternates his residence between town, where he cements his social relations with other elites, and country, where he travels with an entourage from camp to camp, holding "court" by handing down decisions, collecting tribute, and distributing wealth to especially deserving or needy followers.

The chief's second function is to represent the Basseri to other segments of Iranian society. As Barth points out, the Basseri are a distinct subunit of this society, set apart from its peasant and urban segments by their nomadic lifestyle and by deep ethnic divisions. When a Basseri comes into conflict with a farmer, for example, his mobility is a threat to the farmer, just as the farmer's ready access to the court system is a threat to the Basseri; direct negotiation between such different people is difficult. The chief, however, can take up the matter with the farmer's overlords, who are the chief's class equals; and this is the way many such conflicts are resolved.

To summarize, the Basseri subsistence economy centers on the cyclical migration of tent households and camps in search of pasture for small, independently owned family herds. Separate camps, even when they are members of the same oulad, are in competition and distrust and avoid one another. This individualistic economy, however, is limited by a scarcity of pasture land and the necessity of coexisting with peasant farmers under state rule; careful regulation of access to land is necessary to avoid both interpersonal squabbles and overgrazing. A major share of total production is sold in the market, where staple foods and essential materials are obtained.

A paramount chief is needed to maintain order in the countryside, to protect the group from the destructive effects of unimpeded individual exploitation of the environment, and to act as a broker between his subjects and outsiders. In return the chief takes advantage of his central position to maintain exclusive paramountcy over the Basseri, and uses his knowledge of the market system to garner exceptional wealth for himself and his family.

Conclusions

Let us now examine chiefdoms and their evolution in terms of our three key evolutionary processes: intensification, integration, and stratification.

Intensification of the subsistence economy, although important as an underlying process, appears to be much the same in the chiefdom as in the Big Man societies described in Chapter 7. Population density is characteristically high (about 25 persons per square mile) but well

within the range possible for Big Man societies and in some cases—
e.g., the Maori—quite a bit lower. As in simpler societies, the forms
of intensification vary with environmental conditions, ranging from
the short-fallow slash-and-burn cycle of the Trobrianders to the bot-
tomland irrigation agriculture of the Hawaiians. Only the tightly
regulated pasture use of the Basseri is not found in simpler societies.

It is interesting to note that the long-term trend toward a narrower,
simplified, and thus potentially inferior diet that we observed in
Chapters 5-8 is not evident in the chiefdoms we have examined. The
Hawaiians enjoyed a remarkably varied diet, thanks to a diverse environment with many different resource possibilities to which regional peace permitted access. In the Basseri and Trobriand cases external trade in subsistence products was important for dietary variety.

Integration is dramatically more in evidence in the chiefdom than in
simpler societies. Leadership is institutionalized at both the local and
regional levels, and chiefs at both levels are routinely relied on to organize centralized exchange and storage, to construct facilities for the
efficient production of staple products, to organize military operations, to guarantee land use rights, to mediate internal disputes, and
to negotiate or manage external trade relationships.

We can identify the main "causes" of the evolution of centralized
societies as being risk management (Gall and Saxe 1977; Athens 1977),
warfare (cf. Carneiro 1970b), technological complexity (Steward 1955;
Wittfogel 1957), and trade (Sanders 1956; Service 1962). Whether
alone or in combination, it is argued, these prime movers, themselves
an outcome of population growth and intensification, require central
management and thus underlie the evolution of complex societies.
This functionalist logic sees cultural evolution as adaptation, the solution of particular problems brought about by population growth
under particular environmental conditions.

A similar logic is expressed by the nineteenth-century Hawaiian
chief David Malo (1951 [1898]: 187): "The government was supposed
to have one body (kino). As the body of a man is one, provided with a
head, with hands, feet and numerous smaller members, so the government has many parts, but one organization. The corporate body of
the government was the whole nation, including the common people
and chiefs under the king. The king was the real head of the government; the chiefs below the king, the shoulders and chest." As Rathje
and McGuire (1982: 705) point out, this biological analogy also underlies modern functionalism and its analysis of social systems. For
Malo, a Hawaiian chief brought up before missionary contact in 1820,
the ruling chief as the body's head provided the essential direction for

society. To this Polynesian a society without a ruling chief would be as unthinkable as a body without a head.

Are the functionalists correct? Can we explain the evolution of social complexity as a necessary correlate of intensification of the subsistence economy? Not quite, we think. Intensification is indeed necessary, but not sufficient; the crucial matter of *control*, as distinct from management, must also be considered. To put this another way, intensification of the subsistence economy necessitates centralized management, but some forms of management do not necessarily lead to the formation of chiefdoms. It is the particular forms of intensification favoring central control that result in chiefdoms and provide the opportunities for political growth.

Stratification involves the differential control of productive resources, and it is this control above all that distinguishes chiefdoms from simpler societies. Chiefdoms are based on generalized central leadership, as are Big Man societies; but a chief has sufficient institutionalized control over his society's political and economic organizations to be able to restrict leadership to an elite segment. Such control, based on restricted access to critical economic resources, may derive from any of four major conditions:

1. Central storage, instituted originally as a way of handling risk but providing control over capital for use in political affairs (Earle and D'Altroy 1982; D'Altroy and Earle 1985)

2. Large-scale technology, desirable to a local population to minimize production costs but requiring a major capital investment that bonds subsistence producers to the chief (Gilman 1981; Earle 1978)

3. Warfare in naturally circumscribed regions, which requires leadership but allows the victorious chief to control a subjugated population without options for escape (Carneiro 1970b; D. Webster 1975)

4. External trade, which may be necessary to a local population or attractive because of strong external demand, but is not available to most individuals because of the high costs of transport technology (Burton 1975) and the difficulties of intersociety contracts

Once regional control is established, the evolutionary development of chiefdoms toward greater centralization depends on investment opportunities and the costs of controlling or defending whatever investments are made. Some investments, such as irrigation agriculture and marine-based trade with foreign states, offer an exceptionally large potential for control and growth; these characteristically underlie the evolution of states, to which we turn in the next chapter.

ELEVEN

The Archaic State

STATES ARE REGIONALLY ORGANIZED societies whose populations number in the hundreds of thousands or millions and often are economically and ethnically diverse. Whereas chiefdoms vest leadership in generalized regional institutions, in states the increased scope of integration requires specialized regional institutions to perform the special tasks of control and management. The military is responsible for conquest, for defense, and often for internal peace; the bureaucracy is responsible for mobilizing the state's income, for meeting many local managerial responsibilities, and more generally for handling and monitoring information flow; the state religion serves both to organize production and to sanctify state rule. Along with this elaboration of the ruling apparatus comes increasing stratification. Elites are now unrelated by kinship to the populations they govern; their power, underwritten by economic control, is displayed in the conspicuous use of luxury goods and the construction of splendid buildings.

State formation has been a central theoretical concern in anthropology at least since the time of Lewis Henry Morgan (1877). Service (1977) distinguishes two anthropological perspectives on state origins: integration theories and conflict theories. Integration theories derive from cultural ecology (Steward 1955; Service 1962, 1975; Binford 1964) and systems theory more generally (Flannery 1972; Hill 1977; Wright 1977); they view the state as a new level of social integration necessitated by new problems of risk (Gall and Saxe 1977), technological complexity (Wittfogel 1957), and trade (Rathje 1971). Conflict theories emphasize either conquest, whereby one ethnic group comes to dominate others (Ibn Khaldun 1957 [1377]; Carneiro 1967), or class conflict (Engels 1972 [1884]; Adams 1966; Fried 1967); they view the state as a mechanism for maintaining the social, political, and economic domination of one segment over another.

These two theories are not mutually exclusive, and in fact identify two interdependent processes. On the one hand, states are born out of conflict and domination: one ethnic group becomes the ruling elite of a huge empire, and imperial institutions operate to maintain and to strengthen this domination. On the other hand, states develop and function under certain basic conditions that both permit economic control and require central management; local populations are bound economically to the state through a carefully managed dependency that is a consequence of long-term intensification in the subsistence economy.

A further division in theories of cultural development has arisen between unilinear and multilinear evolutionists. Unilinear evolutionists have sought to identify a single line of development reflecting the causal influence of one dominant variable or prime mover, notably technological progress and increasing energy capture (Leslie White 1959) and the managerial requirements of irrigation (Wittfogel 1957). By contrast, multilinear evolutionists (Steward 1955) have seen the development of new levels of complexity taking parallel but distinct lines in accordance with local environmental conditions.

In this chapter we continue to favor a multilinear approach. In Chapter 8 we outlined four opportunities for development, related to risk management, warfare, technology, and trade. In Chapter 10 the emphasis fell on three—risk management, technology, and trade—as conditions both requiring central direction and offering opportunities for control. Now we narrow the emphasis to technology and trade; warfare and risk management remain important areas of state concern, but by themselves are insufficient to provide the financial wherewithal required by the emerging state.

Capital-intensive technology is perhaps the most common basis for the developing political economy of states. Increasing population density requires a level of agricultural intensification that ultimately can be achieved only by large capital improvements such as irrigation systems. Although regional management of irrigation is necessary only for those massive systems constructed well after state formation, irrigation systems even on a fairly small scale permit economic control by elites, who exchange access to irrigated areas for labor or shares of produce. States based on the control of productive technology are typically financed through staples produced on state-controlled improved lands. This form of staple finance, often associated with the

"Asiatic mode of production," formed the financial base of most primary states, including those of Mesopotamia and Egypt. The Inka state, discussed in this chapter, is another good example.

Trade as a source of income for state finance is probably most important on the peripheries of agrarian states. In the eastern Mediterranean the rise of the Mycenean and Athenian states was based on mercantile trade and the large-scale production of export goods by slave labor (Engels 1972 [1884]; Lee 1983; Renfrew 1972). The Aztec state, with a comparatively small bureaucracy in conquered areas, depended for its finance on tribute, often in wealth goods, and the expansion of both long-distance trade and local markets (Berdan 1975; Brumfiel 1980). In this chapter the evolution of medieval France and Japan from simple, staple-financed, chiefdom-like societies into well-financed states is attributed largely to the development of an integrated market system made possible by an increased exploitation of mercantilism and trade.

Although technology and trade are analytically separate sources of wealth, in practice they are usually interrelated. As might be expected, states typically seek out multiple sources of finance to maximize both the amount and the stability of their income. States such as China that were initially dependent on staple finance actively encouraged the development of currency, market exchange, and long-distance trade as new income sources. Indeed there may be a general tendency to replace staple finance with wealth finance because of its greater flexibility, storability, and, most important, mobility (D'Altroy and Earle 1985).

Case 15. France and Japan in the Middle Ages

We now turn away briefly from our ethnographic and archaeological examples to examine some familiar historical materials. Our two examples, medieval France and Japan, are widely separated both spatially and culturally. Yet when the layers of aesthetic, technological, social, and philosophical differences are stripped away—when all that remains is the small set of variables that forms the core of our model of social evolution—we find astonishing similarities between the two societies. This is true even of their rates of change: although developments in Japan took place nearly five centuries later than in France, roughly the same time elapsed in each case between one stage of development and the next. Thus the "first middle age" occupied the tenth and eleventh centuries in France and the fifteenth and sixteenth centuries in Japan, whereas the "second middle age" occupied the twelfth and thirteenth centuries in France and the seventeenth and eighteenth centuries in Japan.

By referring to leaders of the middle ages as "kings" and "emperors" we tend to exaggerate the extent and depth of centralized power available to these leaders. Applying the neutral standards of the previous chapter, we find medieval France and Japan to have been populated with communities ranging from simple to complex chiefdoms, with many areas not integrated beyond the family level or the local group. Under the relentless pressure of continuous population growth and its handservant, the intensification of production, the landscape filled in and the proportion of the countryside that was brought under chiefly control increased, as did the complexity of chiefdoms.

The term "feudalism" is also misleading for at least two reasons. First, we find that many "feudal" institutions, such as the establishment of personal bonds of fealty between lord and vassal, the obligation of military service to the lord, and the granting of estates in land to loyal vassals, are hallmarks of the economic organization of chiefdoms; that is, they are not uniquely "feudal" in and of themselves. Second, some of the uniqueness or idiosyncrasy of medieval society and economy in France and Japan comes from the strong cultural influence of imperial Rome and China, respectively. For example, whereas in chiefdoms the language of social relations, including hierarchical ties, remains rooted in kinship even when actual genealogical distance between individuals may be very great, medieval France and Japan used legalistic language to describe and enforce the various levels of the hierarchy. Yet beneath this largely formal difference the operation of the "feudal" economy, in such central matters as the control of land and capital improvements and the transfer of production to the elites, is essentially that of a chiefdom. A growing awareness of these similarities between medieval society and chiefdoms has led to some illuminating historical reconsiderations, notably of the Viking age in Denmark (Randsborg 1980).

Imperial Precursors

The middle ages of both France and Japan were influenced by contact with external empires. France had been under Roman control for centuries. Japanese rulers were fully aware of the politically developed Chinese state and, perhaps alert to the dangers of a powerful neighbor, had adopted a centralized legal system modeled on the

Chinese one. It was the "ghost" of these externally derived imperial political structures that gave their successors a degree of political structure uncommon in chiefdoms (Asakawa 1965: 196; Hall 1970: 77).

On paper the Merovingian (400-687) and Carolingian (687-900) rulers of France, and the emperors of the Nara (646-794) and Heian (794-1185) periods of Japan, owned all the lands of their respective countries and ruled by decree. The subsequent growth of powerful regional lords who challenged the emperors' supremacy has been generally seen as a kind of "devolution" or "decay" of centralized power (e.g., Lewis 1974: 25-27; Duus 1976: 61; Hall 1970: 75-134), often explained as the inevitable consequence of the greed or inefficiency of the rulers. In this view the reestablishment of centralized rule at the end of the middle ages appears as a stage in a cyclical process of state formation, dissolution, and reformation.

Clearly, however, the centralized states of the late medieval eras in France and Japan were utterly different from the states that preceded them. In earlier times the territories claimed by the so-called emperors were settled by subsistence farming communities at the relatively low population densities characteristic of horticulturalists. Warfare was endemic, and political life centered around war chiefs allied in tenuous federations. In some places population densities were higher; for example, in the ninth century the region around contemporary Paris was settled by 40,000 peasants organized into eight political units (Duby 1968: 12). Such areas were characterized by significant intensification and local centralization, and no doubt paid ample tribute to their rulers. But in both France and Japan these islands of control were surrounded by dangerous, unstable territories that were "owned" by the emperor in name only.

In France slash-and-burn agriculture was practiced in sparsely settled areas, but intensive horticulture was already the more common pattern. The use of pigs, horses, cattle, sheep, and goats was widespread. Short fallows were common, and in some places even more intensive techniques were found: the plow (usually the light wooden *araire*), annual cropping, crop rotation (incorporating legumes), dikes, and manuring (Lynn White 1962: 40-77).

Japan was a hunter-gatherer economy until rice technology was adopted, perhaps around 250 B.C. Dry-rice cultivation coexisted with foraging until about A.D. 300-600, when chiefdoms and archaic states arose in close association with irrigated wet-rice cultivation. Slash-and-burn horticulture is not mentioned for this period, but its persistence to as late as 1960 in isolated regions of Japan suggests it may have been important earlier (Taguchi 1981). By imperial times regions of population concentration showed such signs of intensification as irrigation, manuring, and transplanting, all of which increased rice yields per unit of land (Tsuchiya 1937: 60-78).

This basic pattern—localized population concentrations with intensive production surrounded by large regions of dispersed population and more extensive production—is reflected in other domains of the French and Japanese economies as well. In the central areas economic specialization, markets, and money were of real, if limited, importance; but in peripheral regions roads were poor or nonexistent and subsistence production dominated. In the central areas, too, the new military technology of iron weapons, armor, and warhorses was beginning to create a specialized fighting force, in contrast to the lightly armed fighting forces to which nearly every able-bodied man of the peripheral areas belonged. The new equipment was expensive, and warriors so armed could only be maintained with the income of large estates granted by the emperor.

In short, the medieval eras of Japan and France began in times of authoritarian centralized control of smallish regions of intensifying production surrounded by larger regions not subject to central control and characterized by long fallows, some foraging, intercommunity warfare, and unpredictable political alliances. In each case the earlier extension of imperial control over the peripheral areas left its mark; but the economic base of these areas could not support a state, and when the empires collapsed they were replaced by warring chiefdoms. The developments to which we now turn represent not so much the resurrection of formerly powerful states as the internal evolution of society resulting from the gradual filling in of the countryside and the accompanying socioeconomic changes.

The First Middle Age

A "first middle age" has been identified for France from 900 to 1100 (Bloch 1961: 59-71) and for Japan from 1334 to 1568 (Lewis 1974: 40-48). During this period we find a gradual, continuous unfolding of trends that were already visible in the premedieval era and that came into full bloom in the "second middle age."

The contrast between "developed" and "undeveloped" areas remained marked in the first middle age, but the proportions slowly shifted in favor of developed areas. There was a steady growth of population and a dramatic change in food production. As population expanded, more and more land was brought under cultivation: in France there was an "incessant gnawing of the plough at forest" (Bloch 1961: 60) as the adoption of the heavy steel plow (*charrue*) made it possible to farm the dense dark soils of river valleys like the Loire and the Seine. In both France and Japan regional lords intent on opening up their undeveloped lands offered farmers such incentives as private ownership of plots and low service obligations. In Japan the government and regional lords became active in undertaking major projects (swamp drainage, irrigation works) to create new areas of cultivable land. The resulting destruction of forest became so great that the Japanese state instituted forestry management programs (Tsuchiya 1937: 126).

At the same time the use of existing lands was intensified to raise their productivity. In France a complex set of interrelated changes centered around the heavy steel plow. The plow opened up new lands to annual cropping but required a major investment in draft animals: first oxen, later more expensive but more efficient workhorses. Draft animals were pastured on fallow fields, leaving their manure behind them; this change encouraged the aggregation of households into cooperative groups that rotated their fields together in order to pasture their animals in large fenced areas. A three-field system was devised whereby a household planted one field in winter wheat, planted another in summer crops (generally legumes), and left a third fallow, changing the use of each field every year. This system substantially increased productivity (Lynn White 1962: 40-77).

In Japan a similar intensification took place as irrigation was extended to increasingly marginal lands. Multiple cropping, draining of swamps, and the spread of new crop varieties made for a "great agricultural . . . revival" (Lewis 1974: 53). The threat of famine and intense land hunger are frequently mentioned in commentaries of this period; population continued to grow, and average plot size per household began to decline. For efficiency in wet-rice cultivation households aggregated into self-sufficient cooperative groups that shared labor in periods of peak need (T. Smith 1959: 50-51).

At the beginning of the first middle age, settlement was still almost entirely in homesteads and hamlets spread across the countryside. Bush fallow and the foraging of secondary forest for wild foods were still common. There were few cities or towns, and trade was of minor importance to most people. But local lords were growing in power as their domains filled in with productive households, and their military power defeated imperial efforts to tax and regulate them. There ensued a time of intense warfare in which no large-scale, stable political centralization could take root.

Local lords had much in common with the more powerful chiefs described in Chapters 9 and 10. Kinship was sometimes still important in group formation, but truly tribal peoples disappeared as the warlords' power grew. A lord defended what he considered his territory by alliances if possible and by warfare if necessary. In order to maintain his private army, he allotted his dependents a share of the produce of a section of his territory in return for an oath of personal allegiance and service. Homage to a ruler was still viewed as an act of individual choice.

Population tended to cluster about the lord's residence. From early times it had been common for one house to exceed all others in size and complexity, a house in which the group's common property was stored and cooperative and defensive measures were organized (Mayhew 1973). As population grew, these nuclei gradually became manors surrounded by a peasantry dependent on the manor for protection and security. Beyond the orbit of the manor, however, large depopulated zones existed, sometimes inhabited by scattered hamlets of "free peasants."

A warrior aristocracy arose, characterized by military prowess and strong ties of loyalty to their lord. The values of this class later became rigidified into the high ideals of chivalry (asceticism, fearless defense of lord and honor, strength and skill in battle) that characterized the knights and *samurai* of feudal France and Japan. But in this earlier stage the rights and duties of lords and dependents remained fluid, personal, and negotiable.

The community centering on the lord's manor was self-sufficient. Roads and waterways were only beginning to expand, and markets were just beginning to appear in areas of greater population density. Hall (1970: 113) finds it paradoxical that so much agricultural progress could be made in Japan during a time of economic decentralization and political instability. But there is no paradox if we view the process at the local level rather than from the standpoint of the imperial government. In both countries population pressure was increasing hand in hand with the intensification of food production, and the social division of labor was becoming more complex. In Japan we find workshops of artisans long before towns and cities emerge (Tsuchiya 1937: 82), and in France we find the manor serving to some degree as a center for accumulation and distribution of wealth (Bloch 1961: 236). Hence what seemed to an emperor to be a dismaying loss of control must have seemed on the local level to be a gratifying increase in production, interdependence, and economic and political order.

The Second Middle Age

The second middle age, or "high feudalism," appears in France about 1100-1300 and in Japan from 1568 to 1868, a period preceding and including the Tokugawa shogunate. By this time more powerful rulers were emerging, roads and waterways were being built, and towns and free markets were arising throughout the countryside. Local lords, formerly autonomous, were now compelled to swear fealty to regional overlords, who had superior might and wealth. The power of these regional overlords remained tenuous, however, and they had to shore it up by frequent tours with a retinue through their provinces, accepting food and lodging from local rulers in the so-called "moveable feast" (Bloch 1961: 62; for Japan see Hall 1970: 111). Such tours are common in chiefdoms, as we noted for Hawaii and the Basseri. They are a sign of a leader's weakness when compared to fully developed state rulers, who for the most part confidently reside in palaces and require their subjects to come and pay court to them. Hence we see the second middle age as a period of transition from a society divided into competing chiefdoms to one united as a single state.

Throughout the early part of this period population continued to rise, perhaps at a more rapid rate than before; Japan's population increased by 50 percent (to about 30 million, or about 260 persons per square mile) in the seventeenth century alone, but then leveled off and grew very little thereafter (Hall 1970: 202). The second middle age in both France and Japan has been described as one of great agricultural innovation and "progress" (Duby 1968: 21-22; Duus 1976: 83; Hall 1970: 201-2). Thomas Smith (1959: 87) speaks of a "new attitude toward change, though the reason for it remains obscure." New technologies were perfected and old ones more widely adopted; iron tools were increasingly used, irrigation spread, and new seed varieties were developed and distributed. In Japan the expense of fertilizer became a major cost of production, and commercial fertilizers prepared from fish cakes, fish oil, and human nightsoil were widely available in marketplaces. Average field size continued to decline and labor input per field rose; a kind of "involution" of labor (Geertz 1963; see Chapter 12) took place, as ever greater care was devoted to spacing plants, selecting seedlings, conditioning the soil, and the like. The use of draft animals, double-cropping, and commercial cropping also increased. Cultivation was expanded into previously uncultivated marginal areas, and peasants began to complain of the resulting loss of firewood, manure, and fodder (T. Smith 1959: 95). We recognize all these changes, of course, as integral to the systematic intensification of production in response to population growth, and this would explain the new attitude toward change.

The most dramatic changes during the second middle age, however, took place in the economic, social, and political integration of production and exchange. Lewis (1974: 66) refers to this period as an "age of elaboration and legalism." As items manufactured in towns and artisan guild-communities played a greater role in agriculture, and as it became necessary to put more and more land to its most profitable use by growing single crops for sale rather than multiple crops for subsistence, the market became increasingly important. The great lords (in Japan, *daimyo*) could guarantee the peace of the market and the highway, and could issue money and otherwise support trade.

The whole fabric of medieval society tightened in the second age. Ties of dependence were established through formal rituals, signed legal documents, stricter rules of inheritance, and military conscription. Villages became the key social units beyond the household, defining who could use village lands and also serving as convenient units for taxation. At this point fealty was no longer a matter of choice; nearly everyone was someone's vassal, and what had once been "free peasants" were now "outlaws." Rights to receive rent, taxes, titles, stipends, and shares from the land were elaborately and carefully defined, and the peasants' "fund of rent" (Wolf 1966a) appears to have become continuously more oppressive.

Successful warfare now required large, heavily equipped armies. As the landscape filled in, a kind of "social circumscription" (Carneiro 1970b) made it possible for one faction, by a mixture of threat and compromise, to establish a stable central government uniting all the separate lords. As this process was completed, cities and trade grew rapidly. Manufacture and trade became alternative routes of power and wealth, and even the medieval lords became increasingly profit-oriented. Landless laborers appeared and became migrant wage laborers or "servants" in the homes of land-owning peasants.

Land ownership hence became a matter of great concern; new land surveys were undertaken, legal deeds of ownership accompanied the increasing tendency to buy, sell, and rent land, and peasant uprisings and revolts occurred over issues of land ownership. Some of these issues were increases in taxes and tithes; the frequency of famine (perhaps indicating the inability of the land to sustain further population increases); the replacement of ties of loyalty based on kinship and personal allegiance by legal, impersonal ties enforced by courts and police; and the creation of landless peasants as the feudal protection of land access gave way to an increasingly free market in land.

We can see how the second age spawned a new order. In the place of a "pure feudalism" of autonomous regional lords, a single, powerful unifying ruler emerged. The lord's exclusive position in control of land-based wealth began to fade as wealth was acquired by rising groups of merchants, craftsmen, industrialists, and bureaucrats, all managing their parts of an increasingly complex economy. Leadership came to depend more on control of "exchange" than on the means of production (see Chapter 12). Improvements in transportation, the peace of the internal market, and centralized political power capable of setting foreign policy all increased the importance of trade and commercial production at the expense of the subsistence sector.

In sum, France and Japan in the middle ages developed gradually into states, propelled by pressures and opportunities arising from population increase and the intensification of land use. The growth of a peasantry coincided with the expansion of state-level political structures into tribal areas. At what point does the tribesman paying tribute to a chief become a peasant paying rent to a lord (cf. Bloch 1961: 243)? Although no precise answer is possible, the intensification of labor on the land, and the increase in stratification and bureaucracy at the expense of kinship and personalism are all clearly associated, and a peasantry is the inevitable result.

Case 16. The Inka: An Andean Empire

The Inka empire, Tawantinsuyu, was the largest and administratively most complex polity of the prehistoric New World. The empire, which extended from what is now Chile and Argentina through Peru and Bolivia to Ecuador and Colombia, incorporated about 350,000 square miles and perhaps 8-14 million persons. In contrast to the simpler societies discussed earlier, the scope of political and economic integration of the Inka empire is profound. It exercised power directly over some 85 ethnic groups, themselves originally fragmented into many autonomous units (Rowe 1946: 186-98), and over many diverse environments with special crops and unusual resources.

The rise to power of the Inka was phenomenal. At the end of the Late Intermediate period (around A.D. 1400) the Andean highlands were divided among many warring chiefdoms (Rowe 1946: 274). In the Mantaro valley, north of Cuzco, recent archaeology documents the conditions that typified the highlands prior to Inka expansion (Browman 1970; Matos and Parsons 1979; Earle et al. 1980). The first sedentary villages date to perhaps 800 B.C. and new villages were founded throughout the region as population slowly grew. Villages remained small (5-6 acres), probably with populations in the low hundreds, but their locations shifted through time to higher locations, probably for defense.

Midway through the Late Intermediate period (about 1350) there was a dramatic social change. As population continued to grow, settlements increased rapidly in size and many were now located on ridge tops and hills. For example, the settlement of Tunanmarca, a comparatively large center (53 acres), was located on a high limestone ridge overlooking the Yanamarca valley north of Jauja. Combined with the fortress location, the settlement was surrounded by two concentric fortification walls. The residential zone had an estimated 4,000 house structures that would have housed nearly 10,000 people and a central public plaza with several special buildings. Three smaller contemporaneous settlements located within three miles of Tunanmarca appear to have been politically tied to this center. In all we estimate that the Tunanmarca chiefdom incorporated perhaps 15-20,000 people (Scott and Earle n.d.).

Prior to the Inka conquest, the Mantaro valley and apparently most of the Andean highlands were fragmented into these chiefly polities, which were more or less constantly at war with each other. The Inka were able to build their empire by systematically conquering these formerly independent polities and incorporating their populations and political systems within the empire. How did they do it?

Apparently their unprecedented success is attributable chiefly to their innovative principles of bureaucratic control and indirect rule. The problem was to unify warring chiefdoms by creating a new level of integration. Institutions such as the broad system of labor taxation, though based on existing precedents and ideologies, were transformed to suit the larger and more complex needs of an empire. Essentially the empire was built on the structure and ideology of chiefdoms, but with new hierarchical relationships superimposed. Some examples will be discussed in our analysis of the dual economy of the Andean world.

Prior to the Inka conquest, long-term population growth had caused an intensification of the subsistence economy, violent military conflict, and the initial growth of stratified societies in the Andean highlands. The constant state of war had high economic and psychological costs that made the regional organization and peace of the empire desirable. Warfare was over land; essentially each community fought to protect the land necessary to its very survival. The imperial superstructure imposed regional peace and a system of legal rights of land use in return for labor obligations. The cost of maintaining this system had been significantly lowered by the long-term growth in population density, which lowered administrative costs, and by the increased dependency of the population on intensive agricultural methods (irrigation and terracing) that were easily controlled. Another advantage was the prior evolution of chiefdoms, which permitted the Inka to rule indirectly through existing political systems. Although the Inka conquest must still stand as one of history's most remarkable events, the basic prerequisites for it were in place.

To understand the operation of the Inka empire, we consider the dual economic bases of social and political integration: the subsistence economy, supporting the population of local communities; and the political economy, financing the state and its special interregional institutions. Much of what follows is drawn from the valuable summary descriptions by Rowe (1946), Moore (1958), Wachtel (1977: 60-84), Schaedel (1978), and Murra (1975, 1980 [1956]).

The Environment and the Subsistence Economy

The Andes, home to the Inka empire, are a jagged chain of high mountains a short distance inland from and parallel to the Pacific coast of South America. Generally three environmental zones can be recognized. Along the coast is a dry and barren desert punctuated by green valleys that are fed by streams from the high sierras. The streams were used to irrigate productive agricultural fields near the coast, and rich marine resources added important animal foods. Inland the mountains rise rapidly above the coastal desert and a central sierran zone follows along the Andean range. This sierra contains towering snow-capped peaks, extensive rolling grasslands, and some broad intermontane valleys. The grasslands were used for extensive pasturage, and the rich intermontane valleys for agriculture. To the east the land descends rapidly and is dissected by many steep valleys and cascading streams. Often within 30 miles elevations drop 9,000 feet from the high alpine grasslands to a humid and lush tropical forest. Highland groups lived in the upper reaches of these streams, but the forest environment itself was occupied by distinct tribal groups— such as the Machiguenga (Chapter 3), a mere 100 miles from Cuzco— that were never incorporated into the empire. The widely scattered populations of the tropical forest were dependent typically on foraging and shifting cultivation and were organized at the family or local group level.

In part because of these zonal contrasts, Andean society was quite variable in form. Populations on the coast were densely settled, dependent on large-scale irrigated agriculture and fishing, and characteristically organized as complex states: notably Chimu, with its urban capital of Chanchan (Moseley and Day 1982). Populations in the sierra were less dense, dependent on mixed farming, and characteristically organized as local groups or competitive chiefdoms. From this economic and social diversity the empire organized its massive political superstructure.

In the sierra communities, typified here by the Mantaro valley, archaeologists have documented a sustained and fairly dramatic population increase immediately prior to the Inka conquest (LeBlanc 1981;

Hastorf 1983). Population density under the Inka was about 37 persons per square mile overall in the highlands (LeVine 1985: 450), and locally much higher. Thanks to a mosaic of different soils, precipitation rates, slopes, and elevations (Hastorf 1983), the typical sierra settlement was an island or pocket of very high population surrounded by barren landscape.

The subsistence economy was a mixture of permanent and shifting cultivation of crops and animal husbandry. The crops included maize, potatoes, and quinoa; the animals were chiefly llamas for meat and transport and alpacas for wool. Maize was grown in irrigated fields below 11,000 feet, potatoes and other root crops were grown with shifting cultivation in the uplands to 13,000 feet, and the llamas and alpacas were grazed on the higher-elevation grasslands.

The long-term growth in human population resulted in a selective intensification of agriculture. The shifting cultivation of the uplands is frequently described by early sources as involving a carefully regulated fallow cycle (Rowe 1946). Where feasible, capital improvements for permanent cultivation included irrigation, terracing, and drained field systems (Hastorf and Earle 1985; Donkin 1979). A side effect of this intensification was an increased risk of crop failure as production expanded into valley bottoms, which are susceptible to flooding, and into uplands, which are assailed by hail and frost. In modern times such risks are in part anticipated, and Andean farmers prefer to plant in many diverse locations as a hedge against disaster.

Ethnohistoric studies (Murra 1980 [1956]; D. LaLone 1982) emphasize that the Inka were largely a marketless society. Recent work for the Mantaro (Earle 1985) suggests that exchange was remarkably limited, especially in food. As in the Hawaiian case the extreme environmental diversity in the Andes made a variety of resources available to local populations, thus limiting the need for exchange between communities.

Warfare, as we have seen, was endemic before the conquest. Local leaders questioned by the Spanish about the pre-Inka period described its nature: "Before the Inca, they engaged in wars with each other in order to acquire more lands, and they did not go outside this valley to fight, but it was within the valley, with those from one side of the river which passes through this valley fighting with the Indians from the other side" (Vega 1965 [1582]: 169). Other informants, sounding almost like modern-day anthropologists, interpreted this warfare as caused by increasing population and competition between communities for lands, herds, and women (Toledo 1940 [1570]: 28).

To summarize, the growing population in the Andes created the now familiar problems of agricultural intensification with its associated technology and risk and considerable warfare. As we shall see, these local circumstances prior to the Inka conquest produced the necessary conditions for creating the Inka state.

Social Organization

Andean community organization had two significant levels: the individual household and the *ayllu*, a kinship and territorial group. The individual household was probably a nuclear or minimally extended family composed of a married pair and their children, sometimes joined by a widowed parent, an unmarried sibling, or some other close relative. In contemporary traditional Andean communities this nuclear family household forms the elemental economic unit (Lambert 1977: 3; Mayer 1977: 61). Although we cannot simply extend this pattern back to prehistoric times, highland sites dating to the Inka and immediate pre-Inka periods were typically subdivided into small "patio groups" of several structures opening onto an open work space (Lavallée and Julien 1973; LeBlanc 1981). These groups of structures, typically with one or two buildings and rarely with more than four or five, appear to have been family compounds in which the family's subsistence labors were centered.

The usual division of labor by age and sex permitted the household to approximate a self-sufficient producing and consuming unit. The men were involved in heavy activities such as soil preparation, in many crafts, and in long-distance trading. The women were responsible for other agricultural tasks, food preparation, child rearing, and spinning and weaving (Silverblatt 1978). In agriculture, at least, male and female activities were explicitly complementary. A couple formed a working pair: while the man turned over the soil with a foot plow, the woman broke up the clods; while the man made a planting hole, the woman placed the seeds in the previous hole (Rowe 1946: 213). As long as pasturage was relatively close to the main settlement, the young of both sexes were responsible for tending the herd animals (Murra 1965: 188).

To judge from contemporary traditional Andean communities, a goal of household independence was probably cherished. Contemporary households resist entering into reciprocal relationships with other households lest they prove expensive in terms of future demands on household labor (Lambert 1977: 17). Politically or economically, of course, interhousehold relationships may in some cases be essential for household survival; but such relationships are avoided wherever possible.

The ayllu, a kin group descended from a single defining ancestor, was the main social and economic organization above the household. It ranged in size from a few hundred people to perhaps a thousand (D'Altroy 1984). Membership in the ayllu was once thought to have been patrilineal, but because the group was largely endogamous its members would have been interrelated by numerous overlapping blood and marriage bonds. The ayllu was thus very similar to the Hawaiian *ahupua'a* or community, a medium-sized endogamous corporate group, tightly integrated but regionally quite isolated except where multiple ayllu formed a single larger community.

The ayllu as a corporate group held inalienable rights to community land. For example, all pasturage is said to have been held with undivided group access, and the same was probably true of hunting territories, undeveloped lands, and lands farmed with shifting cultivation. Apparently the ayllu, through its leader, regulated the fallow cycle fairly rigorously, reallocating farmland annually in much the same manner as the chiefs of the Trobriand Islands (Chapter 9). As we have seen, the corporate ownership of land and regulation of farming is a response to the need to intensify shifting agriculture in order to support higher population densities and provide a surplus output.

Rowe (1946: 213) describes the use of cooperative work teams in preparing fields. In a festive spirit, a line of males using foot plows turned over the soil in unison as a line of women broke up the clods. The two facing lines moved across the field together. In any microenvironment, such organized teams were used first to prepare the state fields that each ayllu was responsible for, and then to prepare the fields of the participating households. This ceremonial order establishes symbolically and materially the precedence of the political economy.

A single local community, ideally of one ayllu, typically maintained a generalized subsistence economy that permitted it to be largely self-sufficient, thanks to a diversity of subsistence strategies responding to the diversity of its geographical zones. In the Mantaro valley, for example (Earle et al. 1980), late prehistoric settlements of the Inka and immediate pre-Inka periods were located on the upland slopes and low hills overlooking the river. The upland soils there are ideal for growing potatoes, which provided the starchy staple for the diet. Below the settlements are alluvial bottomlands suitable for intensive maize production, and above the settlements are rolling grasslands used for grazing. Within a few miles of a settlement a community's population had direct and immediate access to a diversity of lands.

During the pre-Inka period community control was probably limited to nearby resources, since hostile neighboring communities would have opposed any attempt to maintain more distant control (Rowe

1946: 274; LeBlanc 1981). Even the restriction to nearby resources, however, would have permitted considerable community self-sufficiency. Communities in different zones would have had access to different resources, and some intercommunity exchange seems probable.

A second type of resource control was exercised by the archipelago community, in which the main community settlement was many days distant from key resource zones like the tropical lowland agricultural areas. The ayllu in effect colonized these resource zones, establishing satellite settlements there and arranging for the long-distance transport of goods by porters or by llama caravans. For the Inka period this type of extended community control has been documented for various locations through the empire (Murra 1972), among them the Mantaro valley communities of the sierra, whose land included lower-elevation zones to the east that produced crops like coca and aji. Although 30 miles or more and high mountains separate this tropical agricultural zone from the valley, we know from historical documents (Vega 1965 [1582]: 168, 172-74; LeVine 1979) that highland communities controlled small villages there.

Can the archipelago community be documented from pre-Inka times or was it an outcome of the conquest? Although complete returns are not in from a number of archaeological projects that are addressing this problem (Hastings 1982; Lynch 1982), what evidence we now have for pre-Inka archipelago communities is at best thin. In times of intercommunity hostility and warfare any such commitment of community resources would surely have been impractical because of the prohibitive costs of defense. It seems reasonable, then, that such communities first made their appearance after the conquest, when the Inka state was in a position to maintain peace and guarantee resource ownership.

Organization above the level of the ayllu is little known and poorly studied. We know that some settlements in the pre-Inka period were quite large and thus probably composed of several ayllu; in the Mantaro valley these large communities were a response to warfare (LeBlanc 1981). During the Inka period there were town-size settlements consisting of several ayllu, and it seems that these may have been related to specific economic and political contingencies of the empire.

Beyond these multi-ayllu communities there was a broader regional formation of ethnic groups with closely related languages, customs, and culture histories. In the upper Mantaro valley, for example, the local ethnic group was the Wanka, and modern communities in the area still identify themselves as Wanka. Prior to the Inka conquest the Wanka did not form a united political group; local communities were politically autonomous and waged war with neighboring Wanka communities (LeBlanc 1981). Although intercommunity exchange and political alliance took place, a community was politically separate in most affairs.

Ethnicity, however, became very important under Inka domination, when the state was divided into ethnically homogeneous provinces. A hierarchy among a province's ayllu, reflecting differences in economic wealth and political relationship to the Inka, was translated into control of the administrative offices of the province's districts and subdistricts. The overall province, however, had no traditional basis beyond general ethnicity, and administrative control was vested in a nonlocal Inka official.

Although the Andean ayllu has often been pictured as egalitarian and organized by principles of kinship and reciprocity, leadership and incipient social differentiation were important at least in some Andean areas. The ayllu leader (*curaca*) was an incipient aristocrat. The position descended in a local patriline, with some flexibility of choice among possible candidates (Rostworoski 1961). Certain specified lands were worked by ayllu members as part of a general obligation to provide for the curaca (Moore 1958: 527), and he apparently also had some rights to local labor and to special resources such as metals and coca (Moore 1958: 39).

In return for control over the community's agricultural and nonagricultural resources and its labor, the curaca was responsible for settling disputes, allocating agricultural lands, and organizing community activities, including local ceremonies and communal labor groups for work on state lands. As a member of the elite, a community leader, and a conductor of ceremonies the curaca is similar to the community chief as discussed in Chapter 10; the main difference lay in his linkage to the state as a local bureaucrat.

The ayllu was primarily organized to solve problems of basic subsistence at both the household and local community levels. In the household resources were pooled in generalized reciprocity; in the ayllu kinship ties were a basis for balanced reciprocal exchanges. On this egalitarian system was imposed a system of social and economic differentiation, with leaders supported primarily by labor contributions from community members. In pre-Inka times the curaca apparently was needed largely for warfare and defense (LeBlanc 1981; Scott and Earle n.d.), but under the Inka the position was transformed.

The Political Economy

The Inka empire was built economically and politically on a base of local communities. It made creative use of existing institutions of finance and control and developed new institutions. The empire arose

out of a social milieu of chiefdoms—socially stratified societies engaged in intense competition for scarce land and other resources
(Toledo 1940 [1570]: 169). The rapid transformation to an empire was
made possible by a shift in goals away from the conquest of land and
the expulsion of defeated populations to the conquest of populations
and the incorporation of their productive capabilities into the financial base of the expanding political system (Rowe 1946: 203).

In many ways the Inka state was like a huge chiefdom. As in Hawaiian chiefdoms, political office was gained by competition among a potential group of hereditary elites, each seeking to marshal support
among different factions. Office carried with it rights to income (Moore
1958: 32), and competition for the office of ruling Inka thus prolifer-
ated into a competition among elite factions for control of desirable
political office. The government was manned by Inka elites in high
position, and thus at least initially there was no separation between
the social elite and the governing bureaucracy.

Nor was religion in any sense an independent institution. The state
religion was represented at administrative settlements throughout the
empire by temple mounds or *ushnu* that stood prominently in the
main plaza and acted as a focus for ceremonial occasions, where they
proclaimed the ruler's divinity and thus his legitimacy. The stability
and fertility of the natural world, being dependent on the super-
natural, were mediated by the ruling Inka. The Inka also tried explicitly to integrate local regions into the empire by moving their main
idols to the capital of Cuzco, where they were placed in state shrines
(Rowe 1946). In our discussion of chiefdoms, we emphasized the
highly generalized nature of the chief as a representative of the social
elite, a political leader, and a divine person; and so it continued to be
in early empires like the Inka, with religious institutions serving as
an important agent of social and political integration (Conrad and
Demarest 1984; cf. Kurtz 1978).

In sharp contrast to Hawaiian chiefdoms, however, the Inka empire
incorporated a vast population of many ethnic groups; and this led to
problems of integration and control that no chiefdom could solve. A
bureaucracy was needed for the ongoing management of state affairs,
and an army to maintain internal peace and repulse external threat:
not a dozen kinsmen and their followers, as in the Hawaiian chiefdoms,
but hundreds or even thousands of specialists linked together in large
hierarchical institutions.

The way state societies develop specialized institutions from earlier
precedents is seen clearly in the economic organization of finance and
production under the Inka state as described by Murra (1980 [1956];

1975). In the pre-Inka period, as we have seen, the curaca financed his position through staple goods grown on lands designated for his use and farmed by commoners as part of their community obligation. On a massive, empire-wide scale this was the financial base for the Inka state.

After conquering a new region the state asserted its ownership of all that region's lands. These lands were then divided into three sections whose produce went respectively to support the state bureaucracy and military, the state religion, and the local community. The community lands remained residually under state ownership, but the right to use them was granted to the community in exchange for its *mit'a* or obligatory labor on state and religious lands and on other state projects such as road maintenance, canal construction, and mining. An ideology of reciprocity was maintained: the use of land, the means of subsistence, was given in exchange for labor in state activities (Wachtel 1977: 66).

The Inka state economy was based on staple finance. Staple foods, including maize, potatoes, and quinoa, were grown on state lands by community labor. Following the harvest the food products were stored in state granaries and used to feed state administrators, military personnel, and others working for the state, including commoners working off their labor obligation. Commoner communities were also obligated to produce craft goods for state use; each family was required to spin wool provided from state herds and to weave a certain amount of cloth, such as one blanket, each year (Murra 1962). (This right to cloth goods may have originated with the community leader, who received products such as shirts and bags woven for him by his support group.) Cloth could then be used as a political currency (D'Altroy and Earle 1985). Thus the state's control over production gave it both products that could be used or consumed immediately by state personnel and convertible, storable wealth to use in later payments.

Although the system of finance through compulsory labor had precedents in the local pre-Inka economy, its scale in the Inka state led to a number of significant changes. One was the introduction of record keeping—not by the introduction of a writing system, as in other early states, but by *khipu*, mnemonic devices with rows of knotted strings used to record the transfer of goods. Local khipu specialists were employed by the state to record all goods going into and coming out of the state's many local storehouses.

Storage was also greatly elaborated under Inka domination. During the pre-Inka period the best evidence for centralized storage complexes is found in the coastal states such as Chimu (Day 1982); in the highlands storage was mainly at the household level (Earle and D'Altroy 1982). The Inka empire, by contrast, needed massive storehouses to hold the staples and craft goods produced for the state. In the Mantaro valley, for example, over 2,000 individual storage units, small silo-like structures, were constructed in orderly rows placed throughout the valley (Earle and D'Altroy 1982). Many of these storage units were placed on the hills directly above the major Inka administrative center of Hatun Xauxa, but an equal number were distributed through the valley in close association with local community settlements. Those on the hills presumably provided for the support of state personnel at Hatun Xauxa, including administrators, state officials on local inspections, and the military. Those in the valley supported state activities in the local communities, including agricultural work, public works projects, and such craft industries as pottery and metal production.

Additionally these state stores would have provided the local resources necessary for supporting military operations, if necessary, and for maintaining local political stability. According to the chronicles, as summarized by Murra (1980 [1956]), they were also frequently to supplement local shortfalls resulting from crop failure. Although reciprocal exchange relationships between families were the first and best way to get through a difficult period, the state provided stored goods as a last resort, thus performing a service that was formerly the curaca's responsibility as both ritual leader and economic manager.

The Inka state also sponsored massive new irrigation and terracing projects, one of which, in the Cochabamba valley in Bolivia (Wachtel 1982; M. LaLone 1985), supported state institutions as far away as Cuzco. The storehouses at this project were maintained by *mitmaq* (ethnic populations removed by the Inka from their native land and resettled in foreign lands), and the land was farmed in rotation by various groups as part of their mit'a labor. Since mitmaq had no traditional claims to the land where they were settled, their rights and broader economic position depended entirely on the state. Mitmaq were not only storehouse workers but farmers of newly developed land and specialists in such crafts as pottery and textiles.

As a continuation of earlier economic arrangements with their curacas, local populations provided the state with craft goods such as cloth, sandals, valuables used as gifts and payments, and probably ceramics. Additionally, villages with special crafts, such as metallurgy and stonemasonry, were required to provide specialists to work for the state. Removed like the mitmaq from their native communities with their traditional system of rights and obligations, these individual specialists were attached to state institutions for which they labored in workshops or work teams.

Among these specialist retainers were the *aclla* or "chosen women," who were weavers attached to the state religious institution (Rowe 1946: 269). Recruited from communities through the empire, these women lived in administrative settlements, where they wove *cumbi*, a particularly fine grade of cloth, and made *chicha*, a kind of beer. Cumbi was a major wealth item in the empire, used especially for political gifts and ceremonial payments. The aclla represented a semi-industrialized form of production, organized for the large-scale manufacture of this highly specific product.

Another category of specialists, called *yana*, worked directly as agricultural laborers and domestic servants for elite patrons and shrines (Murra 1980 [1956]). Some researchers describe the yana as slaves because of their lifetime attachment to an "owner," but apparently they enjoyed many freedoms; and only one of a yana couple's children was required to remain with their father's employer.

The main importance of the mitmaq, the aclla, and the yana is the change they represent in the relations of production. In the characteristic corvée or mit'a system, production is basically organized at the community and household levels, with the products of the labor provided as a rent. By contrast, these new groups were detached from the community and organized by governmental institutions and elites. As described by Murra (1980 [1956]) and Schaedel (1978), this restructuring of production transcends the limits imposed by community production and is a key organizational shift required by state societies to meet their expanded and increasingly specific needs.

Like the Chinese empire, which monopolized the production and sale of salt and iron, the Inka empire raised income by exercising a monopoly over certain important products that were in wide demand. Early chroniclers state that coca, the Andean equivalent of tobacco, was controlled by the state (see Rowe 1946; Moore 1958), which may even have attempted to expand the market demand for it by emphasizing its ceremonial importance in Inka rituals. Similarly all metal mines were owned by the state, being worked as part of a community's labor obligations under the direction of the curaca (Moore 1958: 39).

Indeed, the curaca was a central figure in the operation and finance of the Inka empire. Important in pre-Inka times, at least in highland areas mainly for his leadership in warfare, in Inka times the curaca was selected and supported by the state on the basis of his economic efficacy. The curaca was in a pivotal position: his authority relied both on a local heritage of rights and obligations and on the state's guarantee of support. In the Mantaro valley (D'Altroy 1981) the status of the curaca and the strength of his control were greatly reinforced by imperial incorporation, and local elites accordingly remained strongly disposed to further the state's interests in the region.

Reasons for Inka Imperial Success

A state such as the Inka may be pictured by conflict theorists as operated by ruthless exploiters, or alternatively by functionalist consensus theorists as operated by beneficent managers. It was, and had to be, a little of both, depending as it did on a balance between exploitation and management. Inka rule is best described as rule by enlightened self-interest (Rowe 1946: 273). The empire was financed by mobilizing labor to produce staples and crafts, to construct public works projects, and to support the army; all households of the local community were required to provide corvée labor to these ends. In return the state provided resources and services to the local community that were essential to its subsistence economy, notably orderly access to agricultural land and pastures. Conquest thus established a new set of relations to the means of production that guaranteed the dependence of the local community.

An even greater service provided by the empire to its local communities was that of bringing intercommunity warfare to an end. Among the Wanka, for example, we can document a dramatic improvement in the diet and lifespan of both elite and commoners following the Inka conquest (Earle et al. 1986). The state as grantor of land in return for corvée labor also guaranteed a community's use rights, thus permitting some local communities to extend their resource control vertically and improve the stability and self-sufficiency of their subsistence economy. The state's monopoly over certain goods almost certainly made those goods available to distant communities, often for the first time. And finally, as we have seen, the state's storehouses, though constructed primarily to finance its own activities, provided a residual supply of food for the populace in times of need.

The enlightened self-interest of the Inka empire was characteristic of archaic states, in which the relationship between the subsistence and political economies is carefully balanced. The state continued to depend on the local community for labor and staple products. The community, in return, became dependent on the state. It was of course in the clear economic interest of the state to provide services and resources to strengthen the bond of dependency and to maintain the productive potential of the community, the state's financial base.

Why were the Inka successful in the fifteenth century and not before? Earlier states had existed on the coast of the Central Andes region, notably the artistically renowned Moche state and the Chimu state (Lumbreras 1974); and in the highlands the imperial Wari state had long since established an extensive road system and administrative settlements (Isbell and Schreiber 1978). In part, therefore, the Inka empire can be seen as built on earlier precedent.

But the real key to the Inka's success was a series of developments in the subsistence economy. Long-term population growth through the central Andes had led to a marked escalation in intercommunity warfare and a major intensification of agriculture based on irrigation, terracing, and drained fields. A need for local leaders, mainly for warfare, led to the development of social stratification and chiefdoms throughout the highlands. In turn the high population density, the dependency on capital-intensive agriculture, and the existence of local elites created the ideal opportunity for incorporating these chiefdoms into an imperial state.

Above all, the Inka came along when people were tired of war and ready to appreciate the advantages of peace. The imposition of peace on a region removed the tremendous costs of military preparedness, which included not only the direct costs of maintaining a fighting force and fortifications but the indirect costs of inefficiencies and losses in subsistence production (Schaedel 1978). The restoration of peace and order released a tremendous potential energy surplus, which was channeled by the state into the serving of its own political and social purposes.

Conclusions

Intensification of the subsistence economy is a necessary but insufficient condition for state formation. The necessity of increasing food production, resulting from a consistent growth in population preceding state formation, leads to a filling in of the landscape, capital improvements, carefully managed rotation cycles, clearly demarcated land tenure, intense competition over productive lands, and ultimately a rural population dense enough to support market systems and a specialized urban sector. Without such conditions states cannot exist, except perhaps as satellites tied in through close economic relationships to a major state society. But even where all these conditions obtain, certain measures of economic control and political integration must be taken before a viable state can exist.

Integration on a massive regional or interregional scale is a defining

characteristic of states. Minimally this integration involves a bureaucracy, a military establishment, and an institutionalized state religion.
These institutions assure the state adequate finance, capable economic management, stability, and legitimacy. Over and above these
fundamentally political institutions the establishment of regional
peace by a powerful state permits a rapid increase in economic integration, either through the development of markets and trade, as in
medieval France and Japan, or in the extension of community territories to incorporate diverse production systems, as in the Inka case.

All states are stratified or some more genteel equivalent. They have
to be, because the very institutions of state that are necessary to prevent economic chaos are based on a reliable income for finance. This
income is only possible with economic control, and this control translates into rule by an elite whether socially, politically, or religiously
marked. At the state level stratification appears to be inevitable. The
socialistic and democratic alternatives seem only to decorate a fundamental stratification with an ideology of egalitarianism. As much as
we cringe from this conclusion, the only solution seems a comprehen-
sive simplification of world economic problems made impossible with
pressing populations.

Basic to both state finance and stratification is this element of control. As we have seen, there are two main kinds of control: control
over production, made possible by such technological developments
as irrigation or more weakly by short-fallow, carefully managed farm
lands; and control over distribution (trade), made possible by market
development and the generation of mercantile wealth. In the first instance stratification is defined by two classes: a ruling and landowning elite class, and a producer class of commoners. In the second instance a third class is also present: a merchant class, often attached in
one way or another to the ruling class.

As we have argued throughout the chapter, states can be formed
only where two sets of conditions are present: high population density, with explicit needs for an overarching system of integration; and
opportunities for sufficient economic control to permit the stable finance of regional institutions and to support a ruling class. Where
these two sets of conditions occur together, we find the rapid expansion of the political economy and the beginning of the state.

TWELVE

The Peasant Economy in the Agrarian State

IN CHAPTER 11 we viewed the development of the state from the over-all perspective of the larger system. Here we turn to a more ethnographic view of the state-level economy, focusing on the peasant household and the local community and describing the economy from the ground up. "Peasant society" is a label that applies to a wide range of social systems, each so complex and multitiered that we cannot hope to offer a complete explanation or even a complete typology of peasant economies. Belshaw (1965: 53-58), Halperin and Dow (1977), Potter et al. (1967), C. Smith (1976), and Wolf (1966a) provide admirable overviews and case studies.

Peasant economies are characterized by a relatively high population density and a relatively advanced intensification of production. But so are complex chiefdoms, and we have seen that in stratified societies features of the production system alone no longer serve to distinguish evolutionary states; increasingly the local economy must be understood in the context of a regional economy integrated by market exchange.

In this chapter we review three cases, presented in ascending order of population density and degree of intensification of production. The most meaningful contrast is between the first, a Brazilian *fazenda* (plantation) on which a landlord and certain other local patrons stand as gatekeepers between peasant food producers and the market-dominated political economy, and the other two, villages in China and Java in which the peasant household directly sells its own labor and products on the open market. These three cases represent different points along a continuum of "commercialization" (C. Smith 1976) that is a basic dimension distinguishing types of peasant economy.

We can array state systems along this same continuum. At one end are feudalistic, localized, class-structured societies like complex chiefdoms and early states. Exchanges between strata of society and between ecologically and ethnically distinct regional populations are managed by a mixture of bureaucratic and market mechanisms. Control of land is the basis of wealth and power. At the other end are nation-states, that is, states whose economies are integrated more by the price-fixing competitive market than by hierarchical chains of dependence or bureaucratic controls. Power and wealth in the nation-state depend more on control of the market than on control of land.

In medieval France and Japan (Chapter 11) the food-producing class had only a limited degree of market involvement, and the market itself was nothing more than a localized exchange system controlled by the lord and limited to the area of his political influence. But as those systems evolved into nation-states, markets expanded rapidly. Some ruling elites benefited from the process, others were reluctant to have their monopolistic powers shattered, but there was no halting the market's rapid expansion into the rural hinterlands (see, for example, C. Smith 1976: 356-60).

We find, therefore, an evolutionary development from dependent peasants, bound to a lord who mediates their interaction with other peasants and elites, to independent or "free" peasants who compete directly in a marketplace for access to land, jobs, manufactures, and other essentials of life. In our view this "freeing" of the peasant is a continuation of the evolutionary expansion of the political economy. The economy is now so huge that any effort to move labor and goods through the system by the use of personal, hierarchical chains of command is necessarily less efficient than reliance on the impersonal "free market." In essence the evolution from the complex chiefdom and the archaic state into a market-integrated nation-state is characterized by the increasing dominance of the economy by a competitive, price-fixing market, a dominance made possible by an institutional framework largely devoted to nurturing and protecting the market system.

The peasant household is a highly productive economic unit whose production comes to be divided into four portions, or "funds" (Wolf 1966a). A maintenance fund ("caloric minimum") serves to meet the household's requirements for food. A "replacement fund" assures the continued economic viability of the household economy by providing housing, tools, work animals, seeds, and other necessary and recurrent inputs of production. A "ceremonial fund" consists of expenditures necessary to maintain economically essential ties to other peasant households and to the village community. Finally a "fund of rent," including not only rent but also crop shares, taxes, and other transfers of household production to elites, is paid in return for access to the means of production, especially land, and to security measures under elite control. These four funds can be conceived as meeting obligations at different levels of organization. The maintenance and replacement funds meet the consumption and production requirements of the household. The ceremonial fund meets the social needs of participation in the community. The rent fund supplies necessary payment to the regional elites of the state society.

In evolutionary terms the fund of rent is the final and most burdensome form of intrusion of the political economy into the household economy. It began as a reluctant "gift" from the producer to one or more current Big Men, hardened into the tribute demanded by a powerful chief, and eventually became the legally sanctioned right of landowners and bureaucrats to a share of peasant production. Only the maintenance fund represents food consumed by the peasant household. It is the small and often inadequate proportion of total production that remains after the other three funds have been paid out.

Our case studies will show state influences at work at all levels of the economy: the intensification of production through such methods as irrigation and use of manufactured tools and fertilizers; the regional integration of the economy through markets for labor and produce; and the stratification of the labor force into many varieties both of primary producers and of owners, managers, and bureaucrats.

Two of our cases also illustrate a phenomenon that is beyond the scope of our book but is of great interest as an extension of processes we have examined: namely, the penetration of the world market into the local economy. In Brazil and Java a sort of "dual economy" has been created by the intrusion of cash crops, primarily cane sugar in both cases, into prime agricultural lands that originally supported agrarian populations by starchy staple production. Although the switch to cash cropping denied local populations access to the best lands for food production, it stimulated the development of formerly marginal areas, where state-supported technological investments underwrote huge increases in food production. It also increased the participation of subsistence farmers in a labor market and finally shattered whatever remained of the self-sufficiency of the farm household.

It is difficult to say with any certainty, but the integration of these regions into the world economy may also have encouraged large increases in population by increasing what Chayanov (1966) called the peasant family's capacity for "self-exploitation," i.e., for selling its labor, typically at very low wages. The process goes on. Now it is nation-states that are coming to be seen as component parts of an increasingly unified world economy controlled by powerful but as yet ill-defined international collectivities, in much the way Big Man societies, managed by individualistic, aggressive, alternately hostile and friendly leaders, once groped their way toward the formation of larger collectivities that for better or for worse would transform their way of life.

Case 17. The Brazilian Sharecroppers of Boa Ventura

Our first case illustrates the economy of "dependent" peasants who live under the direct control of landlords. The major production unit of this economy is the nuclear family household, which is linked individualistically to other households and to landlords and other elites through bonds of "friendship" maintained by frequent gift exchanges. Despite a semiarid climate and a terrain unsuited to irrigation, government water works and careful management of land use by landlords have made possible a comparatively dense population that produces staple crops for its own consumption while helping the landlord to raise cattle, cotton, and other products for sale. From our perspective the landlord represents a kind of transitional middleman between the peasant and the state. In more fully commercialized economies such paternalistic middlemen tend to decline in importance, a sign of the growing dominance of the market at all levels of the economy.

The Environment and the Economy

The peasants of Boa Ventura are tenants on a fazenda in Ceará, northeastern Brazil, a region distinguished by a rich, productive humid zone (*littoral*) along the coast and an impoverished, semiarid zone (*sertão*) in the interior (A. Johnson 1971a). Prior to the European conquest the littoral was occupied by warlike horticultural villagers (Tupinamba) who raised root crops and maize in slash-and-burn gardens similar to those described for the Yanomamo in Chapter 5. The interior was sparsely inhabited by hunter-gatherers.

Shortly after the European conquest the humid littoral was taken over for the production of export crops, particularly sugar. This land being now too valuable for food crops, the semiarid sertão was gradually populated by farmers who grew staple foods for sale and reared cattle to provide meat and draft animals to the coast. The labor for these interior fazendas was provided by subsistence-oriented peasant families, who adopted the horticultural methods of their native American predecessors. Even today we find a basic subsistence economy in Ceará that is hardly distinguishable from preconquest horticulture, though it is now overlain by an export economy devoted to the production of cotton, sugar, cacao, cattle, and other commodities.

Over the past century, owing primarily to the construction of large reservoirs and irrigation networks, the population of the sertão has expanded to a density of about 30 persons per square mile. Whether on individual smallholdings or on larger fazendas, people in the sertão prefer to live in nuclear family households with as much autonomy as possible in economic decisions. The vast majority are subsistence farmers practicing bush-fallow horticulture.

The intensification of the subsistence economy and the creation of a market-oriented production system have created a severe shortage of agricultural land. Land is largely owned by an elite class, which manages it to obtain a profit in urban and international markets. Landlords provide houses, water, and land to their tenants, requiring them in return to plant certain crops like cotton, rice, or bananas and sell them at low prices to the landlord. Altogether, by paying shares or giving days of labor to the fazenda, a tenant pays out about 25-30 percent of his total production as rent.

Sharecroppers slash-and-burn their gardens from second growth that has lain fallow for about eight years. Because of the high risk of brush fires during the long dry season, workers carefully clear firebreaks at the borders of their gardens, sweeping them clean with brooms of brush. After the rains begin they plant gardens of intermixed crops such as maize, manioc, beans, squash, sesame, peanuts, and potatoes. In the second year the number of food crops is reduced to make way for tree cotton, which becomes the only crop by the third year and is cultivated for several more years before the field is returned to full fallow.

Sharecroppers obtain virtually all their food from gardens and from backyard animals fed with garden produce. Pigs and goats are raised and when slaughtered are occasions for meat sharing in repayment of earlier gifts. But most protein comes from beans, the "strong food" (*comida forte*) of the region, without which no meal is considered nourishing. An occasional man hunts small birds or rodents with a rifle, but fishing areas are controlled by landlords and are fished by full-time specialists with exclusive contracts. In the broad scheme of things wild foods are of little economic significance to the peasant household.

Tenant households each farm a clearly defined field and decide when, what, and how much to plant, when to weed, etc., with little outside influence. They also take most of the risks of production and thus are true peasants rather than farmworkers; but they can be forced off the land at the landlord's pleasure and thus are dependent peasants in contrast to independent, landowning peasants. In this dependence, and in their personal ties with landlords and other elites, they resemble the farmers found in chiefdoms.

Rainfall in the sertão is unpredictable, crop pests are an ever-present threat, and the land the peasant works belongs to someone else. These insecurities generate economic and social strategies aimed at increasing security even at the expense of some "profit" (A. Johnson 1971b). For example, a sharecropper does not try to plant a single "best crop" in the best land to maximize production, but instead plants a wide mix of crops in as many microenvironments as possible: dry hillsides, fertile riverbanks, the river bottom during the dry season, low-lying humid soils, and the margins of reservoirs (Johnson 1972). Whether the year is wet or dry, the risk-spreading farmer is assured of some food for his larder. This is risk management at the household level, of course, not involving group-level strategies organized by local leaders.

Another strategy is to store a year's supply of food at harvest time, and only then contemplate selling any surplus on the market. This household security strategy has two serious consequences. First, the amounts of food reaching the market fluctuate wildly from year to year; and the resulting insecurity of the food supply for the urban, nonagricultural population can lead to political disturbances. Second, surplus food tends to come on the market all at the same time, after the peasants see how their new crops are progressing and before prices fall with the new harvest; for this reason, the market value of staples like maize and beans begins to fall a month or two before the first harvests of the new year actually reach the marketplace, and peasants are paid less for their food than they might have been under other circumstances.

Perhaps unexpectedly, insecurity does not lead to the complete lack of innovation and experimentation that some observers have attributed to peasant agriculture (Schultz 1964; Wolf 1966a: 16). The sharecroppers are as interested in new crop varieties and techniques as farmers anywhere. Men constantly discuss new crops they have seen during their travels, and try to obtain seeds of those crops for planting. They even perform controlled experiments in their gardens, planting two varieties of seed or using two planting techniques side by side to see which does best. They are not careless of the risk involved, but they minimize risk by restricting innovation to small experimental plots in which crop failure will have little effect on a household's overall production.

In fact, although there are many ideas, methods, and "rules of thumb" that most tenant farmers accept, the degree of individual variation in horticultural practices is remarkable. The reasons for this are various. For one thing, each household has a different composition of producers and consumers, and both of these affect the amount of land a household has under production. For another, people have rather firm opinions about how to farm, even when these opinions differ from those of their neighbors. This leads to no end of disagreement and even disparagement between otherwise friendly tenants. Finally, there are large differences in individual intelligence, skill, and motivation, and these are reflected in wealth and prestige differences between households (cf. Cancian 1972).

Nonetheless the sharecroppers of the sertão live so close to the margin of survival that they visibly lose weight in the months before harvest. The poorest families may be unable to reach their goal of a full year's stored food supply, and may thus experience a shortage of food at a time when their work efforts reach a peak. Children are especially likely to receive less than their share of food (cf. Gross and Underwood 1971), and infant mortality and clinical malnutrition are high. In the frequent years of low rainfall larger numbers of families suffer, and in the periodic droughts all peasant households face basic survival threats. In addition to their risk-spreading strategies in food production at the household level, therefore, they pursue various social means to greater security.

Social Organization

Kinship is less important as a source of social security among Brazilian sharecroppers than among most groups discussed in earlier chapters. Indeed, to hear the sharecroppers tell it, kinsmen are considered unreliable and of little value. Yet groups of kinsmen do form residential clusters here and there on the fazenda (A. Johnson and Bond 1974); and even kinsmen who live some distance apart on the fazenda maintain close exchange ties, whereas non-kin form such ties only with near neighbors.

The emphasis of social relations between tenant households is on "friendship." The importance of friendship in the social organization of peasant communities has been established by Foster (1961) and Wolf (1966b). Foster's model of the "dyadic contract" sets forth the essential features of friendship relations in a Mexican village as follows:

1. Relations are always dyadic. Although friendship between two people inevitably entails relations with the friend's friends, dyadic contracts are structured to minimize such extensions so that friends need not assume responsibility for each other's complete network of social and economic obligations.

2. Short-term exchanges are usually out of balance; that is, one friend "owes" the other. The debt, which shifts back and forth as gift repays gift, is a welcome sign of trust, and any effort to pay the debt off exactly is seen as an effort to end the friendship.

3. Over the long term exchanges should strike a balance so that each friend finds the exchange "fair" (cf. Homans 1958).

4. Friendships are entered into and terminated freely. They are thus fundamentally different from kinship relations, debt-peonage relations, and other structural social relations a person cannot avoid. The economic importance of friendship in peasant society derives especially from this characteristic: one chooses one's network of friends with one's own best interests in mind, and friends who fail to carry their weight can be dropped or avoided.

5. Since friendships are fragile, with weak structural underpinnings, frequent exchanges between friends are necessary to keep the relationship vital and reliable. Most gifts between friends are small, mere tokens of friendship. Overt displays of gratitude are avoided, since they may be interpreted as attempts to end the relationship by paying off the debt with gratitude.

For Brazilian sharecroppers friends add much to the economic security and well-being of the household. Through their separate networks of friends, men and women obtain fresh meat, foods not grown in their own gardens, small temporary loans of cash, days of labor in times of critical labor need, the loan of tools they might not own, and other special favors. They work hard to establish and maintain these friendships, and when their efforts fail the disappointment may be bitter. Close kinsmen and many near neighbors maintain the frequent exchanges that mark friendship ties. But since too many ties can be cumbersome, close regular friendships are limited to two or three per person. Thus in the larger kin clusters not all members exchange equally with each other; some pairs of kin behave like friends, while others treat one another in economic terms more or less like non-kin.

Foster (1961) points out how different this casual, individual-centered relationship is from the rigidly structured, group-centered kinship groups of intensive horticultural societies. Yet friends make important contributions to the economic security and autonomy of households everywhere: in fact, where kin groups control the individual economic lives of their members, "friends" tend to be chosen from among non-kinsmen, to help one another and to serve as buffers against the intrusions of the kin group (A. Johnson and Bond 1974). Although Boa Ventura contains a "village-level" population of about 300, it is a community only in the sense that it is subject to the management policies of the landlord. The significant economic groups in daily life tend to be much smaller than the fazenda community, being limited to hamlet-size clusters of kinsmen and friends in small neighborhoods.

Market Involvement

In addition to the social responses described above, the small landholdings and unpredictable yields of the sharecroppers necessitate some degree of market involvement. The sharecropper often works part time in craft activities to augment his meager income, and sells his agricultural surplus in good years to provide a buffer against bad times to come. Individual exchange ties integrate the fazenda community and extend beyond the fazenda.

A fazenda usually contains an array of economic specialists. Most in demand are blacksmiths to manufacture and repair tools, carpenters to make doors, windows, and furniture, and masons to construct buildings. These specialists are all sharecroppers whose part-time specialties increase their income; they receive lower wages for their work than specialists who live in town, but they enjoy greater security. Specialists who live in town have no gardens to fall back on when customers do not pay for their work, and they need powerful kinsmen or patrons in order to survive. One carpenter who left the fazenda in 1966 to try his luck in town was back in 1967, his savings wiped out in a financially disastrous half-year; having no patron, he had no way to collect what his customers owed him (A. Johnson 1971a: 90-91).

Specialists on other fazendas are also available when needed, but most tenant households obtain the things they do not produce from shops on or off the fazenda. These shops are supplied from markets in town, but peasants rarely go to those markets. Instead they do their buying and selling of crops and commodities through the landlord and shopkeepers, with whom they try to maintain close personal ties very like the "friendship" ties they establish with one another.

Their rare attempts to raise cash crops or to use hired labor invariably fail, either because they cannot control some factor of production (e.g., obtain transportation to market perishable goods) or because their profit margin is too low to sustain the risks of production over any length of time (e.g., losses due to bad weather in one year more than overbalance the profits from a good year).

In such systems the market is in the control of the middlemen and elites. Landlords bulk produce from their own fazendas and ship it to warehouses or other rural bulking centers, where it is stored, partially processed or packaged, and shipped. Markets are primarily places where urban dwellers buy produce that has been bought wholesale from warehouses by middlemen who then break it down into small quantities for resale. Some farmers do sell directly on the market; but they tend to be independent cash-cropping truck farmers, not sharecroppers from outlying fazendas.

Marketing and the Political Economy

Every sharecropper on a fazenda has access to several shops, and since prices on individual items vary it pays to shop around. The best prices are usually found in shops that accept only cash payments and never extend credit. Few sharecroppers, however, can afford to pay cash at all times; credit during scarce times is essential to their survival, and in order to obtain credit a man must become a "loyal customer" (*fregues*) of a single shopkeeper, whose prices are higher because by extending credit he runs a higher risk of nonpayment. Ironically, then, because the cost of credit is embedded in food prices the poorest sharecroppers must pay higher prices for food than their economically more comfortable neighbors, in keeping with the universal principle that "the poor pay more" (Caplovitz 1963). Sharecroppers understand this very well, but cannot forgo the security of being able to obtain beans, manioc flour, kerosene, and cooking oil on credit when there is no cash and the family is hungry.

A similar logic applies in a more general way to relations with the landlord. Dependent peasants do not see the landlord as "the greatest social enemy of the peasant" (Quijano 1967; cf. Feder 1971), but as a potential ally of fundamental importance to their well-being. They actively seek to transform him into a personal patron, since by calling on his personal resources and those of his network of powerful patrons and clients he can provide almost any service a tenant might need.

A change of ownership of a fazenda took place while one of the authors was doing fieldwork in the area. The original owner, known as "The General," had sold the fazenda to a wealthy merchant called Seu

Clovis. Although Seu Clovis espoused democratic ideals and reduced the tenants' obligations (fund of rent), they were dissatisfied with him and almost universally wished the General would return. Why?

One reason was that the General's family was old and prestigious, and he himself had held various high offices; Seu Clovis, although a wealthy man, was of middle-class origins. Not only did tenants share the values of the larger society, which places the old landed aristocracy above the new commercial class, but they understood that the General's superior class status gave him potentially much greater political and economic influence than Clovis.

The patron's *access* to resources, however important, counts less than his willingness to use that access for his clients' benefit. The General demanded higher shares from his sharecroppers than Clovis, but they regarded him as the more generous patron. The General would buy and sell crops in his own "company store," making food and money available on credit to his tenants (albeit at high rates of interest). He provided milk from his cattle and fruits from his irrigated gardens to families with sick children. And he used his influence to intervene in state-level processes on his tenants' behalf; for example, he had an arrested tenant released from jail immediately, and he obtained free government hospitalization for a woman with cancer.

Clovis abolished the company store as "exploitative" and sold his milk and fruit on the market. He lacked the political influence to intervene as the General had. Although the General lost no profit by being generous, since his rent receipts were higher than Clovis's, he was viewed as a stronger and more protective patron. This view was no doubt reinforced by a certain air of class pride and bravado in the General in contrast to Clovis's more unassuming, middle-class demeanor.

Agrarian states are often "feudalistically" organized in ascending chains of patronage (Silverman 1965). At the top powerful patrons control major resources such as government money for irrigation, road building, farm machinery, and social services. This money they allocate to lower-level clients, who are in turn patrons to small landlords and local political leaders. Each lower-level patron distributes the money to his clients, receiving in return their political support for himself and his higher-level patron (Greenfield 1972). The General, for example, regularly told his tenants how to vote, and delivered them in trucks to the polls on voting day. Because Seu Clovis wished to remain apolitical, his manager took over this function (without which the tenants claimed they could not possibly know how to vote), increasing his own political power as a consequence.

The patron-client tie is so important to dependent peasants that it survives despite its numerous inherent contradictions. Patrons are idealized as parental figures, often called "father," who protect and care for their dependents; clients are called "my children" and are expected to be loyal and devoted. But both patron and client openly acknowledge that their tie is basically an instrumental one that works only when both partners strike a "fair" exchange. Sharecroppers on the fazendas of northeastern Brazil point out that they provide labor, votes, or some other item of value to the landlord, and they will move to a new patron if their old one fails to keep up his part of the exchange.

Patrons, like Big Men, must cultivate loyal followings by acts of generosity; yet the patron-client relation is also a power relation in a class-stratified society, and in this regard the patron is not like a Big Man. Behind the familistic phrases and the "fair exchange" expectations lies the ultimate police and military power of the state, a power that can and will be used to maintain differential access to wealth and resources. Dependent peasants do not negotiate whether or not to pay a fund of rent; the only issues are how large that fund will be and what fringe benefits the patron will offer.

In stable agrarian states most peasants see no alternative to this class structure. Thus their dominant world view amounts to a sort of client consciousness that is a typological opposite of proletarian consciousness. The dependent peasant views his dependence as a source of security and strength (Hutchinson 1966). He feels isolated in a society in which democratic ideals and social security protections either do not exist or else do not reach the peasantry. He does not view other tenants as potential allies in a political movement to gain security through unionization and direct political influence over government programs. Rather he sees fellow tenants as being just as powerless and needy as himself, and as potential rivals for the benevolence of their common patron. Beyond his narrow circle of friends and kinsmen he views members of his class with a competitor's wary regard.

Personalism and patronage are important in all peasant economies, but a single individual patron is more important in our present example than in many others. As we shall see, peasant villages often have political and ceremonial institutions that increase peasants' economic security and help to mediate their relations with the state. Or individual peasant households may seek a range of patrons including physicians, pharmacists, notaries, shopkeepers, work-crew bosses, and even well-to-do peasants, spreading their security-seeking efforts widely. But the basic goal always remains that of reducing the impersonal, distant, bureaucratic, and instrumental elements of the political economy of the state (markets, courts, police, tax offices, etc.) to personal, dependable ties with locally known and trusted individual patrons.

Case 18. The Chinese Villagers of Taitou

Prior to the revolution of 1949 Taitou was an agricultural village of about 700 members in Shantung province in northeastern China (Yang 1945). Although a peasant economy, it was very different from the Brazilian fazenda. Two differences are especially worth emphasizing. First, the population density of the Taitou region ranged from 300 to 500 persons per square mile, more than ten times the density of the Brazilian sertão. Figure 9 shows the Chinese landscape densely packed with villages. Second, the peasants of Taitou, as was common but not universal in China prior to 1949, were independent, landowning peasants who rarely lived as tenants on the property of others (cf. J. Buck 1937: 9). The family economy of the Chinese peasant was heavily colored by the need to acquire and manage land carefully under conditions of extreme land scarcity.

The Subsistence Economy

The Taitou region is one of the oldest agricultural areas of China. Even before collectivization virtually all its land had been put to human use. Microenvironmental differences are of great importance, and most families had small holdings scattered among the different zones, including sandy hillsides where sweet potatoes and peanuts were grown on agricultural terraces, flatland fields of heavier soils where millet and wheat were grown, and tiny and expensive plots of wet-rice. According to Yang (1945: 14),

Even within the environs of a single village there is a wide range in the value of soil. The extreme fragmentation prevents ownership of all the land of a given quality by one or a few families and thereby reduces the possibility of complete crop failure for any one family. Since different land is more or less suited to different crops, a family which has land in several places can grow various kinds of food, will always get some return from its land, and, being, therefore, self-sufficient, has less need to trade.

The main staples of the diet were millet, sweet potatoes, wheat, peanuts, and soybeans. To round out the diet barley, maize, and rice were grown in the fields, and cabbage, turnips, onion, garlic, radishes, cucumbers, spinach, string beans, squashes, peas, and melons were grown in small vegetable gardens.

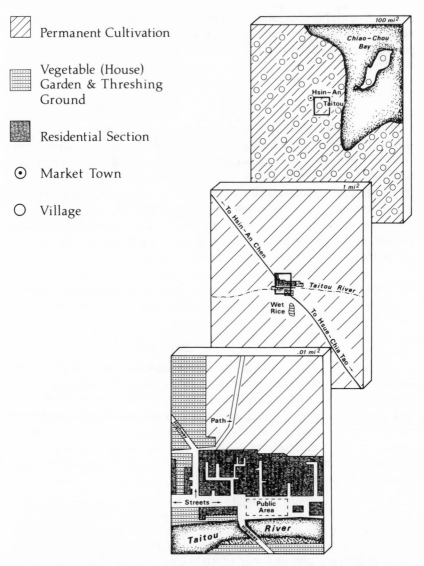

Permanent Cultivation

Vegetable (House)
Garden & Threshing
Ground

Residential Section

⊙ Market Town

○ Village

Fig. 9. Settlement pattern of rural China. The landscape is crowded with villages, each linked to a standard market town. Every inch of land has been used for terraced gardens, rice paddies, paths and roads, and settlements. Each settlement has its blocks of private houses and central public park.

The peasants of Taitou practiced intensive multiple cropping with some seasonal fallowing. The common crop rotation was between winter crops like wheat and barley, and spring crops like sweet potatoes, peanuts, and millet. Every phase of production was accompanied by intensive applications of labor to tease extra produce from the soil. Fertilizer was required for virtually all crops. A family carefully gathered all its animal and human wastes into a compost pit located in the compound of family buildings. When the pit was full, the contents were removed from the compound, covered with mud, and allowed to ferment. Then the compost was sun-dried and ground into fine powder. Ashes from household fires were carefully swept up and added to the compost; even the soot and oxidized bricks from the oven and chimney were periodically ground up and added. Green manure was rarely used, since twigs, brush, and plant stalks were needed as fodder or fuel; but even these materials eventually made their way into the compost heap as dung or ashes. When a field was planted, seeds and fertilizer were carefully mixed by hand to get the proportions right (soybean residue was also added), and then spread by hand on the tilled soil. It was exacting, tedious work, but the people realized that, in Yang's (1945: 17) words, "human labor is cheap and fertilizer and seeds are scarce."

Sweet potatoes, which were a staple of particular importance in poorer families, required heavy labor inputs in all phases of growth. The seedlings were first sprouted in carefully constructed warm, moist beds of sand and then transplanted in heavily fertilized nursery beds, which had to be kept wet. Following the harvest of winter wheat or barley, the field was plowed and carefully ridged. Vines were selected from the nursery beds and transplanted a second time in the ridges, where they were individually hand-watered. They had to be continuously weeded thereafter, and after every rain each plant had to be hand-turned to prevent new roots from growing off the vines into the soil. Ridges required constant repair. Even when the harvest had been completed the sweet potatoes had to be sliced and sun-dried for storage. Yang (1945: 21) comments on the fatigue and muscular pain that accompanied the cultivation of sweet potatoes.

Peasant families generally owned one or two draft animals (mules or oxen) and a number of farm implements: a plow, a harrow, a weeding hoe, a wooden rake, an iron rake, a sickle, a pitchfork, a wheelbarrow. Poorer families did not own all the animals or tools they needed and had to find patrons among the wealthier peasants to loan them what they needed to complete their tool kit. Although many families raised pigs, they rarely ate pork because they needed the money they earned from selling pigs to make purchases essential to the household economy.

Wealth differences between families were substantial, although the villagers themselves tended to deemphasize the extent of stratification in their village. All families had a similar diet, centering on staples like millet and sweet potatoes; but some families were limited to these staples most of the year, whereas others regularly enjoyed wheat bread, fish, and other prized foods. The basic wealth difference was in the amount of land owned. A few successful families owned 20 or more acres of land, many families owned about 10 acres, but the poorest families owned less than two acres. Since the wealthier families tended to be larger, they did not own ten times as much land *per capita* as poor families; nonetheless these figures show a significant degree of stratification in a village of only 700 inhabitants, and interfamily competition was a basic feature of the economy.

Social Organization

The village of Taitou was a compact residential area surrounded by an intensively used landscape (Fig. 9). Its households formed nearly continuous streets of houses, tiny yards, and alleys. At the social center, an attractive open area along the Taitou River, people gathered to pass the time in small tasks of repair or manufacture, in order to hear and repeat the latest news. Surrounding this public area were the houses of the well-to-do, while the poorer neighborhoods tended to be on the village periphery.

The average household contained between five and six members. Wealthy families tended to be larger and lived in large compounds of many rooms. Families were eager to be admired by other villagers for their economic success, and a display of large, sturdy buildings, fine clothing, or fat oxen tied in front of a house evoked neighbors' envy. A wealthy family allowed itself a more varied diet, conducted more elaborate ceremonials, and enjoyed a distinctly superior standard of living.

But a wealthy family that was consuming its wealth rather than saving and investing it was liable to decline. The villagers believed that no family could stay wealthy for as long as four generations; they said of the houses of formerly prestigious families, "Aren't they now only piles of broken bricks and fallen walls?" (Yang 1945: 53). To understand the rise and fall of a peasant family's fortunes, however, we need to examine the social organization of the economy in Taitou.

Although the large family was an ideal, the typical household comprised a single nuclear family or, less commonly, a family extended to include one married son (the "stem family"). We find the expected division of labor by sex into a domestic sphere dominated by women and an external sphere (fields, commerce, politics) dominated by men. The economic complementarity of husband and wife gave the subsistence-oriented household a large measure of economic self-sufficiency, at least in comparison to nonfarming families.

Yet no Taitou household was completely self-sufficient; all had to produce commodities for the market, chiefly peanuts, soybeans, and pigs. The cash obtained for these commodities was needed to pay taxes and to buy food, tools, and other essential goods and services. For farming families life in Taitou involved considerable give and take. They neither made nor repaired their own tools; they also had to pay day laborers during certain phases of the agricultural cycle. Women bought raw cotton in the market and spun it into thread; but they had to pay specialists to dye this thread and weave it into cloth, which they then cut and sewed into clothing for their families. Other specialists in the village included one carpenter, three soy oil pressers, five or six masons, a schoolteacher, and various public officials.

A far more complex division of labor was found in the larger system of villages of which Taitou was a part. In China a unit of major social and economic significance beyond the village was the "standard market area" (Skinner 1964). Taitou and some 20 other villages all did business in one standard market town (Hsinanchen), located about two-thirds of a mile from the village along a dusty road. Hsinanchen was much larger than any of these peasant villages and had large buildings and broad avenues lined with shops and restaurants. There were drugstores, blacksmiths, silversmiths, bakeries, hardware stores, wine-makers, carpenter shops, a bookstore, and many inns and restaurants. On regular market days a large market opened and villagers poured into town. These market days were coordinated with market days in other standard markets in the region so that itinerant tinkers and peddlers could move from one market to the next in succession without missing a market day (Yang 1945: 90-202; cf. Skinner 1964). A network of paths joined the region as an economic entity.

Villagers from Taitou and other villages visited the market town regularly. In addition to buying and selling, villagers established important economic ties. Men obtained credit from shopkeepers and tradesmen that was essential for maintaining their economic production. While drinking in teashops or wineshops, they learned how the regional economy was doing and considered how to gear their own efforts accordingly. Even men who had nothing to buy or sell at the market still traveled there every few days, carrying empty baskets, just to be seen and to keep their lines of credit and communication open.

The regional elites made the standard market town their center of operations. Whereas peasants seldom traveled beyond the borders of their standard market area, elites maintained economic and social ties with elites in other market towns. Higher-level markets and administrative centers were further concentrations of more powerful elites. This central-place hierarchy was almost exclusively economic rather than political; no chain of patronage was associated with it. "In the contract-oriented and highly commercialized traditional Chinese society, 'patron-client' relationships were almost insignificant" (Myron Cohen 1984).

Although peasant farm families predominated in Taitou, many opportunities for income existed in part-time or full-time specialty occupations off the farm. It was the ability of a large, united family to tap these additional sources of wealth and convert them into holdings of farmland that laid the basis for economic stratification in Taitou. Not only was this economic truth plain to all, but powerful ideals, deeply ingrained in childhood and reinforced through ceremonies, religious teachings, mottoes, and folktales, supported family loyalty and unity. Yet most married sons asked for their share of the family estate soon after marriage and set up independent households of their own.

Even though this is a general characteristic of peasant social organization (Myron Cohen 1970: xx-xxiv), it is not always easy to explain why some families remain united and flourish while the majority do not. It is instructive, however, to examine how the decision to set up an independent household was typically arrived at in Taitou. Before marriage a young man worked exclusively for the household, turning all his earnings over to his father and receiving a small allowance at his father's discretion. His parents selected his wife for him; indeed he and his wife might meet face to face for the first time at their wedding. A room was provided for them in the parents' house, where his new wife came under the economic direction of her mother-in-law. The son continued to turn all his income over to his father, and to follow his father's wishes in his choice of career and in any business dealings.

The parents welcomed the daughter-in-law as a source of labor that would increase the wealth and prestige of the household, but they also said "a son is lost when he is married" (Yang 1945: 58). The son's loyalty to his natal family eroded as he became increasingly devoted to his wife and their children. The daughter-in-law encouraged this change. As an outsider she felt no great loyalty to her husband's family; indeed she might wonder pointedly whether her husband's economic contribution to the family exceeded his brothers', and whether, when the family estate was finally divided, her husband would receive a fair share of the wealth he had helped to generate. Yang (1945: 80) speaks of her attitude as "menacing" the communal spirit. Furthermore, daughters and daughters-in-law did not come under the same communal financial control as sons. Daughters were entitled to work at odd jobs to earn money for themselves, and a diligent daughter might accumulate 30 to 50 dollars by the time she was married. After marriage she was entitled to invest this money in chickens, loans, or other enterprises and to keep the profits for herself. Out of these funds she might buy special foods and other gifts for her husband and children, and her savings became a financial base for establishing a separate household when the family estate was divided among the sons (Myron Cohen 1968).

A daughter-in-law was typically treated as a drudge by her mother-in-law, and suffered along with everyone else in the family from her father-in-law's authoritarianism and stinginess. Small wonder that she was always after her husband to leave his parents' household and set up his own. In most families the pressure built to the breaking point and the sons demanded their share of the family estate. This destroyed the basis of the father's absolute economic authority, although some forms of aid and strategic cooperation might continue among family members.

There are two main reasons why large families could gain more wealth than small families. One was the frugality imposed on all members by strict parents. Families seeking to better their lot grimly refused to spend even small sums. Yang (1945: 130) recalls a father lecturing his family in these terms: "Listen, children, there is nothing in this world that can be won easily. A piece of bread must be earned by one day's sweat. You cannot buy any piece of land unless you save all that you can spare through two or three years. The desire for better food, better dress, a good time, or the easy way will lead but to the ruin of our family."

The other advantage of a large family lay in its division of labor. Unless a family already had a great deal of land, it did not need all its sons' labor to run the farm. One son might be enough, freeing other sons to become merchants, tradesmen, artisans, petty officials, or farmworkers. Supporting his extended family as far as possible from its farm income, the father could invest the additional income earned by his sons in new pieces of land. As the family's landholdings increased, it rose to prominence in the village.

A large family, then, was often one whose members were sufficiently motivated by pride and ambition to undertake great individual sacrifice, both in material terms and in lost nuclear family autonomy. Although for a time the successful family's wealth would help it to remain intact, in due course there would be strong pressures to spend the money on a higher standard of living rather than more land, or to divide the family property among the sons and let them decide how to spend their share. Few families could hold up under these pressures. Only occasionally did one hear of middle-aged sons, themselves fathers of large families, still living in their aged father's house, turning their earnings over to him, and accepting his direction in economic matters.

The common pattern was for large families to dissolve into separate nuclear units, which remained tied together by strong loyalties and typically cooperated and lived together in neighborhoods. Owing to patrilocal residence these hamlet-sized groups of families had the same surname and formed "clans" that had certain social security functions. In strong clans well-to-do families helped their less fortunate clansmen, and widows, orphans, the elderly, and the sick were given money and food if they had been moral people and loyal members of the clan. In addition, unrelated neighbors formed dyadic contracts that made everyday life more convenient and enjoyable, joining in one another's ceremonies and doing favors like loaning small sums of money without interest.

The village as a whole was united in common cause for limited purposes. At the time of Yang's study, the main such purpose was defense against outlaws, who were prevalent because the central government was weak. Villagers built and manned barricades, organized armed night patrols, and provided armed battalions that joined similar battalions from neighboring villages in fighting off bandit attacks. Villagers also pooled resources to hire a full-time "crop watcher," who guarded the fields against animal pests and thieves.

Public standards were enforced mainly by gossip, threats of "loss of face" (prestige), and ostracism. Villagers rarely instituted legal proceedings against one another and did not report one another to government officials for lawbreaking. The posture of the village toward the outside was in this sense defensive. As far as possible, disputes were resolved by village leaders, men who commanded respect by virtue of their economic success and "proper" behavior. Government officeholders in the village, by contrast, tended to have low status; to obtain villagers' cooperation in such government projects as road building and canal repair, they first had to win the support of village leaders.

These leaders appear also to have provided minor forms of patronage. Not only did wealthy clan members help their poorer kin, as we have seen, but wealthy families hired workers, loaned money, rented land, and in other ways provided resources poor families needed whether they were kin or not. Poor families had to be respectful, honest, and hardworking in order to obtain such resources, but expected to be treated respectfully and generously in return. For example, unless a family that hired farmworkers on a regular basis provided them with good meals and other favors, it might have trouble hiring workers the next time they were needed.

Our discussion of Taitou refers exclusively to the immediately pre-Maoist era in the Chinese countryside. Since then much has happened. Riding a wave of optimism and hope for a future classless society, the Maoists attempted to convert China's family-based economic and social relations to a system characterized by collectivization and direct state control. They failed; and their failure bears out one message of this book, namely that self-serving individuals and families, far from being the recent products of a depraved capitalism, are the fundamental economic unit in all societies. To create an economy and a society without regard to the desires of this basic economic unit is impossible, which is why we see in today's China an attempt to reconstruct many of the aspects of traditional peasant life.

Case 19. The Javanese Villagers of Kali Loro

Our final case study is of the village of Kali Loro in central Java (B. White 1976). It is a fitting last case for our book, since the economy of Kali Loro reflects extremes of population density and intensification of production that are rarely exceeded in nonindustrial economies. Since this degree of intensification, as we shall see, is due partly to the encroachment of a world industrial market in Java over the past century or more, this case also illustrates the changes that take place in an agrarian economy as it becomes incorporated in the world market.

The Subsistence Economy

Kali Loro is a "village complex" located 20 miles northwest of the coastal city of Yogya Karta on a narrow plain between the Progo River and the Menorah Mountains. Although the dense planting of house gardens with fruit trees and a diversity of crops gives the countryside the appearance of a jungle, the landscape here has been radically transformed over hundreds of years of dense human settlement, and there remain virtually no wild or natural habitats in the region. Since the early nineteenth century the population of Java has grown at a steady rate of between 1 and 2 percent a year. Low as this rate seems by modern standards (it is roughly the current rate of growth in the United States), Java's population increased from 5 million in 1815 to about 80 million in 1975 and many areas now have population densities well over 1,000 persons per square mile. As may be imagined, rural Javanese today experience great pressure on land resources and most achieve only a marginal subsistence.

In the circumstances, why has Java's population continued to grow? One reason is that the Javanese value large families: White found that the women of Kali Loro want on the average to rear five children to maturity. Because of the frequency of infant and child mortality many poor women fail to achieve this number and are disappointed, but large families are common. To heighten the paradox, the people of Kali Loro frequently complain about the growth of population and the resulting extreme scarcity of land—yet they continue to desire and produce large families.

In Kali Loro population is so dense (about 1,850 persons per square mile) that the village areas—consisting of houses with small gardens—occupy nearly as much land as the wet-rice fields. Wet-rice land is so scarce that there is only enough to provide the average adult with about 40 days of work a year. This has led some economists to postulate a "hidden unemployment" among rural workers. White shows, however, that there is no hidden unemployment in Kali Loro, where even children's labor is a valued and necessary part of the peasant household's adaptation to extreme land scarcity.

Hence the paradox shifts focus. Instead of wondering why parents continue to have large families where population is overabundant, we now wonder why, in an economy seemingly oversupplied with people, labor is so scarce that households strive to increase their labor supply by having many children.

To resolve this paradox we must first understand that although opportunities for work in rice cultivation are scarce, there are many other ways of earning income. Most of these are less profitable than agriculture and by themselves do not generate even a marginal subsistence income; but in the peasant household, short of land and without alternatives, even starvation wages are preferable to no wages at all. Hence "we should speak not of unemployment ('no work to do'), nor of underemployment ('not enough work to do'), but of low labour-productivity or labour-efficiency, which for a landless or near-landless household means 'a lot of work to do, with very low returns'" (B. White 1976: 91).

Everybody wants land for wet-rice cultivation, and most people manage to own some. But in order to reach the "poverty line" of bare subsistence a household must farm at least 0.2 hectare of wet-rice, and most families in Kali Loro fall below this minimum. In order to bring their household income up to a more comfortable level (*cukupan*, or "enough"), families employ a variety of strategies.

Foremost among these is, and has been for generations, labor-intensive gardening on small plots. Geertz (1963) describes the Javanese as applying "hair-splitting techniques" to extract ever larger quantities of rice from the same land, a process he calls "agricultural involution." By careful and frequent weeding, the application of manure by hand, the laborious preparation of seedbeds, the careful transplanting of seedlings, the grading of plots to equalize the distribution of water in a field, and other techniques, the diligent farmer can get the most out of his small plot. By expanding Java's irrigation networks the government has greatly increased the amount of land available for this kind of cultivation.

Another strategy for dealing with the scarcity of wet-rice land is to plant a garden on the family's home site. Such gardens typically produce as much food per hour of labor as rice fields. They are used to grow as many as 50 distinct cultigens including root crops, tree crops, and items of utilitarian value such as firewood and wrapping leaves. They add diversity to the diet and increase household security. But they are unirrigated and cannot be intensified to yield as much food per hectare as wet-rice yields.

Wet-rice and garden activities together only account for a small proportion of an adult's time (about 2.5 hours per day for men and 0.5 hour per day for women). The average adult's workday in Kali Loro is filled up by a wide variety of additional activities:

1. Livestock are kept in a stable next to the house. Because very little pasture land is available, grazing animals like sheep or cattle require heavy labor inputs in raising fodder. Fodder must also be raised or bought for draft animals. It is in such demand, and therefore so expensive, that some families in Kali Loro cannot afford to maintain a draft team and have been forced to operate their plows with human labor.

2. Various opportunities for wage labor in and out of agriculture are exploited seasonally.

TABLE 8
Kali Loro Time Allocation
(Hours per day)[a]

Activity	Men		Women	
Food production				
Hunting	0.0		0.0	
Fishing	0.0		0.0	
Collecting	0.0		0.0	
Agriculture	2.6		0.5	
Livestock	1.3		0.1	
		3.9		0.6
Food preparation	0.1		2.7	
Food consumption[b]	—		—	
Commercial activities				
Collecting natural products	0.3		0.4	
Cash cropping	0.4		2.3	
Manufacturing	1.4		1.0	
Wage labor	0.7		1.7	
Marketing	1.1		0.2	
		3.9		5.6
Housework				
Housekeeping	0.1		1.0	
Water and fuel	0.2		0.1	
		0.3		1.1
Manufacture	0.0		0.0	
Social				
Socializing, visiting[b]	—		—	
Child care	0.4		1.0	
Public events[b]	—		—	
		0.4		1.0
Individual	—		—	
TOTAL	8.6		11.0	

SOURCE: B. White 1976: 209.
[a] Workdays, including weekends and holidays.
[b] Social and recreational activities not reported.

3. Many handicrafts and foods are produced in the home for sale in the marketplace.

As Table 8 shows, none of these activities is dominant. Rather the average household is characterized by an "occupational multiplicity" that allows its members to keep working and earning additional income even when, for seasonal or other reasons, any single source of employment fails. Considering that the averages in Table 8 take into account all days including holidays and periods of illness, it is striking that the average workday, including necessary home-centered activities like child care and food preparation, is 8.6 hours for adult men and 11.0 hours for adult women.

Tasks are not all valued equally, however. Some pay much better than others (agricultural labor pays several times better than other kinds of work), whereas some pay so little that they barely cover the worker's subsistence needs and add little to the family's cash supply. But low-paying tasks have the advantage of being available when agricultural employment is scarce and alternatives for labor are few, and many can be performed by people with few skills, including children.

Social Organization

As in other peasant societies, households tend to be self-contained nuclear families. In Kali Loro households average 4.6 members and are clustered together in 26 villages of approximately 300 members each. Most social contacts between households take place within the village or between members of immediately neighboring villages. Villagers who live more than two miles apart are generally strangers.

Changes in the domestic cycle over time greatly influence the economic status of the family. Newly married couples strive to set up independent housekeeping as soon as possible, although this is made difficult by the scarcity of land. As they begin to have children they enter what White calls the "early expansion" phase, when large amounts of parental time, including fathers' time, is devoted to child care. With hungry mouths to feed, mere subsistence is a struggle and capital accumulation is nearly impossible.

As children grow up the family moves into a "late expansion" phase. Older siblings take over the care of children, freeing the parents for directly productive labor: food- or income-producing work increases by over 25 percent above levels in "early expansion" households. With children helping to feed themselves and with older members free to seek productive work, "late expansion" families are able to save money and invest in land, houses, and capital goods.

As in other peasant societies where land is owned and inherited, however, a struggle arises when children prepare to marry. Children demand a share of the household's wealth in order to establish their independent households, but parents are reluctant to give up their control of their children's income and fear that children who have established separate households will not support them in their old age. Parents seek to retain a secure hold on their children's production by holding on to their land and having their married children work for them as sharecroppers.

Even parents of large families fear they do not have enough children to look after them when they are too old to work. Their fear reflects the tenuousness of kinship ties in peasant society, where security depends as much on friendship as on kinship.

Still another security mechanism is found in Kali Loro: the *slametan*, a series of exchanges of gifts and services organized by the ceremonial system. Although larger families have larger networks and participate more fully in the slametan, even small families with marginal incomes spend remarkably large sums in these exchanges, which account for an average of 15 percent of total household expenses for the village as a whole. In return villagers become part of a security network on which disadvantaged families can call, sustaining an ethic of "shared poverty" (Boeke 1953) in which ceremonial exchanges act to some extent as a leveling mechanism (Wolf 1957), equalizing life chances for all community members.

The pressure to equalize life chances intensifies with population density; those villages with the greatest abundance of land per capita are those with the most unequal distribution of land. In the more densely settled communities practices such as sharecropping, harvest sharing, and cooperative labor exchanges help equalize household incomes.

Nonetheless economic stratification within villages persists. Some families are landless; others have exceptionally large holdings. The ownership of prized wet-rice land is particularly skewed: 37 percent of the villagers own none, whereas the wealthiest 6 percent own more than 50 percent of wet-rice acreage. Many "landless" households have access to wet-rice land by renting or sharecropping, and 90 percent of the villagers own at least some garden land; but unequal access to resources is the rule.

As a result we find patron-client ties between wealthy and poor families. Clients work their patron's lands or care for his animals at lower than average wages in exchange for an acknowledged status as a quasi-member of the patron's family, a status that entitles them to protection and aid. Wage labor for patrons, whether agricultural or not, is a highly desired source of income even in families with wet-rice plots of their own.

When colonial enterprises converted much of Java's best land from rice to sugarcane and other export crops, peasants were forced to intensify their production of rice on inferior lands, including new lands made available for cultivation by government irrigation projects. At the same time, colonialism opened up opportunities for wage labor and craft manufactures destined for the world market. It is not clear exactly how these developments affected population growth, but this much seems certain: the proportion of peasant family income derived from subsistence farming has declined as population and occupational multiplicity have risen.

To return by way of summary to White's main argument, we can see that it is "rational" for a husband and wife in Kali Loro to want many children. Although young children are a hardship, older children make a major labor contribution to all areas of household production. Households with older children are more "efficient" to the degree that older children produce more income than they consume, and larger families produce a greater surplus above subsistence needs that can be invested to increase income and security. Where land is extremely scarce and alternatives to agriculture yield even smaller returns than the overworked fields themselves, every effort is made to increase family income by the exploitation of family labor.

The extraordinarily long working day of the Javanese adult is an index of the scarcity of opportunities for productive work. The people of Kali Loro correctly attribute this scarcity to population growth, yet they are victims of their own "tragedy of the commons" (Chapter 10). Any household that strives to serve the common good by limiting its births achieves nothing but the disadvantage of being labor-short in a highly competitive economy in which more labor means better living for the family.

Conclusions

The agrarian state and the peasant communities that are its rural foundation are the final elaboration of the evolution of the economy as we have examined it in this book. The analysis can undoubtedly be carried farther, into industrial and even postindustrial society, but it would take another book of this length to do justice to so complex a subject.

Given a nonindustrial technology, peasant economies generally make maximum use of the land, with little or no fallowing and virtually no dependence on wild foods. But, more importantly, peasants are integrated into large, hierarchically structured economic systems; and in this sense, despite a significant measure of household subsistence autonomy as compared to modern families, they are the least self-sufficient of all the peoples examined in this book.

Even Boa Ventura, at much lower population densities than Taitou and Kali Loro, represents a high degree of intensification of agriculture: many hours of labor and much husbandry of resources are needed to keep the household going. But Taitou and Kali Loro are our most dramatic and poignant examples of intensification: the "involuted" application of family labor to tiny plots of sweet potatoes and wet-rice, with each plant hand-tended through every demanding step of production; the expropriation of all available land to human purposes; the need to use all resources, even human feces and the soot of a brick oven, to replenish the land and coax any small additional measure of food from it; and the scattering of efforts among several very small plots, each in a different microecological zone, to minimize risks of crop failure and maximize the diversity of foods in the diet.

And yet for all this hard work and careful management, peasant economies provide a less satisfactory subsistence than others we have examined. Although many economic systems may be exposed to sudden, unpredictable disasters that result in famine and death, only among peasants do we find a substantial portion of the population persistently fluctuating, not between feast and famine, but between barely adequate diets and serious undernourishment. The larger economy may provide them with opportunities to bolster their economic security, but competition is extreme and the net gain in security is meager and costly.

The peasant household is in one sense self-sufficient: the needs of the political economy have grown beyond the limits of effectiveness of extended corporate kin groups. These large but comparatively intimate social units, such as the clans of the Central Enga and the Trobriand Islands, have fallen away as their risk-spreading, technological, defensive, and trade functions have passed onto still larger, more distant institutions such as armies, markets, and bureaucratic agencies.

What is left for the peasant family are dyadic friendship ties designed to ensure that short-term shortfalls will be made up by gifts and other aid from friends. That very poor families, such as those in Kali Loro, will spend up to 15 percent of their household budget on gifts, feasts, and other social expenses is not a sign of economic foolishness, but a measure of the importance of exchange ties to neighbors and of full membership in the village community.

A household's freedom to choose its friends, however, may not compensate for the loss of obligatory, institutionalized ties to a strong kin group. A peasant family's best chance to increase its security is to acquire more land. Since land is in inadequate supply, this can only mean intense competition with one's neighbors; and the best way to win in such competition is to make the individual sacrifices necessary to remain a large, economically unified family with autocratic control over all labor and income. But this is in a way to repudiate ties with neighbors in favor of the family.

In the final analysis the economic survival of a peasant family is its own concern; the role of hamlet and village-level groups is now very circumscribed, and the impersonal, competitive market has taken on many functions formerly handled by chiefs, lords, and bureaucrats. For one thing, the market is an efficient means of storage. The peasant whose winter wheat is abundant can sell his surplus on the market and "store" it in money. If his summer rice should fail, he can retrieve his winter surplus by spending his money for someone else's rice. Since money is compact and does not spoil, it is more efficiently stored than food.

An efficient market not only evens out individual ups and downs, but helps in dealing with large-scale catastrophes like drought or crop disease; the fully commercial markets of China, for example, can move foodstuffs long distances over good roads in response to high market demand in a given location. An efficient market also undermines distinctions of class, ethnicity, kinship, and patronage; in market transactions, only money talks. Economic specialization and class stratification persist and even proliferate, but the free flow of business in the market leads to much individual mobility (C. Smith 1976: 360-67).

In purely economic terms the market makes for efficiency in two main ways. First, it makes prices more responsive to supply and demand. When demand is high, prices rise; producers are motivated to increase production, and intermediaries are eager to move goods to their destination, reducing waste and spoilage and minimizing warehousing costs. Second, the market encourages specialization and economies of scale. The larger the market area, the larger the demand, even for unusual and seldom needed items. Capital investments in technology and a trained workforce become possible on an unprecedented scale.

At the level of the household economy, however, the market has costs as well as benefits. Land that may be bought may also be sold. The necessary corollary of upward mobility is downward mobility. That the market brings strangers into economic transactions, without mediation by Big Men or chiefs, makes for more efficient pricing and a more efficient flow of goods, but strangers have no moral obligations to one another and hence no built-in recourse against fraud. And the fact that a peasant can buy tools, seeds, and fertilizers in the market more cheaply than he can make them himself means that the very food production on which family subsistence is based is dependent on the market and will suffer if some unexpected occurrence, such as a war, cuts off essential market supplies.

Peasants are uncomfortable with these negative aspects of the free market and take steps to offset them. Thus just as the impoverished sharecroppers of Boa Ventura willingly buy exclusively from shopkeepers who charge higher prices, peasants in general seek to establish ties of friendship and patronage with people who occupy vital roles in the market network (Belshaw 1965: 78-81). The peasant's emphasis on acquiring land also goes against the market principle, since only land allows direct subsistence from household labor and frees the household from the necessity of selling its labor on the market. Thus also the peasant practice of planting many different crops in as many soil types as possible, though it results in an "inefficient" miniaturization and scattering of plots, makes it possible to produce a secure and diversified diet without reference to the market. The storage of foods in the home, despite a free market where food can be sold for money that can be saved and invested, is yet another peasant practice aimed at self-sufficiency. Still others, such as the *jajmani* system in India (Dumont 1970) and *pratik* in Haiti (Mintz 1961), could be enumerated at length.

One reason peasants do not trust the market is that they do not control it. In advanced state systems elites acquire their wealth less from land or tribute than from commerce and taxes, less from control of the means of production than from control of the "means of exchange" (C. Smith 1976). In this sense China and Java, where peasants hold legal title to their lands, are evolutionarily beyond northeastern Brazil with its "backward" system of patronage and large estates filled with dependent peasants. In an evolved nation-state the elite have no objection to peasants' owning their land because the power base of society has shifted from production to exchange, from the land to the market; it is no longer in the countryside but in the market towns and the commercial metropoles that the rich and powerful concentrate their activities.

These elites are beyond reach of the peasant household. The peasant takes advantage of the market, but does not feel personally involved in its creation and maintenance. Elites control where markets are allowed to situate, who will be licensed to sell, which commodities may be sold freely and which are to be monopolistically regulated; they are responsible for law and order, roads, food inspection, and the standardization of weights and measures. To the peasant household these are simply part of the environment, like soil fertility and rainfall, to be exploited but not understood and certainly not controlled. Hence the peasant's indifference to the state and his suspicion of outsiders generally.

The peasant's ties to the elite are direct and local, taking the form where possible of personal patron-client ties to wealthy, well-connected people who will help in emergencies and intercede when necessary with the bureaucracy. As in the case of land, however, households are in competition with one another for the limited supplies of patronage, and whatever links they do achieve with patrons always depend on mutual advantage for their continued existence. Other advantages peasants seek in return for their favors and services to the wealthy are access through rental or sharecropping to subsistence plots at the patron's disposal and connections to elites in the larger integrated economy, such as the standard market area, who may provide employment or other opportunities not locally available.

Because of such advantages, whether enjoyed or just aspired to, peasants view personal dependence on a patron as a source of strength and thus, paradoxically, of freedom. This "client consciousness" puzzles observers from more fully commercialized economies, who equate freedom with a free market and see any patron-client relationship as smacking of exploitation. Historically, however, class consciousness or proletarian consciousness takes root in rural areas only after a large class of migrant workers who are both landless and without patrons has been created by the further commercialization of agriculture, a process that goes beyond the scope of this discussion (see A. Johnson 1975b). Here we simply observe that a patron-client relationship, despite its class inequality, is still an effort to build trust and loyalty into vertical economic relations, whereas the market, which takes over management of the vertical flows of labor, crops, handicrafts, raw materials, and money, is, at its most efficient, an impersonal "invisible hand" that knows no loyalty and is unmoved by human suffering.

Overall we find the peasant family to be highly vulnerable in a land-scarce, competitive, densely populated economy. Although the family carries most of the risks of production, it enjoys little profit. Why? Primarily because such labor-intensive methods of production as we have encountered in this chapter produce low returns to labor; secondarily because elites and government agencies are too powerful, and too removed from local control, to feel any pressure to return much of the wealth they extract from the agrarian sector. The capacity for intensification depends to a considerable extent on services provided by the state, but these merely serve to keep production levels up and avoid mass starvation, not to relieve individual families of the burden of scarcity. Significantly, a married couple's greatest fear is that they will be abandoned in their old age by offspring whose own battle with scarcity is too all-consuming to leave them the time and energy to care for their aged parents.

THIRTEEN

The Economy of the Regional Polity

WE HAVE IDENTIFIED the following evolutionary typology: chiefdom, state, nation-state. From the simplest chiefdom to the most complex nation-state the scale of political integration increases dramatically. The polity grows from the small local group of the Trobriand hamlet cluster (with perhaps 500 people) to the empire or nation-state with a population in the millions. As the scale of political integration expands, new institutions become necessary. Kinship and personal loyalty are replaced by impersonal bureaucracies and markets. Religion, formerly centered in the household and the community, becomes an arm of the state. The whole nature of life changes from the inward-looking, environmentally constrained local community to the broadly organized, market-integrated nation-state.

The following are the salient characteristics of the regional polities discussed in Chapters 9-12:

1. Population density is characteristically high, although there is a wide range from the Basseri's 2 persons per square mile to Taitou's 400 and Kali Loro's 1,850.

2. Environmental conditions are diverse but usually include either (a) rich resources like irrigated lands or bottom alluvium, or (b) opportunities for trade resulting from capital-intensive transport (often water-based) or an established outside market for the polity's main product. In both situations the critical factor is the ease with which intensive production can be controlled.

3. The technology is typically intensive agriculture that requires either major capital improvements (irrigation canals, flood control, terraces, drainage, etc.) or careful management of a shifting cultivation cycle. Where trade is important, it requires either heavy capital investment in transport technologies (canoes and ships) or the careful man-

agement of external diplomatic relationships. Control is comparatively easy to attain in all four cases.

4. The settlement pattern is hierarchical, with a center providing economic, political, and religious services for outlying settlements. Cities are common in state societies but not universal.

5. Social organization is hierarchical and regionally organized. Local groups continue to function as important community organizations, but central and regional political, social, and religious institutions dominate. Most importantly, there are class divisions and a demarcated ruling elite.

6. Territoriality is a matter of private ownership by elites, institutional ownership as a means of finance, and use rights given to commoners in return for a portion of their labor or production. The state acts to guarantee the land tenure rights of both elites and commoners.

7. Warfare is common but directed outside the regional polity. Conquest extends a polity and its resource base; and warfare offers an opportunity for advancement to the individual warrior, who may receive land from new conquests in return for his military service. Within the polity warfare is outlawed and force is monopolized by military specialists.

8. Ceremonialism, in addition to its local functions, becomes critically important for legitimizing social stratification. In particular, ceremonies assert the divinity of rulers and the sanctity of their leadership.

9. Leadership is institutionalized in defined offices open only to certain members of the elite. Succession to office involves considerable maneuvering and is not always peaceful. There is intense competition for high positions in the social hierarchy.

In Chapter 8 we found that the primary variables relating to the initial formation of local corporate groups and intergroup collectivities were risk management, capital investments in technology, warfare, and trade. The contributions of these variables, however, were not equal; warfare and risk management were most important, with technology and trade playing limited roles. In the formation of stratified societies (Chapters 9-12) we find that the same four processes remain important, but in differing proportions at different levels of economic integration and stratification.

Warfare plays an important role, since conquest is the way to expand a society and increase its resource base. Yet warfare does not create the basis for economic integration beyond the local group. Rather it represents the weakness of the political system at its extremes, where the pressure for economic integration is too weak to

TABLE 9
Economic Processes in Political Evolution

Local group	Regional chiefdom	Early state	Nation-state
WARFARE	warfare	warfare	warfare
risk management	RISK MANAGEMENT	(risk management)	(risk management)
technology	technology	TECHNOLOGY	(technology)
trade	trade	trade	TRADE

offset the competitive stresses between groups. Hence military considerations do not *explain* the regional integration of political systems, since without other bases for integration the same competitive pressures that cause warfare between large groups can also cause fissioning and warfare within such groups.

For the simple chiefdom, as in the Trobriand case, risk management is of great significance, with transport technology and trade of real, though secondary, importance. But as we reach complex chiefdoms like those of Hawaii and early states like the Inka, the highest levels of political integration are concerned with technology, often irrigation; risk management is handled routinely at subsidiary regional levels by staple food storage and the promise of redistribution. And finally, in nations like Brazil, China, and Java (Indonesia), the economy is integrated at the highest level by markets and trade, although risk management (e.g., famine relief) and technology (e.g., irrigation networks, reservoirs) are at least partially dependent on national and elite support. In Table 9 capital letters are used to represent the dominant process or explanatory principle at each evolutionary level, and parentheses to represent processes that are largely supported by lower-level political structures.

As in Chapters 4 and 8, we shall review the evolutionary process from the ecological, structural, and economic points of view.

The Ecological Perspective

The dominant ecological process continues to be the intensification of human use of the environment. At this stage the diet of all societies depends on a limited number of starchy staples, for carbohydrates are a much more efficient source of food energy than animal or other plant products. Peoples like the Basseri and the Trobriand fishing villagers, who formerly consumed mostly animal foods, now trade their animal products for the starchy staples that are their major sources of food energy; a similar process is evident in the Chinese household that raises pigs but cannot afford to eat them.

The first response of a horticultural people to a shortage of fertile land is to expand into virgin territory, where slash-and-burn techniques with long fallows are the least expensive means of maintaining soil fertility and hence high crop yields. With growing population pressure sacrifices become necessary. Less desirable plots may be cleared where steep slopes, poor drainage, or the like have discouraged earlier clearing; or the fallow period may be shortened and a new crop planted before the soils become fully restored by natural regeneration. All such responses to population pressure involve declines in labor efficiency, either because special labor increases are necessary, as in drainage canals or terraces, or because the resulting yields are lower per hour of labor owing to declines in soil fertility.

As yields decline, households and kin groups may invest extra labor in the family fields; animal and vegetable manure may be added; mounds, pits, and terraces may be built to overcome specific limits to the production of special crops like sweet potatoes, yams, and taro; dikes and channels may be constructed to manage the flow of water through a field; seedbeds may be dug where seedlings can mature properly before being transplanted to the harsher habitat of a field; fences may be erected to manage the access of animals to a field, which can be beneficial if it occurs at the right time in the agricultural cycle; and labor may be invested in feeding work animals to supplement household labor in food production and food processing.

Another way to offset declining yields is to invest family labor in the political economy, in the form of work for chiefs, lords, or patrons or work as a hired hand. By such means the household may gain access to richer soils owned by the elites, to irrigation water, to fertilizer from the lord's stables, and to tools such as axes, plows, and hoes produced by the chief's craftsmen or by specialists for sale on the market. Such aids to intensification are not produced directly by the household but must be paid for with labor or food, whether given as rent or sold for money in the marketplace. Thus the ultimate outcome of intensification is to undercut the household's independence and to bind it economically to regional elites and the market system.

It may well be that certain technological innovations increase the labor efficiency of a household in the short run, as when an irrigation ditch or a steel plow opens up hitherto unused lands. Technological innovation, however, by raising carrying capacity, then encourages population growth. This seemingly inevitable rise in population eventually consumes the added resources, and the intensification of labor inputs resumes. The cycle of technological innovation and population growth, identified originally in Chapter 1, seems to accelerate, inevitably locking the peasant into dependency.

In Japan, for example, we see that medieval development was closely tied to the spread of wet-rice cultivation, which depended on an expanded irrigation network for its fullest development. But this economic development was matched and perhaps even exceeded by step-by-step population increases. In the later stages of development in medieval Japan, therefore, we found an increasing use of fertilizer, seed selection, transplanting, and other labor-intensive techniques going hand in hand with the government's expansion of irrigation. In medieval France we found a comparable complex change involving the spread of the steel plow, the rotation of fields through cereal crops, nitrogen-fixing crops, and pasturage in a regular cycle, improved seed varieties, the breeding of more efficient work animals, and other examples of intensification, all evolving as a system. At every stage the peasant's labor burden and his bondage increased.

In Taitou in China and Kali Loro in Java we find extremes of labor intensification, stretching the limits of the peasant family's capacity for "self-exploitation." In the Taitou and Kali Loro cases it is necessary to make every square foot of agricultural land productive. No organic matter that can serve as manure is wasted. Small imperfections in a field are corrected by diligent effort, each plant receives special attention, and all family members contribute somehow to production. Peasants are understandably preoccupied with the exact boundaries between properties, since many plots are so small that a shift of even a few inches in the property line can have a significant impact on a family's food production.

The Structural Perspective

As we learned from the Central Enga case (Chapter 7), complex political integration is brought about not by intensification as such but by the kinds of technology and social organization on which intensification is based.

As society becomes more complex the family and the hamlet retain their central position in the subsistence economy, the family striving to remain self-sufficient as a unit of production and consumption, the hamlet continuing as a support group in which certain forms of cooperation—such as house-building and fencing—are important. In this regard the main trend in regional polities, which we examine in more detail below, is the increasing dependence of the family-hamlet group on resources external to its control, such as irrigation water, rented land, and trade goods.

The local group also persists as a recognizable unit throughout the evolution from chiefdom to state. In the Trobriand case, for example, the chief or his designated "garden magician" carefully regulates the use of scarce garden land for the local group as a unit. The chief also stores his yams in full public view at the center of the village, where they attest the group's prestige.

The village-level group is also common among peasants, especially the ceremonially defined and integrated village with public material exchanges like the Javanese *slametan* or the Mesoamerican *cargo* (Cancian 1965). That the cost of such exchanges is very high underscores the importance of this "ceremonial fund" to the impoverished peasant community in its struggle for survival in an exploitative and threatening world.

Yet these low-level structures must co-exist alongside the emerging political, social, and economic institutions of the regional polity. Unquestionably these institutions confer some benefits on the household, for example by providing water for its fields, legal safeguards for its rights in land, and markets for its surplus goods or labor. Intensification and corresponding resource competition create conditions likely to result in the tragedy of the commons. The regional organization of the Basseri or the Inka create a management of resources at a level necessary to limit disastrous resource depletion and intraregional conflicts. In the nation-states, some attempt to regulate market fluctuations is often imposed.

These services by elites "on behalf of" commoners, however, should not obscure the more fundamental issue of control. Essentially the economic opportunities open to the commoner household have narrowed to situations in which its autonomy is relinquished. Economic circumscription translates into political control.

As political control incorporates greater numbers of local groups, it necessarily becomes increasingly impersonal. For the simple chiefdom kinship continues to provide the language of property relations, debt and credit, and labor allocation. Rank is still justified on grounds of genealogical closeness to a common ancestor. And even warfare is fought "for the group," to displace other groups so that one's own group may enjoy access to their territories.

By contrast, the economic functions of the complex chiefdom and the state center more on the use of staple finance (i.e., taxation in kind) to underwrite public works. The highest elites are now far re-

moved from local populations, and the fiction of kinship links between elites and commoners is rarely maintained. Elites fight wars to
gain control of new regions and their wealth. The commoner family
may welcome the pacification of the countryside and the stabilization
of storage and famine relief, but increasingly the workings of the system are beyond its ken. It responds by taking refuge in the political
passivity and indifference to the state that are universally characteristic of peasants.

As the state evolves into the modern nation, much of the management of risk and technological investment becomes further imper-
sonalized in the market. Here again there are benefits to the household. Money serves to store value and can be more or less freely
exchanged for food, tools, or whatever else may be needed. Food
need not rot in local granaries when distant buyers will pay high
prices for it; the specialized production of many crafts and foods is
possible when large populations generate a significant demand for
even unusual items; and in general, labor, technological inputs, and
other factors of production are sensitive to small changes in supply
and demand and hence move quickly to the places where they are
most needed.

On the other hand, the market ends the domestic group's secure attachment to the basic means of livelihood. Land can now be sold,
kinsmen can abandon their social obligations to seek their fortune in
the marketplace, and money may be spent, lost, or stolen. The result
is a more efficient market but a more isolated family unit, one whose
uneasy personal ties to the elite are its best remaining source of social
security.

The efficiency of the market is such that in the nation-state, for the
first time in history, there comes into being a landless, uprooted, exposed class of "free labor," in the form of mobile family-level groups
reminiscent of the forager groups with which we began our study.
Free laborers move opportunistically from job to job, own little, store
little, and depend for daily subsistence on the momentary opportunities offered by a fluctuating environment. They maintain extended
family ties and networks of friendship, but must be prepared to move
immediately when local resources give out.

Yet these free laborers forage not in a natural environment free from
territorial constraints, but in a densely settled, "cultivated" environment occupied by tenacious, intensely competitive families who are
desperately afraid of losing the bits of earth they happen to control at
the moment. Free laborers' only access to a means of livelihood is
through a labor market that is run by no particular human being and

has no human feeling or responsibility for their well-being. Thus the free market creates a class that is free not only from feudal domination and the rigid controls of small community life, but also from the traditional security mechanisms that could save it from destitution.

The Economic Perspective

In chiefdoms and states individual decision-making is central to the lives of both elites and commoners. Elites, acting in their own self-interest and in the interest of the ruling institutions they represent, carefully seek out opportunities for investment and expanded economic and political control. Just as a capitalist entrepreneur explores various ways of maximizing his profits with the implicit goal of improving his status and his standard of living, so elites in state or chiefdom seeking to expand their income to maintain political domination calculate carefully, if not always accurately, the costs and benefits of different military adventures and public works projects (see for example Luttwak's [1976] economic analysis in *The Grand Strategy of the Roman Empire*).

Commoners are equally concerned with making effective decisions. The peasant assesses many opportunities and costs in deciding what to grow and how to market his produce; in the Guatemalan peasant market system for example, which Tax (1953) has labeled "Penny Capitalism," each decision is based on a surprisingly complete knowledge of different costs and market prices. The analogy to the forager is again relevant. The commoner searches out each opportunity for income based on a calculation of benefits and moves from market to market, from craft to craft, from crop to crop, as immediate prospects change.

One indicator of how individuals economize is the manner in which they allocate their limited time to the many alternatives available to them. In Figure 10, we can get a sense of how time is spent differently in societies at different levels of population density and commercialization, based on data in Tables 5, 7, and 8. The Machiguenga time pattern emphasizes household self-sufficiency: the bulk of work time is spent procuring wild foods, producing garden foods for home consumption, and manufacturing items for household use. In New Guinea, the emphasis remains high on subsistence agriculture, but wild foods do not occupy much time, and neither does manufacture for household use; instead, the emphasis has shifted to commercial activities of cash cropping and wage labor. This reflects the recent commercialization of the Highland New Guinea economy, but in the

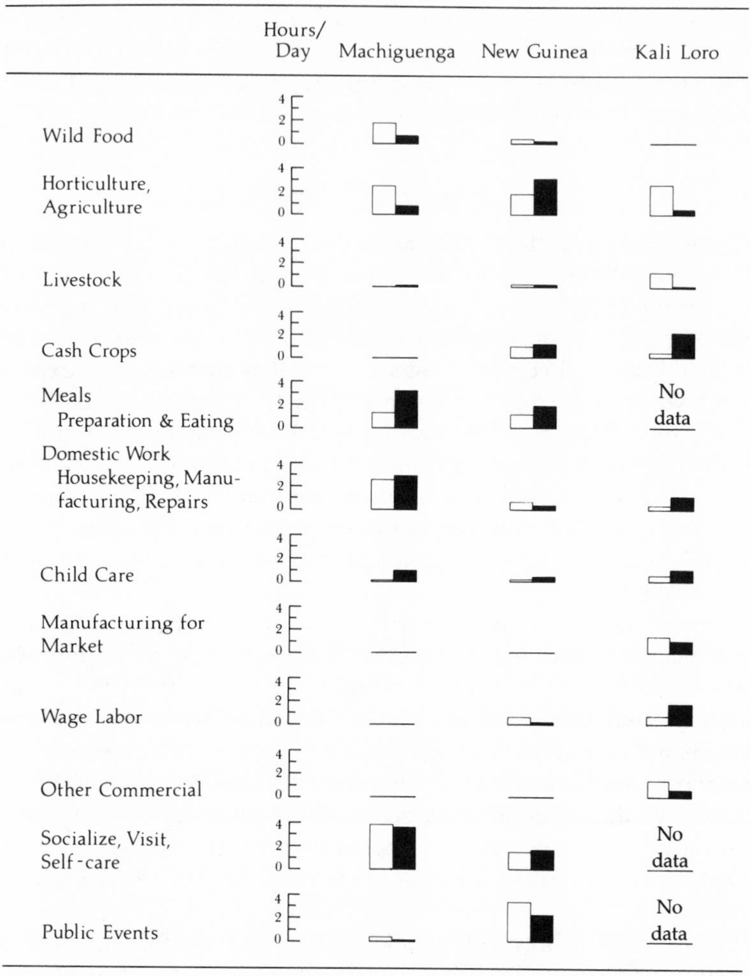

	Hours/ Day	Machiguenga	New Guinea	Kali Loro
Wild Food	4 2 0			
Horticulture, Agriculture	4 2 0			
Livestock	4 2 0			
Cash Crops	4 2 0			
Meals Preparation & Eating	4 2 0			No data
Domestic Work Housekeeping, Manufacturing, Repairs	4 2 0			
Child Care	4 2 0			
Manufacturing for Market	4 2 0			
Wage Labor	4 2 0			
Other Commercial	4 2 0			
Socialize, Visit, Self-care	4 2 0			No data
Public Events	4 2 0			No data

Fig. 10. Time use in the family-level society, the local group, and the regional polity. The white bars represent male working hours, the black bars female. The Machiguenga data are for a 13-hour day, the New Guinea data for a 12-hour day, and the Kali Loro data for all waking hours, including weekends and holidays.

past, production for public feasts and gift exchanges played a large role in time allocation. In comparing the New Guinea and Machiguenga patterns, we can also note that leisure or "discretionary" time is allocated differently, with the Machiguenga remaining individualistic and relatively private, in contrast to the attendance at public events in the New Guinea case.

Although the Kali Loro data do not include discretionary time, we can see that in subsistence food production the emphasis has shifted exclusively to agriculture and livestock. Furthermore, commercial activities have taken on major significance, and a wider range of activities is represented, a range that was referred to as "occupational multiplicity."

The division of labor by sex also follows certain common patterns. With the shift from the extensive horticulture of the Machiguenga to the intensive horticulture of the New Guinea group, there is a strong shift toward women's work in agriculture. This then shifts back to the men in Kali Loro. Some of the men's time in the New Guinea example has shifted to public events, on which the family's position in the political economy depends. We cannot judge whether the public involvement of men has also declined in the Kali Loro example, but we may note that in Kali Loro men, and women even more so, spend long hours every day in productive activities, longer than in the two other groups, and indeed probably longer than the vast majority of people elsewhere who live at lower population densities and in less commercialized agrarian economies.

A further example of economic behavior in the regional polity is the "tragedy of the commons," perhaps best thought of as the inevitable result of individual economizing decisions regarding use of a common resource. Why should a Basseri herder withhold his herds from common pasture that is being degraded when another herder will gladly take the opportunity for his own hungry animals? And why should a Javanese family restrain its own production of children in an overpopulated landscape when its standard of living may only be raised by the contribution of hardworking children? The problems that arise from the economizing behaviors of the family require structural solutions such as the land management provided by the Basseri chief.

Both elites and commoners in stratified societies rely heavily on dyadic contracts to form personal networks that can be used to obtain a wide range of necessary resources. Although such contracts further the free pursuit of individual self-interest, they lack the long-term stability and reliability of more structured ties based on social status and reinforced by jural rules.

In a similar way peasants seek personal, feudal-type ties of loyalty with large landowners, shopkeepers, government officers, physicians, and other local elites; and these elites seek similar ties for themselves with higher-ranking elites, so that everyone in the society is theoretically reachable through vertical networking. For the peasant these ties provide access to state institutions and security, for the elite a loyal labor force and a political support base.

In a political sense the peasant is less a "follower" than a "subject." The peasant does control his loyalty, and even in a densely populated countryside elites value the loyalty of hard workers who will make their lands and herds produce to their potential. Otherwise, however, the peasant has little political clout except in times of rapid change or the weakening of the state due to outside pressures, when peasant movements may sporadically mobilize the innate wish for greater control. Nor does the weakening of the state offer the peasant any real hope, since it entails the disruption of food distribution arrangements, public works, and the market, and such disruptions must be felt as losses by the peasant household. With the self-sufficiency of the family-level group no longer even a memory, the peasant household's notorious political passivity or "fatalism" comes at least partially from a realistic recognition that the security of the domestic economy, such as it is, depends on a smoothly operating political economy.

FOURTEEN

Conclusion

THE EVOLUTION OF HUMAN SOCIETY is most dramatically seen in the growth of the polity from the camps and hamlets of family-level society to the empires and nations of state society (Table 10). The family camp of a hunter-gatherer society, with perhaps 10 to 30 members, is an informal, biologically based group whose members are genetically close; the village community, with a few hundred members, is composed of people more distantly related and liable to fission along extended-family lines; the chiefdom, with several thousand members, includes many people unknown to each other personally who share little other than a leader and a ceremonial identity in common; and the state, with its millions of faceless persons, encompasses disparate and foreign elements with separate ethnic identities, histories, and interests. The evolution of complex societies dramatically changes social life, as cultural identification and economic interdependence supersede biological bonding as the fiber of social cohesion.

In this chapter we review the evidence presented in earlier chapters concerning the evolution of nonindustrial society. We begin by summarizing the patterned changes that we found in the basic variables of population, technology, settlement, social organization, and political relations as we moved from level to level along the evolutionary trajectory. We then highlight the basic processes that account for these patterns.

The Evolution of Human Societies

We have identified the following levels of cultural evolution: the family, the local group, the Big Man collectivity, the chiefdom, the archaic state, and the nation-state. These labels do not signify perfectly discrete levels or plateaus, to one or another of which all known cul-

TABLE 10
The Size of Communities and Polities in Evolutionary Perspective

Polity type and case	Size of community	Size of polity
Camp		
Shoshone	30	30
San	20	20
Hamlet		
Machiguenga	25	25
Nganasan	30	30
Local group		
Yanomamo	150-250	150-500
Taremiut	150-300	150-300
Tsembaga	200	200
Turkana	20-25	100-200
Big Man collectivity		
NW Coast fishers	500-800	500-800
Central Enga	350	350
Kirghiz	20-35	1,800
Chiefdom		
Trobriand Islanders	200-400	1,000
Hawaii Islanders	300-400	30,000-100,000
Basseri	200-500	16,000
State		
Inka	±400	14,000,000
Brazil	±300	80,000,000+
China	±300	600,000,000+
Java (Indonesia)	±300	100,000,000+

NOTE: The cases of medieval France and Japan (Chapter 11) are excluded because they cover a long period over which population size and political integration changed dramatically.

tures must be assigned; rather they designate stations along a continuum at which it is convenient to stop and make comparisons with previous stations. "Chiefdom," for example, is a convenient abstraction for a culture that is still evolving from (and contains elements of) the Big Man collectivity or the local group, and for one that may be well along the road to becoming a state. Since the evolutionary continuum represents a transformation of many variables at once, local conditions and history produce many variants that appear "more evolved" in some respects and "less evolved" in others when compared to their neighbors on the continuum.

Let us now review the elements of our evolutionary typology in the light of what we have learned from our case studies.

The Family. The basic feature of the family-level society is the tendency for population to reach a uniform distribution across the landscape in order to minimize procurement costs. Theoretically a uniform distribution would put each individual at the center of an area that was the same size for all, assuming that resources in the landscape were uniformly distributed. In reality resources are not uniformly distributed, and people in a family-level society tend to cluster around resources that are concentrated seasonally or permanently.

But the major deviation from a perfectly uniform distribution is, of course, the inevitable clustering of individuals in family groups. As a unit for the nurturance of children during their long period of dependency, made more efficient by a complementary division of labor by sex and strengthened by the security obtained from sharing among food producers, the family is a unique and seemingly irreplaceable institution (Campbell 1966: 276-81). So resilient and adaptive is the family group that it has survived the most momentous changes in the economy and society, changes that in some cases reach to the heart of the family economy. Family groups remain the basis of the subsistence economy, as primary units of production and consumption, at all the evolutionary levels we have discussed.

Technology at the family level supports the family's economic autonomy. A household can fashion, usually from locally available materials, the tools its members use for daily production, including bows and arrows, spears, digging implements, bags, traps, clothing, and shelter. Some persons may gain reputations as especially skilled workers in wood or stone, and some materials like salt or obsidian may be traded great distances; but neither specialization nor trade leads to significant social stratification. The family group is able to obtain everything it needs either on its own or else through egalitarian exchanges with similar family groups.

The social organization of the family group is based on the nurturance and trust generated in the daily give and take of family life. Camps and hamlets are the mature phase of the nuclear family. Grownup children, when they marry, are generally free to affiliate with either the husband's family or the wife's; often they move from one to the other opportunistically, but feelings of special closeness on one side or the other usually solidify into more permanent alliances. With kinship reckoning bilateral and kin groups flexible, social structure is readily manipulated for personal ends.

Social control occurs naturally in the family setting. The bully, thief, or idler is ostracized and must either learn to behave acceptably or stand alone. Individual violence occurs, but organized warfare is virtually absent. The low degree of competition in family-level societies is a function of low population density and small group size. A family is free to move in order to avoid a fight. Starvation does not result simply from being excluded from a rich resource base, although it can happen to any family unlucky enough to experience a sudden, unforeseeable, and protracted diminution in its supply of wild food.

Few authorities exist who can coordinate the behavior of many families at once. Common understandings may arise giving individual families priority of access to resources within a home range, but it is courtesy and perhaps some anticipation of aggressive defense of a home range, not political control by suprafamilial leaders, that prevents intrusion by an outgroup. Agreements partitioning foraging territories, such as those the Nganasan make each spring before leaving their winter hamlets, are of this sort. When a leader is given temporary control of food production, as in the Shoshone rabbit drive, it is always by mutual agreement of participants who recognize the immediate benefits of cooperation. The family group's control of food processing, consumption, and exchange is not affected, and the arrangement ends when the food supply has been harvested or has moved on.

The camp, however, is not the boundary of social interaction. Through marriage and exchange partnerships households create networks of reciprocal obligations that reach out broadly through the region. These regional personalized networks, presaging the extensive networks seen at the next two social levels, improve security by giving families access in time of economic hardship to the home ranges of neighboring camps. They also facilitate the trading of surplus goods for material unavailable on the home range, such as high-quality stone for axes.

The Local Group. Under certain circumstances individual families cannot afford to hive off from a local group at will. The local group does not arise naturally out of human sociability. Family groups do meet in larger aggregates, as we have seen, and these are occasions of intense socializing, dancing, and feasting that give great pleasure to their participants; but these aggregates are too large to be maintained on a permanent basis by the natural trust that keeps nuclear and extended families together. Without the threat of force or famine, local groups naturally fission into family groups.

Some local groups are held together by the necessity of multifamily capital investments and food-sharing arrangements. Others form when the density of population relative to resources is great enough to provoke warfare on a more or less continuing basis. This is not as radical a shift as may at first appear. The "nonviolence" of such groups as the !Kung San, the Machiguenga, or the Nunamiut Eskimos is based on firm social sanctions against aggression and self-promotion. Underlying such sanctions lie jealousies and perceived injustices that do not boil over into violence only because of the long "cooling-off" periods afforded by family isolation in the food quest and the possibility of responding to a dispute by leaving.

With the more frequent encounters between members of separate family groups that inevitably accompany population growth, competitive frictions intensify and violent attacks and reprisals occur. Family groups respond by forming local groups for effective attack and defense, and by establishing ties outside the local group to provide allies in battle and avenues of escape in defeat.

Within the local group the economy remains largely in the hands of the family. Technology also remains family-centered in our cases, except among the Tareumiut whalers. Local groups are inherently unstable because their constituent families become dissatisfied with present arrangements and form new alliances. Local group leaders are experts at conciliating and mediating, at "speaking well" in the common interest of the group and convincing autonomous and skeptical householders that their well-being lies with the group.

The Big Man Collectivity. Without allies the local group is vulnerable; in the vicinity are neighboring groups eager to attack and seize land or other property. Thus as population density continues to increase and warfare intensifies, local groups tend to form alliances, and these alliances develop into interdependent collectivities managed by Big Men.

In other Big Man systems, especially in hunter-gatherer and pastoral economies, the group leader's role as manager is even more significant than his military role. He may underwrite major investments in technology and food storage, as on the Northwest Coast, or take charge of large-scale risk-spreading arrangements, as among the Kirghiz. Through careful management of resources, including the mobilization of labor to produce, harvest, and store food and the subsequent distribution of surplus product, Big Men help make it possible for a larger population to exist than could be supported on the same resource base in a less organized economic system.

A Big Man stands for the political unity of the local group, and represents the group's economic interests in the complex reciprocity of the intergroup collectivity. He may organize certain elements of technology, such as fishing boats, weirs, and storage cellars, that extend the range, yield, and security of the local group's food production. He may also organize war expeditions, but his more critical function is to maintain alliances that protect his group's access to the means of production on which they depend.

His political power is conditional and transitory, depending on his ability to maintain a flow of benefits to his followers. He does this through intricate chains of debt and credit that depend on a sense of trust rooted in a history of satisfactory exchanges and, in a broader sense, on prestige. His power, such as it is, depends on his ability to call in debts at times when his local group's needs are high and his debtors are well provisioned. But since his ties are to greater and lesser Big Men whose own competitive position is continually being tested, and since younger men are always trying to rise to eminence, the Big Man is continually in danger of losing his following. Competitive divisions and hostilities erupt, shifting alliances and limiting the Big Man's ability to unite groups under his own firm hand.

As in the local group, ceremonial life in the intergroup collectivity consists largely of public displays of alliance and group interest beyond the family group. Insignia and performances proclaim the place of individuals within groups and the rights of those groups in productive property. Intergroup ceremonies also afford occasions for ordinary people and group leaders to create and reinforce exchange networks, negotiate transactions, and trade information about their specific locales. Above all, they provide an opportunity for Big Men to display the size, strength, and importance of their group for the world to see, and thus to improve their group's position in the competitive struggle for resources in an increasingly crowded world.

The Chiefdom. The chiefdom unites the local groups of a given region within one political institution dominated by an aristocratic leader or chief. Descended from gods and invested with special powers, the chief has the final say in all matters involving the group, including ceremonies, adjudications, war, and diplomacy. The chief is much like the Big Man in status and duties, but his domain is larger and more stable.

The stability of the chiefdom depends not on myth or mystification, but on the real economic benefits that stability confers. Among the factors making for stability is the development of elites, who come to power as a result of their services to the chief, use their power to acquire productive resources, and with the chief's help, retain their power for many years and even from generation to generation. A positive feedback arises as the elite's control of productive resources generates income that can be used by the chief to further strengthen his control. Ceremonial occasions legitimize social inequality, provide social glue for a polity divided by conflicting interests (one local group vs. another; elites vs. commoners), and generally uphold the sanctity of the society's structure against pressures to fragment and fight.

The State. The state differs from the chiefdom mainly in its larger scale, larger and more diverse total population, and more rigid stratifi-cation. These elaborations in scale and internal differentiations create further difficulties of integration as local, class, and special-interest groups proliferate. Integration at this level is beyond the informal con-trol of a hereditary elite. It requires a state bureaucracy, a state reli-gion, a judiciary, and a police force.

Many archaic states are characterized by a system of staple finance, often referred to as the "Asiatic mode of production." Control at the state level is based on state ownership and improvement of land. Con-tinued population growth calls forth major technological improve-ments to agriculture, including land reclamation and irrigation, that require centralized control. And even where they do not literally *re-quire* centralized control, as in the case of irrigation systems simple enough to be built and managed by the local community, they cannot escape it; community members now depend on irrigation for their livelihood, and with the filling in of the landscape by population growth they have nowhere to go.

Such communities are ripe for control in Carneiro's (1970b) sense of circumscription: although they can manage the irrigation system, they cannot defend it from the state's exactions. The militarily power-ful chief, and later the state, "protect" the community's access to its essential technology, but exact labor or produce in return. A similar kind of circumscription occurred in medieval times, when families who preferred to remain outside the manorial polity were gradually transformed from "free peasants" into "outlaws."

The Nation-State. In the continuing evolution of the nation-state and of modern multinational economic communities we can recognize trends identified in previous stages. With the nation-state, trade and markets rise to such prominence as to set the institutional framework of the entire economy. To be sure, the state also provides defense, risk management, and large-scale technological investment. But economic transactions and relationships are increasingly determined by the market.

With each step in the growth of the political economy we have iden-tified new mechanisms by which the individual household—the core of the subsistence economy—becomes integrated into ever-larger ag-gregates of households. The family-level economy, based primarily on bonds of kinship and friendship—that is, reciprocity and trust—gives way first to modest hierarchies based on reputation and publicly demonstrated "prestige," later to official hierarchies still rooted in the language of kinship, and still later to bureaucracies in which the pre-tense of kinship between ruler and ruled is finally abandoned, though it remains faintly perceptible in patron-client relationships of various sorts.

In the nation-state the latest stage of depersonalization of the economy is achieved, with the "invisible hand" of the market allocating goods and services without regard for social relationships, ethics, or need, but only according to the buyer's purchasing power and the seller's costs. The state supports this system by guaranteeing ownership in property and protecting the market against disruption and manipulation. The apparatus of the state may also be used to maintain political stability by protecting individuals or groups from the full impact of market forces: by controlling food prices, for example, or providing famine relief. From the strict market perspective these interventions are "imperfections" that distort the free market. But we may also see them as reflecting values that are important in the subsistence sector but are ill-served by the impersonal workings of the market.

The Individual in Society

Much economic behavior is best understood in terms of the dialectic of the individual *versus* society (Murphy 1971). The "subsistence economy" has its origins in a hypothesized natural, individualistic family or camp society that evolved under pressures of natural selection during two million years of foraging. The family, or its extension as a camp or hamlet, forms a "natural" unit with strong biological attachments cementing social relations. That is, natural biogenetic rewards, actualized in innumerable nurturant exchanges, keep an individual attached to the group despite its "costs." This view is essentially consistent with "kin selection theory" in sociobiology.

The "self," as defining the unit of "self-interest" in economic theory, is normally thought of as the individual, but a biological definition of self can with proper caution be extended to include the nuclear family and the kin cluster of the camp and hamlet. This may help us understand the remarkable continuity of the household and the kin cluster as the basic economic units through all levels of our analysis: the !Kung camp, the Machiguenga hamlet, the Eskimo men's house, the Enga lineage, the Chinese family, each an informal support group for the individual in a broad range of social, economic, and political affairs. There are no doubt cognitive and affective limits on the size of these groups, on the number of people we can know and depend on personally. "The magic number 25" (Birdsell 1968b; Williams 1981) comes to mind: families tend to have five members and hamlets tend to have five families, or 25 members altogether.

Even within the family and the camp or hamlet, however, a certain contradiction between individual self-interests is evident. Parents commonly want to keep their grown children close for labor, emotional support, and old age security; whereas the offspring want to establish themselves in independent, "adult" lives with their own families. Every family member must sacrifice his self-interest to that of other members now and again; that is what a family is. An implicit calculation of self-interest or fairness, however, remains critical to the household group. The ability of the family to stay together, especially past the maturity and marriage of the children, reflects the clear economic advantage to individuals of interdependence and is often heavily reinforced by symbolic means.

The same dialectic that characterizes the household extends to the camp or hamlet and its periodic fragmentation into its composite families. The camp of the forager is a family that has grown up and maintained its bonds, rooted in personal affect but continued largely for pragmatic reasons. The group stays together at certain times of the year when cooperation is desired and the costs of aggregation are low, only to fragment again as the ratio of costs to benefits rises. The resulting pattern of aggregation-dispersion is the most visible reflection of the dialectic of conflicting interests in families and multifamily groups.

This simple dialectic becomes more complex when economic interdependence extends beyond the family group, i.e., when the political economy comes into being. Indeed, the political economy would *never* have come into being were it not for the evolution of "cultural" rewards and punishments that supplement or override biological rewards.

The dialectic of individual vs. society or subsistence vs. political economy is a human universal. Wolf's four "funds," for example (see Chapter 12), describe universal allocations of total household production. Roughly speaking, the maintenance fund ("caloric minimum") and the replacement fund are the core of the subsistence economy, whereas the ceremonial fund and the fund of rent are fundamental to the political economy. In a family-level society the preponderant allocation is to the maintenance and replacement funds, feeding the family and keeping up its clothing and tools. The equivalents of the ceremonial fund and the fund of rent are minor shares of total household production, normally less than 10 percent, spent in order to have friendly relations beyond the camp or hamlet group. The goal is to maintain the family's security by averaging risks in subsistence procurement by sharing food and exchanging rights of access to needed resources.

In more complex societies the basic goals of individuals in the subsistence economy are remarkably unchanged: the household, whether in the !Kung camp, the Enga village, or the Brazilian fazenda, seeks to be a self-sufficient, economically independent unit producing all it needs. But if the goals do not change, the possibilities for achieving them change dramatically. The evolving political economy mobilizes an increasing percentage of the household's production as part of the ceremonial fund and the fund of rent. In a peasant economy such as Boa Ventura, approximately one-third of the household's output must be used to meet the community's ceremonial requirements and the rental agreement with the landowner. The mobilization of subsistence products from the peasant household is, of course, the necessary financial base for the complex institutions of the chiefdom and the state.

As we have seen, at each evolutionary stage existing organizational units are embedded within new, higher-order unifying structures. Hamlets are made up of families, local groups of hamlets, regional chiefdoms of local groups, and states of regional chiefdoms. The earlier levels continue to operate but with modified functions. Thus the local group of a stateless society, which had formerly been a unit of defense, is transformed into a unit of taxation and administration as it becomes incorporated into the state (as with the Inka ayllu in Chapter 11). The new levels of integration are carefully modeled on the preexisting order and characteristically preserve the ideology of the lower levels in order to facilitate the very difficult task of unifying formerly separate groups and reconciling their conflicting interests. Thus the chiefdom and even the archaic state espouse ideologies of reciprocity and redistribution appropriate to the more intimate social life of the family and the local community.

To sustain economic integration beyond the capacity of the biological bonds that underpin the familistic group, it is necessary to extend the individual's sense of "self-interest" to broader social units. This extension of self is based on symbols. The evolution of the political economy represents the elaboration of this symbolizing capacity in order to overcome the individualistic, competitive, divisive, centrifugal stresses that continually threaten to defeat efforts to cooperate beyond the family level.

It is for this reason that Leslie White's (1959: 241) dismissal of the prestige economy as merely "a social game" is such a thoroughgoing error. The prestige economy, which in certain social situations *is* the political economy, is one dramatic manifestation of the general principle that economic integration beyond the family level is stabilized through the elaboration of public symbols. This process begins in the symbolic extension of familistic relations via classificatory kinship, expands in the ceremonial constructions of collective consciousness, as in totemism (Durkheim 1947: 463-74), and finds spectacular expression in Big Man feasts. Far from being economically irrelevant, these symbolically rich events and institutions create channels of economic support, identify economically important groups, and advertise such crucial information as property ownership, group strength, and alliance formation.

In the same light, such familiar symbolic expressions as the regalia and ritual that sanctify class superiority and surplus extraction in complex societies are central to the political economy. And for the ultimate in abstract symbolism, purified of all taint of individuals and their family or local group affiliations, we need only look at modern money: standing for all things of value, it links everyone within a nation's borders (and even beyond) in its ceaseless flow from hand to anonymous hand.

Why is symbolization at the political level able to overcome the persistent "selfishness" of individuals and their families? To put this question more precisely, what is it that *changes* about an individual family's self-interest that allows the political economy to evolve?

We have assumed from the outset that the family level is the level of economic integration with which humans are biologically most comfortable and for which they are biologically best equipped. Households in family-level societies are amazingly independent and self-reliant from a modern perspective. They meet nature on its own terms. They are capable of mustering all the necessities and most of the desirables in life through their own efforts. And they like it that way: other people just get in the way.

The evolution of the political economy is accompanied by the steady erosion of the autonomy of the family. Each higher-order integration of the economy not only introduces new ("emergent") cultural processes, but also entails a further reduction in the sphere of free action of the family in the natural environment. More and more of a family's time and energy must be spent in the sociocultural environment; in a word, the economy is increasingly socialized.

The power of symbolization has been to create and maintain a sense of unity (i.e., of shared self-interest) beyond the limits of the normal biologically and emotionally based unity of the family. Since symbols alone can create only a temporary group cohesion, real, material benefits must be offered by the political process if it is not to disintegrate under the pressures of family self-interest. The boundary of the

TABLE 11

Characteristics of the Family-Level Society, the Local Group, and the Regional Polity

Characteristic	Family-level society	Local group	Regional polity
Population density	less than 1 per square mile	more than 1 per square mile	more than 10 per square mile
Environment	scattered resources, meager and unpredictable	seasonally concentrated resources and/or transformable resources	concentrated, controllable resources and/or trade opportunities
Technology	individual procurement and tools, wild foods	capital improvements, domesticated and/or stored foods	major capital technology
Settlement	camp/hamlet	village/hamlet cluster with dance ground	settlement hierarchy
Social organization	family and bilateral networks	corporate and defensive groups	social stratification and regional institutions
Territoriality	customary use	defended local territories, group ownership	elite and/or institutional ownership, state-guaranteed private ownership
Warfare	controlled aggression, impulsive homicide	group aggression/ defense, personal fierceness, ceremonial regulation	conquest warfare, military specialists, internal peace
Ceremonialism	family ritual, ad hoc ceremonies	group and intergroup ceremonies	ceremonies of legitimization
Leadership	ad hoc leadership, institutionalized modesty	Big Man competition and display, differential success	hereditary elites, institutionalized leadership, status rivalry
Determinant variables	risk	for agriculturalists, warfare; for hunter-gatherers and pastoralists, risk, technology, and/or trade	technology and/or trade

polity, where avoidance, negative reciprocity, and warfare—that is, economic disintegration and chaos—occur, is the point at which the costs of group life to the family exceed its benefits. What are the changes that allow this boundary to encompass ever larger and more diverse populations?

The primary motor of change in the subsistence economy is the positive feedback relationship between population growth and technology. World prehistory and history record long-term, consistent population growth and technological growth. As Table 11 shows, a correlate of increasing social complexity is higher population density. This relationship is not a simple one; the marginality of the environment and the mode of subsistence both radically affect population density at different levels. Thus, for example, very high densities among New Guinean horticulturalists accompany a social organization structurally comparable to the low-density hunter-gatherer groups of the Northwest Coast.

The critical factor seems to be how population density affects the benefits of group membership to the household (or, conversely, the costs of not being a group member). The steady increase in population density that underlies cultural evolution creates problems that only the group can solve, or, at least, that the group can solve most efficiently. To put this another way, increasing population density causes the benefits of group membership for the family to rise correspondingly.

In nonstratified society, population growth presses on resources and requires a number of specific cultural solutions resulting in higher degrees of cooperation (risk spreading, technology, mutual defense, and trade). These group solutions relieve the population pressure, allowing population to rise until renewed pressure produces further cooperative solutions; and this positive feedback goes on ad infinitum. The negative feedback in this developmental process is also strong: greater cooperation runs against the perceived self-interest of the family group, which is rooted in biology, and each step in group formation is accordingly resisted as an erosion of family autonomy and control. This is no doubt why the pace of cultural evolution in the subsistence economy is slow, as evidenced by the very low rates of population growth up to the Neolithic period.

Ultimately, however, the long-term intensification of the subsistence economy gives rise to opportunities for investment and control that permit further increases in population but only at the cost of institutionalized leadership and social stratification. It is at this point that the subsistence economy gives way to the political economy as the main locus of evolution; at this point that the conservatism of the household is at last overcome as a major force; at this point that the basic issue changes from "What behavior best serves the immediate self-interest of the family group?" to "What behavior provides the greatest opportunities for investment and control?" The political economy is in place, and the way is clear for the rapid expansion in the size of human populations and the scale of their economies that has taken place since Neolithic times.

Bibliography

Bibliography

Adams, J. 1973. The Gitksan potlatch. Toronto: Holt, Rinehart & Winston.

Adams, R. 1966. The evolution of urban society. Chicago: University of Chicago Press.

AMRF. 1979. Studies on the epidemiology and treatment of hydatid disease in the Turkana District of Kenya. Third annual report. Nairobi: African Medical and Research Foundation.

Anduze, P. 1960. Shailili-Ko: Descubrimiento de las fuentes del Orinoco. Caracas: Talleres Graficos Ilustraciones.

Arvelo-Jimenez, N. 1984. The political feasibility of tribal autonomy in Amazonia. Manuscript. Instituto Venezolano de Investigaciones Cientificas, Caracas.

Asakawa, K. 1965. Land and society in medieval Japan. Tokyo: Japan Society for the Promotion of Science.

Athens, J. 1977. Theory building and the study of evolutionary process. In For theory building in archaeology, ed. L. Binford, pp. 353-84. New York: Academic Press.

———. 1984. Prehistoric taro pondfield agriculture in Hawaii: modeling constraints for its development. Unpublished manuscript.

Atlas. 1979. Atlas de Venezuela. 2d ed. Caracas: Ministerio del Ambiente y de los Recursos Naturales Renovables.

Austen, L. 1945. Cultural changes in Kiriwina. *Oceania* 16: 15-60.

Baksh, M. 1984. Cultural ecology and change of the Machiguenga Indians of the Peruvian Amazon. Unpublished Ph.D. dissertation, Anthropology, University of California, Los Angeles.

Barnett, H. 1968. The nature and function of the potlatch. Eugene: Department of Anthropology, University of Oregon.

Barth, F. 1964. Nomads of south Persia. London: Allen & Unwin.

Bartholomew, G., and J. Birdsell. 1953. Ecology and the protohominids. *American Anthropologist* 55: 481-98.

Beckerman, S. 1979. The abundance of protein in Amazonia: a reply to Gross. *American Anthropologist* 81: 533-60.

———. 1980. Fishing and hunting by the Bari of Columbia. *In* Studies in hunting and fishing in the neotropics, ed. R. Hames, pp. 67-109. Bennington, Vt.: Working Papers on South American Indians 2.

———. 1983. Does the swidden ape the jungle? *Human Ecology* 11: 1-12.

Belshaw, C. 1955. In search of wealth. American Anthropological Association Memoir 80.

———. 1965. Traditional exchange and modern markets. Englewood Cliffs, N.J.: Prentice-Hall.

Benedict, R. 1934. Patterns of culture. Boston: Houghton Mifflin.

Berdan, F. 1975. Trade, tribute and market in the Aztec Empire. Unpublished Ph.D. dissertation, Anthropology, University of Texas, Austin.

Bergman, R. 1974. Shipibo subsistence in the Upper Amazon rainforest. Unpublished Ph.D. dissertation, Geography, University of Wisconsin, Madison.

Berlin, E., and E. Markell. 1977. An assessment of the nutritional and health status of an Aguaruna Jivaro community, Amazonas, Peru. *Ecology of Food and Nutrition* 6: 69-81.

Bernal, I. 1969. The Olmec world, tr. D. Heyden and F. Horcasitas. Berkeley: University of California Press.

Best, E. 1924. The Maori. Polynesian Society Memoir.

———. 1925. Maori culture. Bulletin 9. Wellington, N.Z.: Dominion Museum.

Bettinger, R. 1978. Alternative adaptive strategies in the prehistoric Great Basin. *Journal of Anthropological Research* 34: 27-46.

———. 1982. Aboriginal exchange and territoriality in Owens Valley, California. *In* Contexts for prehistoric exchange, ed. J. Ericson and T. Earle, pp. 103-27. New York: Academic Press.

Binford, L. 1964. A consideration of archaeological research design. *American Antiquity* 29: 425-41.

———. 1968. Post-Pleistocene adaptations. *In* New perspectives in archaeology, ed. S. Binford and L. Binford, pp. 313-41. Chicago: Aldine.

———. 1980. Willow smoke and dogs' tails: hunter-gatherer settlement systems and archaeological site formation. *American Antiquity* 45: 4-20.

Binford, L., and S. Binford. 1966. A preliminary analysis of functional variability in the Mousterian of Levallois facies. *American Anthropologist* 68: 238-95.

Binford, S. 1968. Early Upper Pleistocene adaptations in the Levant. *American Anthropologist* 70: 707-17.

Biocca, E. 1971. Yanoama: the narrative of a white girl kidnapped by Amazon Indians. New York: Dutton.

Birdsell, J. 1953. Some environmental and cultural factors influencing the structure of Australian aboriginal populations. *American Naturalist* 87: 171-207.

———. 1968a. Comment on infanticide. *In* Man the hunter, ed. R. Lee and I. DeVore, p. 243. Chicago: Aldine.

———. 1968b. Some predictions for the Pleistocene based on equilibrium systems among recent hunter-gatherers. *In* Man the hunter, ed. R. Lee and I. DeVore, pp. 229-40. Chicago: Aldine.

Bloch, M. 1961. Feudal society. Chicago: University of Chicago Press.

Boas, F. 1898. Final report on the northwestern tribes of Canada. B.A.A.S. Report, pp. 628-88.

―――. 1910. Kwakiutl tales. New York: Columbia University Press.

―――. 1921. Ethnology of the Kwakiutl. 35th annual report of the Bureau of American Ethnology to the Secretary of the Smithsonian Institution. Washington, D.C.: U.S. Government Printing Office.

―――. 1949 [1896]. The limitations of the comparative method of anthropology. *In* Race, language, and culture, pp. 270-80. New York: Macmillan.

―――. 1949 [1920]. Methods of ethnology. *In* Race, language, and culture, pp. 281-89. New York: Macmillan.

―――. 1966. Kwakiutl ethnography, ed. Helen Codere. Chicago: University of Chicago Press.

Boeke, J. 1953. Economics and economic policy of dual societies, as exemplified by Indonesia. New York: Institute of Pacific Relations.

Bohannan, P. 1955. Some principles of exchange and investment among the Tiv. *American Anthropologist* 57: 60-69.

Boserup, E. 1965. The conditions of agricultural growth. Chicago: Aldine.

Browman, D. 1970. Early Peruvian peasants: culture history of a central highlands valley. Unpublished Ph.D. dissertation, Anthropology, Harvard University.

Brown, P. 1972. The Chimbu: a study of change in the New Guinea highlands. Cambridge, Mass.: Schenkman.

Brown, P., and A. Podolefsky. 1976. Population density, agricultural intensity, land tenure and group size in the New Guinea highlands. *Ethnology* 15: 211-38.

Brumfiel, E. 1980. Specialization, market exchange, and the Aztec state. *Current Anthropology* 21: 459-78.

Brumfiel, E., and T. Earle. 1986. Specialization, exchange, and complex societies. Cambridge: Cambridge University Press.

Brush, S. 1976. Man's use of an Andean ecosystem. *Human Ecology* 4: 147-66.

Buchbinder, G. 1973. Maring microadaptation: a study of demographic, nutritional, genetic and phenotypic variation in a highland New Guinea population. Unpublished Ph.D. dissertation, Anthropology, Columbia University.

Buck, J. 1937. Land utilization in China. Shanghai: Commercial Press.

Buck, P. (Te Rangi Hiroa). 1932. Ethnology of Tongareva. Honolulu: Bernice P. Bishop Museum Bulletin 92.

―――. 1934. Mangaian society. Honolulu: Bernice P. Bishop Museum Bulletin 122.

―――. 1938. Ethnology of Mangareva. Honolulu: Bernice P. Bishop Museum Bulletin 157.

Burling, R. 1962. Maximization theories and the study of economic anthropology. *American Anthropologist* 64: 802-21.

Burrows, E. 1936. The ethnology of Futuna. Honolulu: Bernice P. Bishop Museum Bulletin 138.

———. 1937. The ethnology of Uvea. Honolulu: Bernice P. Bishop Museum Bulletin 145.

Burton, R. 1975. Why do the Trobriands have chiefs? *Man* 10: 544-58.

Byers, D. 1967. The prehistory of the Tehuacan Valley: environment and subsistence. Austin: University of Texas Press.

Campbell, B. 1966. Human evolution. Chicago: Aldine.

Cancian, F. 1965. Economics and prestige in a Maya community. Stanford, Calif.: Stanford University Press.

———. 1972. Change and uncertainty in a peasant economy. Stanford, Calif.: Stanford University Press.

Caplovitz, D. 1963. The poor pay more. Glencoe, Ill.: Free Press.

Carneiro, R. 1960. Slash-and-burn agriculture: a closer look at its implications for settlement patterns. *In* Men and culture, ed. A. Wallace, pp. 229-34. Philadelphia: University of Pennsylvania Press.

———. 1967. The evolution of society: selections from Herbert Spencer's *Principles of Sociology.* Chicago: University of Chicago Press.

———. 1970a. Scale analysis, evolutionary sequences, and the rating of cultures. *In* A handbook of method in cultural anthropology, ed. R. Naroll and R. Cohen, pp. 834-71. New York: Columbia University Press.

———. 1970b. A theory of the origin of the state. *Science* 169: 733-38.

———. 1981. The chiefdom as precursor to the state. *In* The transition to statehood in the New World, ed. G. Jones and R. Kautz, pp. 37-79. Cambridge: Cambridge University Press.

Chagnon, N. 1968a. Yanomamo: the fierce people. New York: Holt, Rinehart & Winston.

———. 1968b. Yanomamo social organization and warfare. *In* War: the anthropology of armed conflicts and aggression, ed. M. Fried, M. Harris, and R. Murphy, pp. 109-59. New York: Natural History Press.

———. 1980. Highland New Guinea models in the South American lowlands. *In* Studies in hunting and fishing in the neotropics, ed. R. Hames. Bennington, Vt.: *Working Papers on South American Indians* 2: 111-30.

———. 1983. Yanomamo: the fierce people. 3d ed. New York: Holt, Rinehart & Winston.

Chagnon, N., and R. Hames. 1979. Protein deficiency and tribal warfare in Amazonia: new data. *Science* 203: 910-13.

Chance, N. 1966. The Eskimo of north Alaska. New York: Holt, Rinehart & Winston.

Chayanov, A. 1966 [1925]. The theory of peasant economy. Homewood, Ill.: Irwin.

Chibnik, M. 1981. The evolution of cultural rules. *Journal of Anthropological Research* 37: 256-68.

Childe, V. 1936. Man makes himself. London: Watts.

———. 1942. What happened in history? Baltimore: Penguin.

———. 1951. Social evolution. London: Watts.

Christenson, A. 1980. Change in the human niche in response to population growth. *In* Modeling change in prehistoric subsistence economies, ed. T. Earle and A. Christenson, pp. 31-72. New York: Academic Press.

Clarke, W. 1966. From extensive to intensive shifting cultivation: a succession from New Guinea. *Ethnology* 5: 347-59.

———. 1971. Place and people: an ecology of a New Guinean community. Berkeley: University of California Press.

———. 1982. Comment. *In* Individual or group advantage, ed. J. Peoples. *Current Anthropology* 23: 301.

Coale, A. 1974. The history of the human population. *Science* 231(3): 40-51.

Codere, H. 1950. Fighting with property. Seattle: University of Washington Press.

Cohen, Mark. 1977. The food crisis in prehistory. New Haven, Conn.: Yale University Press.

Cohen, Myron. 1968. A case study of Chinese family economy and development. *Journal of Asian Studies* 3: 161-80.

———. 1970. Introduction. *In* A. Smith, Village life in China, pp. ix-xxvi. Boston: Little, Brown.

———. 1984. Personal communication.

Conkey, M. 1978. Style and information in cultural evolution: toward a predictive model for the Paleolithic. *In* Social archaeology: beyond subsistence and dating, ed. C. Redman et al., pp. 61-85. New York: Academic Press.

Conrad, G., and A. Demarest. 1984. Religion and empire: the dynamics of Aztec and Inca expansion. Cambridge: Cambridge University Press.

Cordy, R. 1974. Cultural adaptation and evolution in Hawaii: a suggested new sequence. *Journal of Polynesian Society* 83: 180-91.

———. 1981. A study of prehistoric social change: the development of complex societies in the Hawaiian islands. New York: Academic Press.

Dalton, G. 1961. Economic theory and primitive society. *American Anthropologist* 63: 1-25.

———. 1977. Aboriginal economies in stateless societies. *In* Exchange systems in prehistory, ed. T. Earle and J. Ericson, pp. 191-212. New York: Academic Press.

D'Altroy, T. 1981. Empire growth and consolidation: the Xanxa region of Peru under the Incas. Unpublished Ph.D. dissertation, Anthropology, University of California, Los Angeles.

———. 1984. Personal correspondence.

D'Altroy, T., and T. Earle. 1985. Staple finance, wealth finance, and storage in the Inca political economy. *Current Anthropology* 26: 187-206.

D'Aquili, E. 1972. The biopsychological determinants of culture. Module in Anthropology 13. Reading, Mass.: Addison-Wesley.

Day, K. 1982. Storage and labor service: a production and management design for the Andean area. *In* Chanchan: Andean desert city, ed. M. Moseley and K. Day, pp. 333-49. Albuquerque: University of New Mexico Press.

DeVore, I., and K. Hall. 1965. Baboon ecology. *In* Primate behavior, ed. I. DeVore, pp. 20-52. New York: Holt, Rinehart & Winston.

Donald, L., and D. Mitchell. 1975. Some correlates of local group rank among the Southern Kwakiutl. *Ethnology* 14: 325-46.

Donkin, R. 1979. Agricultural terracing in the aboriginal New World. New York: Viking Fund Publication in Anthropology 56.

Driver, H. 1969. Indians of North America, 2d ed. Chicago: University of Chicago Press.

Drucker, P. 1955. Indians of the northwest coast. Anthropological Handbook No. 10. New York: American Museum of Natural History.

————. 1965.Cultures of the north Pacific coast. San Francisco: Chandler.

Drucker, P., and R. Heizer. 1967. To make my name good: a reexamination of the Southern Kwakiutl potlatch. Berkeley: University of California Press.

Duby, G. 1968. Rural economy and country life in the medieval west. Columbia: University of South Carolina Press.

Dumont, L. 1970. Homo hierarchicus. Chicago: University of Chicago Press.

Durham, W. 1982. Interaction of genetic and cultural evolution: models and examples. *Human Ecology* 10: 289-323.

Durkheim, E. 1947 [1912]. The elementary forms of the religious life. Glencoe, Ill.: Free Press.

Duus, P. 1976. Feudalism in Japan. New York: Knopf.

Dyson-Hudson, R., and J. McCabe. 1985. South Turkana nomadism: coping with an unpredictably varying environment. New Haven, Conn.: HRAFLEX.

Earle, T. 1976. A nearest-neighbor analysis of two formative settlement systems. *In* The early Mesoamerican village, ed. K. Flannery, pp. 196-223. New York: Academic Press.

————. 1977. A reappraisal of redistribution: complex Hawaiian chiefdoms. *In* Exchange systems in prehistory, ed. T. Earle and J. Ericson, pp. 213-29. New York: Academic Press.

————. 1978. Economic and social organization of a complex chiefdom: the Halelea district, Kauai, Hawaii. Ann Arbor: Museum of Anthropology, University of Michigan, Anthropological Paper 63.

————. 1980a. A model of subsistence change. *In* Modeling change in prehistoric subsistence economies, ed. T. Earle and A. Christenson, pp. 1-29. New York: Academic Press.

————. 1980b. Prehistoric irrigation in the Hawaiian islands: an evaluation of evolutionary significance. *Archaeology and Physical Anthropology in Oceania* 15: 1-28.

————. 1982. The ecology and politics of primitive valuables. *In* Culture and ecology: eclectic perspectives, ed. J. Kennedy and R. Edgerton, pp. 65-83. Washington, D.C.: American Anthropological Association Special Publication 15.

————. 1985. Commodity exchange and markets in the Inka state: recent archaeological evidence. *In* Markets and marketing, ed. S. Plattner, pp. 369-97. Monographs in Economic Anthropology 4.

Earle, T., and T. D'Altroy. 1982. Storage facilities and state finance in the upper Mantaro valley, Peru. *In* Contexts for prehistoric exchange, ed. J. Ericson and T. Earle, pp. 265-90. New York: Academic Press.

Earle, T., T. D'Altroy, C. LeBlanc, C. Hastorf, and T. Levine. 1980. Changing settlement patterns in the upper Mantaro valley, Peru. *Journal of New World Archaeology* 4(1).

Earle, T., T. D'Altroy, C. Hastorf, C. Scott, C. Costin, G. Russell, and E. Sandefur. 1986. The impact of Inka conquest on the Wanka domestic economy. Los Angeles: University of California, Institute of Archaeology Monograph.

Ellis, W. 1853. Polynesian researches during a residence of nearly eight years in the Society and Sandwich Islands. London: H. G. Bohn.

———. 1963 [1827]. Journal of William Ellis. Honolulu: Advertiser Publishing.

Emerson, A. 1960. The evolution of adaptation in population systems. *In* The evolution of life, *Vol. 1 of* Evolution after Darwin, ed. Sol Tax, pp. 307-48. Chicago: University of Chicago Press.

Engels, F. 1972 [1884]. The origin of the family, private property, and the state. New York: International Publishers.

Ericson, J. 1977. Egalitarian exchange systems in California: a preliminary view. *In* Exchange systems in prehistory, ed. T. Earle and J. Ericson, pp. 109-26. New York: Academic Press.

Evans-Pritchard, E. 1940. The Nuer. Oxford: Oxford University Press.

Feder, E. 1971. The rape of the peasantry. New York: Doubleday.

Feinman, G., and J. Neitzel. 1984. Too many types: an overview of prestate sedentary societies in the Americas. *In* Advances in archaeological method and theory 7, ed. M. Schiffer, pp. 39-102. New York: Academic Press.

Firth, R. 1929. Primitive economics of the New Zealand Maori. London: Routledge.

Flannery, K. 1969. Origins and ecological effects of early domestication in Iran and the Near East. *In* The domestication and exploitation of plants and animals, ed. P. Ucko and G. Dimbleby, pp. 73-100. Chicago: Aldine.

———. 1972. The cultural evolution of civilizations. *Annual Review of Ecology and Systematics* 3: 399-426.

Foster, G. 1961. The dyadic contract: a model for the social structure of a Mexican peasant village. *American Anthropologist* 63: 1173-92.

Fowler, D. 1966. Great Basin social organization. *In* The current status of anthropological research in the Great Basin: 1964, ed. W. D'Azevedo et al., pp. 57-74. Reno: Desert Research Institute.

Fried, M. 1967. The evolution of political society. New York: Random House.

Friedman, J. 1974. Marxism, structuralism and vulgar materialism. *Man* 9: 444-69.

Frisch, R. 1978. Population, food intake and fertility. *Science* 199: 22-30.

Frisch, R., G. Wyshak, and L. Vincent. 1980. Delayed menarche and amenorrhea in ballet dancers. *New England Journal of Medicine* 303: 17-19.

Gall, P., and A. Saxe. 1977. Ecological evolution of culture: the state as predator in succession theory. *In* Exchange systems in prehistory, ed. T. Earle and J. Ericson, pp. 255-68. New York: Academic Press.

Geertz, C. 1963. Agricultural involution. Berkeley: University of California Press.

Gilman, A. 1981. The development of stratification in Bronze Age Europe. *Current Anthropology* 22: 1-23.

————. 1984. Explaining the Upper Paleolithic revolution. *In* Marxist perspectives in archaeology, ed. M. Spriggs, pp. 115-26. Cambridge: Cambridge University Press.

Godelier, M. 1977. Perspectives in Marxist anthropology. Cambridge: Cambridge University Press.

Goldman, I. 1970. Ancient Polynesian society. Chicago: University of Chicago Press.

Goldschmidt, W. 1959. Man's way: a preface to the understanding of human society. New York: Holt, Rinehart & Winston.

Goodfellow, D. 1968. The applicability of economic theory to so-called primitive communities. *In* Economic anthropology, ed. E. LeClair and H. Schneider, pp. 55-65. New York: Holt, Rinehart & Winston.

Goody, J. 1971. Technology, tradition and the state in Africa. Cambridge: Cambridge University Press.

Greenfield, S. 1972. Charwomen, cesspools, and road building: an examination of patronage, clientage, and political power in southeastern Minas Gerais. *In* Structure and process in Latin America: patronage, clientage, and power systems, ed. A. Strickon and S. Greenfield, pp. 71-100. Albuquerque: University of New Mexico Press.

Gross, D. 1975. Protein capture and cultural development in the Amazon basin. *American Anthropologist* 77: 526-49.

Gross, D., and B. Underwood. 1971. Technological change and caloric costs: sisal agriculture in northeastern Brazil. *American Anthropologist* 73: 725-40.

Grossman, L. 1984. Peasants, subsistence ecology, and development in the highlands of Papua New Guinea. Princeton, N.J.: Princeton University Press.

Gubser, N. 1965. The Nunamiut Eskimos. New Haven, Conn.: Yale University Press.

Gulliver, P. 1951. A preliminary survey of the Turkana. Kenya: Colonial Social Science Research Council.

————. 1955. The family herds. London: Routledge & Kegan Paul.

————. 1975. Nomadic movements: causes and implications. *In* Pastoralism in tropical Africa, ed. T. Monod, pp. 369-84. London: Oxford University Press.

Gunther, E. 1972. Indian life on the northwest coast of North America. Chicago: University of Chicago Press.

Hall, J. 1970. Japan: from prehistory to modern times. New York: Dell.

Halperin, R., and J. Dow, eds. 1977. Peasant livelihood: studies in economic anthropology and cultural ecology. New York: St. Martin's Press.

Hames, R. 1982. Comment on A. Johnson, "Reductionism in Cultural Ecology: the Amazon Case." *Current Anthropology* 23: 421-22.

Hamilton, W. 1963. The evolution of altruistic behavior. *American Naturalist* 97: 354-56.

Handy, E. 1923. The native cultures in the Marquesas. Honolulu: Bernice P. Bishop Museum Bulletin 9.

Handy, E., and E. Pukui. 1958. The Polynesian family system in Ka'u, Hawaii. Rutland, Vt.: Tuttle.

Hardesty, D. 1977. Ecological anthropology. New York: Wiley.

Hardin, G. 1968. The tragedy of the commons. *Science* 162: 1243-48.

Harris, M. 1959. The economy has no surplus? *American Anthropologist* 61: 185-99.

——. 1974. Cows, pigs, wars, and witches: the riddles of culture. New York: Random House.

——. 1977. Cannibals and kings: the origins of cultures. New York: Random House.

——. 1979. Cultural materialism. New York: Random House.

Hastings, C. 1982. Implications of Andean verticality in the evolution of political complexity: a view from the margins. Unpublished paper presented at the 81st annual meeting of the American Anthropological Association, Dec. 4-7, Washington, D.C.

Hastorf, C. 1983. Prehistoric agricultural intensification and political development in the Jauja region of central Peru. Unpublished Ph.D. dissertation, Anthropology, University of California, Los Angeles.

Hastorf, C., and T. Earle. 1985. Intensive agriculture and the geography of political change in the upper Mantaro region of central Peru. *In* Prehistoric intensive agriculture in the tropics, ed. I. Farrington, pp. 569-95. Oxford: British Archaeological Reports, International Series 232.

Hawkes, K., and J. O'Connell. 1981. Affluent hunters? Some comments in light of the Alyawara case. *American Anthropologist* 83: 622-26.

Hayden, B. 1981a. Subsistence and ecological adaptations of modern hunter/gatherers. *In* Omnivorous primates, ed. R. Harding and G. Teleki, pp. 344-421. New York: Columbia University Press.

——. 1981b. Research and development in the stone age: technological transitions among hunter-gatherers. *Current Anthropology* 22: 529-48.

Heider, K. 1970. The Dugum Dani. New York: Viking Fund Publication in Anthropology 49.

Herskovits, M. 1952. Economic anthropology. New York: Norton.

Hill, J. 1977. Systems theory and the explanation of change. *In* Explanation of prehistoric change, ed. J. Hill, pp. 59-103. Albuquerque: University of New Mexico Press.

Hipsley, E., and N. Kirk. 1965. Studies of dietary intake and the expenditure of energy by New Guineans. Noumea, New Caledonia: South Pacific Commission Technical Paper 147.

Hitchcock, R. 1978. The traditional response to drought in Botswana. *In* Proceedings of the symposium on drought in Botswana, ed. M. Hinchey, pp. 91-97. Gaborone: The Botswana Society.

Holmberg, A. 1969. Nomads of the longbow. Garden City, N.Y.: Natural History Press.

Homans, G. 1958. Social behavior as exchange. *American Journal of Sociology* 62: 597-606.

———. 1967. The nature of social science. New York: Harcourt, Brace.

Hommon, R. 1976. The formation of primitive states in pre-contact Hawaii. Unpublished Ph.D. dissertation, Anthropology, University of Arizona.

———. 1981. A model of the pre-contact history of the island of Kaho'olawe. Paper delivered at the 46th annual meeting of the Society for American Archaeology, San Diego.

Howell, N. 1979. Demography of the Dobe !Kung. New York: Academic Press.

Hutchinson, B. 1966. The patron-dependent relationship in Brazil. *Sociologia Ruralis* 6(1).

Ibn Khaldun. 1956 [1377]. The mugaddimah: an introduction to history. New York: Pantheon.

Ingold, T. 1980. Hunters, pastoralists and ranchers. Reindeer economies and their transformations. Cambridge: Cambridge University Press.

Isaac, G. 1978. The food-sharing behavior of protohuman hominids. *Scientific American* 238: 90-108.

Isbell, W., and K. Schreiber. 1978. Was Hurai a state? *American Antiquity* 43: 372-89.

Jackson, J. 1975. Recent ethnography of indigenous northern lowland South America. *Annual Review of Anthropology* 4: 307-40.

Jochim, M. 1984. Paleolithic complexity and salmon productivity. Paper presented at the 49th annual meeting of the Society of American Archaeology, Portland, Ore.

Johnson, A. 1971a. Sharecroppers of the Sertão. Stanford, Calif.: Stanford University Press.

———. 1971b. Security and risk-taking among poor peasants: a Brazilian case. *In* Studies in economic anthropology, ed. G. Dalton, pp. 143-78. Washington, D.C.: American Anthropological Association.

———. 1972. Individuality and experimentation in traditional agriculture. *Human Ecology* 1: 149-59.

———. 1975a. Time allocation in a Machiguenga community. *Ethnology* 14: 301-10.

———. 1975b. Landlords, patrons, and "proletarian consciousness" in rural Latin America. *In* Ideology and social change in Latin America, ed. J. Nash and J. Corradi, pp. 78-96. New York: City University of New York.

———. 1980. The limits of formalism in agricultural decision research. *In* Agricultural decision making, ed. P. Bartlett, pp. 19-43. New York: Academic Press.

———. 1982. Reductionism in cultural ecology: the Amazon case. *Current Anthropology* 23: 413-28.

———. 1983. Machiguenga gardens. *In* Adaptive responses of native Amazonians, ed. R. Hames and W. Vickers, pp. 29-63. New York: Academic Press.

Johnson, A., and C. Behrens. 1982. Nutritional criteria in Machiguenga food production decisions: a linear programming analysis. *Human Ecology* 10: 167-89.

Johnson, A., and G. Bond. 1974. Kinship, friendship, and exchange in two communities: a comparative analysis of norms and behavior. *Journal of Anthropological Research* 30: 55-68.

Johnson, O. 1978. Domestic organization and interpersonal relations among the Machiguenga Indians of the Peruvian Amazon. Unpublished Ph.D. dissertation, Anthropology, Columbia University.

Jones, W. 1959. Manioc in Africa. Stanford, Calif.: Stanford University Press.

Keesing, R. 1975. Kin groups and social structure. New York: Holt, Rinehart & Winston.

———. 1983. 'Elota's story: the life and times of a Solomon Islands Big Man. New York: Holt, Rinehart & Winston.

Kikuchi, W. 1976. Prehistoric Hawaiian fishponds. *Science* 193: 295-99.

Kirch, P. 1974. The chronology of early Hawaiian settlement. *Archaeology and Physical Anthropology in Oceania* 9: 110-19.

———. 1977. Valley agricultural systems in prehistoric Hawaii: an archaeological consideration. *Asian Perspectives* 20: 246-80.

———. 1980. Polynesian prehistory: cultural adaptation in island ecosystems. *American Scientist* 68: 39-48.

———. 1982. Advances in Polynesian prehistory: three decades in review. *In* Advances in world archaeology, Vol. 1, pp. 51-97. New York: Academic Press.

———. 1983. Man's role in modifying tropical and subtropical Polynesian ecosystems. *Archaeology and Physical Anthropology in Oceania* 18: 26-31.

———. 1984. The evolution of Polynesian chiefdoms. Cambridge: Cambridge University Press.

———. 1985. Intensive agriculture in prehistoric Hawaii: the wet and the dry. *In* Prehistoric intensive agriculture in the tropics, ed. I. Farrington, pp. 435-54. Oxford: British Archaeological Reports, International Series 232.

Kirch, P., and M. Kelley. 1975. Prehistory and ecology in a windward Hawaiian valley: Halawa Valley, Molokai. Honolulu: Pacific Anthropological Records 24.

Kirchhoff, P. 1955. The principles of clanship in human society. *Davidson Anthropological Society Journal* 1: 1-11.

Konner, M. 1982. The tangled wing: biological constraints on the human spirit. New York: Harper & Row.

Konner, M., and C. Worthman. 1980. Nursing frequency, gonadal function and birth spacing among !Kung hunter-gatherers. *Science* 207: 788-91.

Kroeber, A. 1939. Cultural and natural areas of native North America. Berkeley: University of California Publications in American Archaeology and Ethnology 38.

Kurtz, D. 1978. The legitimation of the Aztec state. *In* The early state, ed. H. Claessen and P. Skalnik, pp. 169-89. The Hague: Mouton.

LaLone, D. 1982. The Inca as a nonmarket economy: supply on command versus supply and demand. *In* Contexts for prehistoric exchange, ed. J. Ericson and T. Earle, pp. 292-316. New York: Academic Press.

LaLone, M. 1985. Indian land tenure in southern Cuzco, Peru: from Inca to colonial patterns. Unpublished Ph.D. dissertation, Anthropology, University of California, Los Angeles.

Lambert, B. 1973. Bilaterality in the Andes. *In* Andean kinship and marriage, ed. R. Bolton and E. Mayer, pp. 1-27. Washington, D.C.: American Anthropological Association.

Lathrap, D. 1970. The upper Amazon. New York: Praeger.

Lavallée, D., and M. Julien. 1973. Les établissements Asto à l'époque préhispanique. Lima: Travaux de l'Institut Français d'Etudes Andiens 15.

Leach, J., and E. Leach, eds. 1983. The Kula: new perspectives on Massim exchange. Cambridge: Cambridge University Press.

Leacock, E., and R. Lee. 1982. Politics and history in band society. Cambridge: Cambridge University Press.

LeBlanc, C. 1981. Late prehispanic Huanca settlement patterns in the Yanamarca Valley, Peru. Unpublished Ph.D. dissertation, Anthropology, University of California, Los Angeles.

LeClair, E. 1962. Economic theory and economic anthropology. *American Anthropologist* 64: 1179-1203.

Lee, R. 1976. !Kung spatial organization. *In* Kalahari hunter-gatherers, ed. R. Lee and I. DeVore, pp. 73-97. Cambridge: Harvard University Press.

———. 1979. The !Kung San. Cambridge: Cambridge University Press.

———. 1983. Greeks and Victorians: a re-examination of Engels' theory of the Athenian polis. Unpublished manuscript.

———. 1984. The Dobe !Kung. New York: Holt, Rinehart & Winston.

Lee, R., and I. DeVore. 1968. Man the hunter. Chicago: Aldine.

———, eds. 1976. Kalahari hunter-gatherers. Cambridge: Harvard University Press.

Legros, D. 1977. Chance, necessity, and mode of production: a Marxist critique of cultural evolutionism. *American Anthropologist* 79: 26-41.

LeVine, T. 1979. Prehispanic political and economic change in highland Peru: an ethnohistorical study of the Mantaro Valley. Unpublished Masters thesis, Archaeology, University of California, Los Angeles.

———. 1985. Inka administration in the Central Highlands: a comparative study. Unpublished Ph.D. dissertation, Archaeology, University of California, Los Angeles.

Lewis, A. 1974. Knights and samurai: feudalism in northern France and Japan. London: Temple Smith.

Lowman, C. 1980. Environment, society and health: ecological bases of community growth and decline in the Maring region of Papua New Guinea. Unpublished Ph.D. dissertation, Anthropology, Columbia University.

———. 1984. Personal correspondence.

Lumbreras, L. 1974. The people and cultures of ancient Peru. Washington, D.C.: Smithsonian.

Luttwak. E. 1976. The grand strategy of the Roman empire from the first century A.D. to the third. Baltimore: Johns Hopkins University Press.

Lynch, T. 1982. Personal correspondence.

MacNeish, R. 1964. Ancient Mesoamerican civilization. *Science* 143: 531-37.

———. 1970. Social implications of changes in population and settlement pattern of the 12,000 years of prehistory in the Tehuacan Valley of Mexico. *In* Population and economics, ed. P. Deprez, pp. 215-50. Winnipeg: University of Manitoba Press.

Maine, H. 1870. Ancient law. London: John Murray.

Malinowski, B. 1922. Argonauts of the western Pacific. New York: Dutton.

———. 1929. The sexual life of savages in northwestern Melanesia. London: Routledge.

———. 1935. Coral gardens and their magic. London: Allen & Unwin.

Malo, D. 1951 [1898]. Hawaiian antiquities, 2d ed. Honolulu: Bernice P. Bishop Museum Special Publication 2.

Malthus, T. 1798. An essay on the principle of population. London: Johnson.

Marshall, J. 1957. Ecology of the !Kung bushmen of the Kalahari. Unpublished senior honors thesis, Anthropology, Harvard University.

Marshall, L. 1976. Sharing, talking, and giving. *In* Kalahari hunter-gatherers, ed. R. Lee and I. DeVore, pp. 349-71. Cambridge: Harvard University Press.

Matos M., R., and J. Parsons. 1979. Poblamiento prehispanico en la cuenca del Mantaro. *In* Arqueologia peruana, ed. R. Matos M., pp. 157-71. Lima: Centro de Proyeccion Cristiana.

Mauss, M. 1967 [1925]. The gift: forms and functions of exchange in archaic societies. New York: Norton.

Mayer, E. 1977. Beyond the nuclear family. *In* Andean kinship and marriage, ed. R. Bolton and E. Mayer, pp. 60-80. Washington, D.C.: American Anthropological Association.

Mayhew, A. 1973. Rural settlement and farming in Germany. New York: Harper & Row.

Mead, M. 1930. Social organization of Manu'a. Honolulu: Bernice P. Bishop Museum Bulletin 76.

Meggers, B. 1954. Environmental limitation on the development of culture. *American Anthropologist* 56: 801-24.

Meggitt, M. 1964. Male-female relations in the highlands of New Guinea. *American Anthropologist* 66: 202-24.

———. 1965. The lineage system of the Mae Enga of New Guinea. New York: Barnes & Noble.

———. 1967. The pattern of leadership among the Mae Enga. *Anthropological Forum* 2: 20-35.

———. 1972. System and subsystem: the Te exchange cycle among the Mae Enga. *Human Ecology* 1: 111-23.

———. 1974. Pigs are our hearts!: the Te exchange cycle among the Mae Enga of New Guinea. *Oceania* 44: 165-203.

————. 1977. Blood is their argument: warfare among the Mae Enga tribesmen of the New Guinea highlands. Palo Alto, Calif.: Mayfield.

————. 1984. Personal correspondence.

Meillassoux, C. 1972. From reproduction to production. *Economy and Society* 1: 93-105.

Migliazza, E. 1972. Yanomama grammar and intelligibility. Unpublished Ph.D. dissertation, Indiana University.

Mintz, S. 1961. Pratik: Haitian personal economic relations. *In* Proceedings of the 1961 annual spring meeting of the American Ethnological Society.

Moore, S. 1958. Power and property in Inca Peru. New York: Columbia University Press.

Moran, E. 1979. Human adaptability. North Scituate, Mass.: Duxbury.

Morgan, L. 1877. Ancient society. Chicago: Kerr.

Morrell, M. 1985. The Gitksan-Wet'suwet'en fishery in the Skeena river system. Hazelton, B.C.: Gitksan-Wet'suwet'en Tribal Council.

Moseley, M., and K. Day. 1982. Chan Chan: Andean desert city. Albuquerque: University of New Mexico Press.

Murphy, R. 1971. The dialectics of social life. New York: Basic Books.

————. 1979. Lineage and lineality in lowland South America. *In* Brazil: anthropological perspectives, ed. M. Margolis and W. Carter, pp. 217-24. New York: Columbia University Press.

Murra, J. 1962. Cloth and its functions in the Inca state. *American Anthropologist* 64: 710-28.

————. 1965. Herds and herders in the Inca state. *In* Man, culture, and animals, ed. A. Leeds and A. Vayda, pp. 185-215. Washington, D.C.: American Association for the Advancement of Science.

————. 1972. El "control vertical" de un máximo de piso ecológicos en la economía de las sociedades andinas. *In* Visita de la provincia de León de Huánuco en 1562, ed. J. Murra, 2: 429-76. Huánuco, Peru: Universidad Nacional Hermilio Valdizán.

————. 1975. Formaciones económicas y políticas del mundo andino. Lima: Instituto de Estudios Peruanos.

————. 1980 [1956]. The economic organization of the Inka state. Greenwich, Conn.: JAI Press.

Nelson, E. 1899. The Eskimo about Bering Strait. Bureau of American Ethnology Annual Report 18. Washington, D.C.: U.S. Government Printing Office.

Netting, R. 1968. Hill farmers of Nigeria: cultural ecology of the Kofyar of the Jos Plateau. Seattle: University of Washington Press.

————. 1977. Cultural ecology. Menlo Park, Calif.: Cummings.

Newman, P. 1957. An intergroup collectivity among the Nootka. Masters thesis, Anthropology, University of Washington.

Oberg, K. 1973. The social economy of the Tlingit Indians. Seattle: University of Washington Press.

Odum, E. 1971. Fundamentals of ecology, 3d ed. Philadelphia: Saunders.

Oliver, D. 1974. Ancient Tahitian society. Honolulu: University of Hawaii Press.

Oswalt, W. 1979. Eskimos and explorers. Novato, Calif.: Chandler & Sharp.

Patton, M. 1981. The ecology of hydatid disease in Turkana District, Kenya. Masters thesis, Geography, University of Minnesota.

————. 1982. *Idia* to *ekile nawi*: career development of the Turkana pastoralist. Unpublished manuscript, Geography, University of California, Los Angeles.

People of Ksan. 1980. Gathering what the Great Nature provided: food traditions of the Gitksan. Seattle: University of Washington Press.

Peoples, J. 1982. Individual or group advantage? A reinterpretation of the Maring ritual cycle. *Current Anthropology* 23: 291-310.

Pianka, E. 1974. Evolutionary biology. New York: Harper & Row.

Pinkerton, E. 1985. Personal communication.

Polanyi, K. 1957. The economy as instituted process. *In* Trade and market in the early empires, ed. K. Polanyi, C. Arensberg, and H. Pearson, pp. 243-70. New York: Free Press.

Popov, A. 1966. The Nganasan. Bloomington: Indiana University Publications.

Potter, J., M. Diaz, and G. Foster, eds. 1967. Peasant society: a reader. Boston: Little, Brown.

Potts, R. 1984. Home bases and early hominids. *American Scientist* 72: 338-47.

Powell, H. 1960. Competitive leadership in Trobriand political organization. *Journal of the Royal Anthropological Institute* 90: 118-45.

————. 1969. Territoriality, hierarchy and kinship in Kiriwina. *Man* 4: 580-604.

Quijano, A. 1967. Contemporary peasant movements. *In* Elites in Latin America, ed. S. Lipset and A. Solari. New York: Oxford University Press.

Ramos, A. 1972. The social system of the Sanuma of northern Brazil. Ph.D. dissertation, University of Wisconsin, Madison.

Randsborg, K. 1980. The Viking age in Denmark. London: Duckworth.

Rappaport, R. 1967. Pigs for the ancestors. New Haven, Conn.: Yale University Press.

————. 1971. Nature, culture, and ecological anthropology. *In* Man, culture, and society, ed. L. Shapiro, pp. 237-67. New York: Oxford University Press.

Rathje, W. 1971. The origin and development of lowland Classic Maya civilization. *American Antiquity* 36: 275-85.

Rathje, W., and R. McGuire. 1982. Rich man . . . poor man. *American Behavioral Scientist* 25: 705-15.

Reidhead, V. 1980. The economics of subsistence change: a test of an optimization model. *In* Modeling change in prehistoric subsistence economies, ed. T. Earle and A. Christenson, pp. 141-86. New York: Academic Press.

Renfrew, C. 1972. The emergence of civilization: the Cyclades and the Aegean in the third millennium B.C. London: Methuen.

————. 1973. Monuments, mobilization, and social organization in neolithic Wessex. *In* The explanation of culture change, ed. C. Renfrew, pp. 539-58. London: Duckworth.

Rick, J. 1978. Prehistoric hunters of the high Andes. New York: Academic Press.
————. 1984. Structure and style at an early base camp in Junin, Peru. Paper presented at the 49th annual meeting of the Society for American Archaeology, Portland, Ore.

Rosendahl, P. 1972. Aboriginal agriculture and residence patterns in upland Lapakahi, island of Hawaii. Unpublished Ph.D. dissertation, Anthropology, University of Hawaii, Honolulu.

Rosman, A., and P. Rubel. 1971. Feasting with mine enemy: rank and exchange among Northwest Coast societies. New York: Columbia University Press.

Rostworoski de Diez Canseco, M. 1961. Curacas y sucesiones (Costa Norte). Lima: Imprenta Minerva.

Rowe, J. 1946. Inca culture at the time of the Spanish Conquest. *In* Handbook of South American Indians, ed. J. Steward, 2: 183-300. Bureau of American Ethnology Bulletin 143. Washington, D.C.: U.S. Government Printing Office.

Sackett, J. 1984. Personal communication.

Sackschewsky, M. 1970. The clan meeting in Enga society. *In* Exploring Enga culture, ed. P. Brennan, pp. 51-101. Wapenamanda, New Guinea: Kristen Press.

Sahlins, M. 1958. Social stratification in Polynesia. Seattle: University of Washington Press.
————. 1963. Poor man, rich man, big man, chief: political types in Melanesia and Polynesia. *Comparative Studies in Society and History* 5: 285-303.
————. 1968. Notes on the original affluent society. *In* Man the hunter, ed. R. Lee and I. DeVore, pp. 85-89. Chicago: Aldine.
————. 1972. Stone age economics. Chicago: Aldine.

Sanchez, P. 1976. Properties and management of soils in the tropics. New York: Wiley-Interscience.

Sanders, W. 1956. The Central Mexican symbiotic region. *In* Prehistoric settlement patterns in the New World, ed. G. Willey. New York: Viking Fund Publication in Anthropology 23: 115-27.

Sanders, W., J. Parsons, and R. Santley. 1979. The basin of Mexico: ecological processes in the evolution of a civilization. New York: Academic Press.

Sauer, C. 1948. Geography of South America. *In* Handbook of South American Indians, ed. J. Steward, 6: 319-44. Washington, D.C.: Bureau of American Ethnology.

Schaedel, R. 1978. Early state of the Inkas. *In* The early state, ed. H. Claessen and P. Skalnik, pp. 289-320. The Hague: Mouton.

Schultz, T. 1964. Transforming traditional agriculture. New Haven, Conn.: Yale University Press.

Scott, C., and T. Earle. n.d. The development of chiefdoms among the Wanka ethnic group in the central highlands of Peru. Unpublished manuscript.

Service, E. 1962. Primitive social organization. New York: Random House.
————. 1975. Origins of the state and civilization. New York: Norton.

————. 1977. Classical and modern theories of the origins of government. *In* Origins of the state, ed. R. Cohen and E. Service, pp. 21-34. Philadelphia: ISHI.

Shahrani, M. 1979. The Kirghiz and Wakhi of Afghanistan: adaptation to closed frontiers. Seattle: University of Washington Press.

Sherratt, A. 1981. Plough and pastoralism: aspects of the secondary products revolution. *In* Pattern of the past, ed. I. Hodder, G. Isaac, and N. Hammond, pp. 261-305. Cambridge: Cambridge University Press.

Silberbauer, G. 1981. Hunter and habitat in the central Kalahari desert. Cambridge: Cambridge University Press.

Silverblatt, I. 1978. Andean women in the Inca empire. *Feminist Studies* 4: 37-61.

Silverman, S. 1965. Patronage and community-national relationships in central Italy. *Ethnology* 4: 172-89.

Skinner, G. 1964. Marketing and social structure in rural China, Part 1. *Journal of Asian Studies* 24: 3-43.

Smith, B. 1978. Mississippian settlement patterns. New York: Academic Press.

Smith, C. 1976. Exchange systems and the spatial distribution of elites: the organization of stratification in agrarian societies. *In* Regional analysis, ed. C. Smith, *Vol. 2*, Social systems, pp. 309-74. New York: Academic Press.

Smith, T. 1959. The agrarian origins of modern Japan. Stanford, Calif.: Stanford University Press.

Smole, W. 1976. The Yanoama Indians: a cultural geography. Austin: University of Texas Press.

Spencer, R. 1959. The North Alaskan Eskimo. Washington, D.C.: Smithsonian.

Steinvorth-Goetz, I. 1969. Uriji jami! Life and belief of the forest Waika in the Upper Orinoco. Caracas: Asociacion Cultural Humboldt.

Steward, J. 1930. Irrigation without agriculture. Ann Arbor: Papers of the Michigan Society of Science, Arts, and Letters 12: 144-56.

————. 1936. The economic and social basis of primitive bands. *In* Essays on anthropology in honor of Alfred Louis Kroeber, ed. R. Lowie, pp. 311-50. Berkeley: University of California Press.

————. 1938. Basin-Plateau aboriginal sociopolitical groups. Washington, D.C.: Bureau of American Ethnology.

————. 1955. Theory of culture change. Urbana: University of Illinois Press.

————. 1977. The foundations of Basin-Plateau Shoshonean society. *In* Evolution and ecology, ed. J. Steward and R. Murphy, pp. 366-406. Urbana: University of Illinois Press.

Stewart, H. 1977. Indian fishing: early methods on the Northwest Coast. Vancouver: J. J. Douglas.

Suttles, W. 1968. Coping with abundance: subsistence on the Northwest Coast. *In* Man the hunter, ed. R. Lee and I. DeVore, pp. 56-68. Chicago: Aldine.

Taagapera, R. 1981. Super-cancer of the biosphere: a world population growth model. Paper delivered to the Jacob Marschak Interdisciplinary Colloquium on Mathematics in the Behavioral Sciences, University of California, Los Angeles, April 10, 1981.

Taguchi, M. 1981. Swidden agriculture in Japan. Unpublished manuscript, Anthropology, University of California, Los Angeles.

Tanaka, J. 1976. Subsistence ecology of central Kalahari San. *In* Kalahari hunter-gatherers, ed. R. Lee and I. DeVore, pp. 98-119. Cambridge, Mass.: Harvard University Press.

Tax, S. 1953. Penny capitalism. Institute of Social Anthropology Publication 16. Washington, D.C.: Smithsonian.

Taylor, K. 1974. Sanuma fauna: prohibitions and classifications. Instituto Caribe de Antropologia y Sociologia, Monografia 18. Caracas: Fundacion La Salle de Ciencias Naturales.

Thomas, D. H. 1972. Western Shoshone ecology: settlement patterns and beyond. *In* Great Basin cultural ecology: a symposium, ed. D. Fowler, pp. 135-53. Reno: Desert Research Institute Publication in the Social Sciences 8.

———. 1973. An empirical test for Steward's model of Great Basin settlement patterns. *American Antiquity* 38: 155-76.

———. 1983a. The archaeology of Monitor Valley: 1. Epistemology. Anthropological Papers 58(1). New York: American Museum of Natural History.

———. 1983b. On Steward's models of Shoshonean sociopolitical organization: a great bias in the Basin. *In* The development of political organization in North America, ed. E. Tooker, pp. 59-68. Washington, D.C.: American Ethnological Society.

Thomas, E. 1959. The harmless people. New York: Knopf.

Thompson, R. 1983. Modern vegetation and climate. *In* The archaeology of Monitor Valley, ed. D. Thomas, pp. 99-106. New York: American Museum of Natural History.

Toledo. 1940 [1570]. Informacion hecha por orden de Don Francisco de Toledo. . . . *In* Don Francisco de Toledo, supremo organizador del Peru, su vida, su obra, ed. R. Levillier, 2: 14-37. Buenos Aires: Espasa-Calpe.

Truswell, S., and J. Hansen. 1976. Medical research among the !Kung. *In* Kalahari hunter-gatherers, ed. R. Lee and I. DeVore, pp. 166-94. Cambridge, Mass.: Harvard University Press.

Tsuchiya, T. 1937. An economic history of Japan. Tokyo: Asiatic Society of Japan Transactions 15.

Tylor, E. 1913 [1871]. Primitive culture. London: John Murray.

Uberoi, J. 1962. Politics of the Kula ring. Manchester: Manchester University Press.

Vayda, A. 1961. A reexamination of Northwest Coast economic systems. New York: New York Academy of Sciences Transactions 23: 618-24.

———. 1967. Pomo trade feasts. *In* Tribal and peasant economies, ed. G. Dalton, pp. 494-500. Garden City, N.Y.: Natural History Press.

Vayda, A., A. Leeds, and D. Smith. 1961. The place of pigs in Melanesian subsistence. *In* Proceedings of the American Ethnological Society, ed. V. Garfield, pp. 69-77. Seattle: University of Washington Press.

Vega, A. 1965 [1582]. La descripción que se hizo en la provincia de Xauxa. . . . Relaciones Geográficas de Indias, Biblioteca de Autores Españoles 183: 166-75. Madrid: Ediciones Atlas.

Veja. 1982. O país afinal comprou o sonho do jari. *Veja* (Rio de Janeiro), 13 de Janeiro, pp. 68-73.

Wachtel, N. 1977. The vision of the vanquished, tr. B. and S. Reynolds. Hassocks, Sussex: Harvester Press.

———. 1982. The *mitimas* of the Cochabamba Valley: the colonization policy of Huayna Capac. *In* The Inca and Aztec states, 1400-1800, ed. G. Collier, R. Rosaldo, and J. Wirth. New York: Academic Press.

Waddell, E. 1972. The mound builders. Seattle: University of Washington Press.

Wagley, C. 1976. Amazon town: a study of man in the tropics. New York: Oxford University Press.

Webster, C., and P. Wilson. 1966. Agriculture in the tropics. London: Longmans.

Webster, D. 1975. Warfare and the evolution of the state. *American Antiquity* 40: 464-70.

Webster, S. 1971. An indigenous Quechua community in exploitation of multiple ecological zones. *In* Actas y Memorias del XXXIX Congreso Internacional de America, 3: 174-83.

Weiner, A. 1976. Women of value, men of renown. Austin: University of Texas Press.

———. 1983. "A world of made is not a world of born": doing Kula in Kiriwina. *In* The Kula: new perspectives on Massim exchange, ed. J. and E. Leach, pp. 147-70. Cambridge: Cambridge University Press.

Wenke, R. 1980. Patterns in prehistory. Oxford: Oxford University Press.

White, B. 1976. Production and reproduction in a Javanese village. Bogor, Indonesia: Agricultural Development Council.

White, Leslie. 1959. The evolution of culture. New York: McGraw-Hill.

White, Lynn, Jr. 1962. Medieval technology and social change. Oxford: Oxford University Press.

Wiessner, P. 1977. Hxaro: a regional system of reciprocity for reducing risk among the !Kung San. Unpublished Ph.D. dissertation, Anthropology, University of Michigan, Ann Arbor.

———. 1982. Beyond willow smoke and dogs' tails: a comment on Binford's analysis of hunter-gatherer settlement systems. *American Antiquity* 47: 171-78.

Wilbert, J. 1966. Indios de la region Orinoco-Ventuari. Caracas: Fundacion La Salle de Ciencias Naturales.

———. 1972. Survivors of El Dorado. New York: Praeger.

Wilkinson, R. 1973. Poverty and progress: an ecological perspective on economic development. New York: Praeger.

Willey, G. 1953. Prehistoric settlement patterns in the Virú Valley. Washington, D.C.: Bureau of American Ethnology Bulletin.

Williams, B. 1974. A model of band society. Washington, D.C.: Memoir of the Society for American Archaeology 29.

————. 1981. A critical review of models in sociobiology. *Annual Review of Anthropology* 10: 163-92.

Wilmsen, E. 1978. Seasonal effects on dietary intake in Kalahari San. *Federation of American Societies for Experimental Biology Proceedings* 37: 65-71.

Winterhalder, B., and E. Smith, eds. 1981. Hunter-gatherer foraging strategies. Chicago: University of Chicago Press.

Wittfogel, K. 1957. Oriental despotism. New Haven, Conn.: Yale University Press.

Wobst, M. 1976. Locational relationships in Paleolithic society. *Journal of Human Evolution* 5: 49-58.

Wolf, E. 1957. Closed-corporate peasant communities in Meso-America and central Java. *Southwestern Journal of Anthropology* 13: 1-13.

————. 1966a. Peasants. Englewood Cliffs, N.J.: Prentice-Hall.

————. 1966b. Kinship, friendship, and patron-client relations in complex societies. *In* The social anthropology of complex societies, ed. M. Banton. London: Tavistock.

Wright, H. 1977. Toward an explanation of the origin of the state. *In* Explanation of prehistoric change, ed. J. Hill, pp. 215-30. Albuquerque: University of New Mexico Press.

————. 1984. Prestate political formations. *In* On the evolution of complex societies, ed. T. Earle, pp. 41-77. Malibu, Calif.: Undena Publications.

Yang, M. 1945. A Chinese village: Taitou, Shantung province. New York: Columbia University Press.

Yellen, J. 1976. Settlement pattern of the !Kung: an archaeological perspective. *In* Kalahari hunter-gatherers, ed. R. Lee and I. DeVore, pp. 47-72. Cambridge: Harvard University Press.

————. 1977. Archaeological approaches to the present. New York: Academic Press.

Yellen, J., and H. Harpending. 1972. Hunter-gatherer populations and archaeological inference. *World Archaeology* 4: 244-53.

Yellen, J., and R. Lee. 1976. The Dobe-/Du/da environment: background to a hunting and gathering way of life. *In* Kalahari hunter-gatherers, ed. R. Lee and I. DeVore, pp. 27-46. Cambridge: Harvard University Press.

Yengoyan, A. 1972. Ritual and exchange in aboriginal Australia: an adaptive interpretation of male initiation rites. *In* Social exchange and interaction, ed. E. Wilmsen, pp. 5-9. Ann Arbor: Museum of Anthropology, University of Michigan, Anthropological Papers 46.

Zerries, O., and M. Schuster. 1974. Mahekodotedi. Munich: Klaus Renner Verlag.

Index

Library of Congress Cataloging-in-Publication Data

Johnson, Allen W.
  The evolution of human societies.

  Bibliography: p.
  Includes index.
  1. Social evolution. 2. Ethnology. I. Earle,
Timothy   II. Title.
GN360.J65 1987   306   86-23204
ISBN 0-8047-1339-1